B&T
2/12/83
39.50

ABSCISSION

Leaves   H. O. Miller

ABSCISSION

FREDRICK T. ADDICOTT

With illustrations by
ALICE B. ADDICOTT

University of California Press
Berkeley · Los Angeles · London

University of California Press
Berkeley and Los Angeles, California
University of California Press, Ltd.
London, England
© 1982 by
The Regents of the University of California
Original line drawings: Figs. 2:1, 2, 5, 10, 18, 19; 3:3, 5, 12, 13, 14, 15, 16, 17, 21; 4:1; 5:22; 6:8; 7:38; 8:4; and 9:6. © 1981 by Alice Baldwin Addicott. All rights reserved.
Printed in the United States of America

1 2 3 4 5 6 7 8 9

Library of Congress Cataloging in Publication Data

Addicott, Fredrick T.
  Abscission.

  Bibliography: p.
  Includes index.
  1. Abscission (Botany) I. Title.
QK763.A32   581.3   81-4065
ISBN 0-520-04288-3   AACR2

Contents

PREFACE xiii
ACKNOWLEDGMENTS xv
ABBREVIATIONS AND CHEMICAL NAMES xvii

Prologue 1
A. ABSCISSION IN LEGEND AND SONG 1
B. EARLY SCIENTIFIC WRITING ON ABSCISSION 2
C. THE GERMAN SCHOOL 1860–1930 2
D. THE PRESENT IMPETUS 5

1. *Introduction* 7
A. SCOPE OF ABSCISSION PHENOMENA 7
B. SOME ABSCISSION HABITS OF HIGHER PLANTS 9
   1. Abscission Habits of Leaves 9
   2. Branch Abscission Habits 10
   3. Flowers, Fruits, and Seeds 10
C. ABSCISSION AS A CORRELATION PHENOMENON 11
   1. The Concept of Correlative Plant Behavior 11
   2. Inhibition and Senescence as Factors in Abscission 12
      a. Inhibition of Abscission 12
      b. Abortion and Senescence in Abscission 13
   3. Organ Competition and Abscission 13
   4. Role of Abscission in Plant Homeostasis 15
D. BENEFITS OF ABSCISSION 16
E. THE ABSOLUTE REQUIREMENTS FOR ABSCISSION 19
F. ABSCISSION TERMINOLOGY 20

2. *Anatomy of Abscission*   22
   - A. THE ABSCISSION ZONE   23
   - B. SEPARATION LAYER   26
   - C. PATTERNS OF SEPARATION   27
   - D. PROTECTIVE LAYERS   28
   - E. DEHISCENCE   31
     1. Fruit Dehiscence   32
        a. General   32
        b. Ontogeny of Dehiscence Zones   32
        c. Circumscissile Dehiscence   34
        d. Anther Dehiscence   37
   - F. MECHANICAL FACTORS IN ABSCISSION AND DEHISCENCE   38
     1. Abscission   38
        a. Tissue Tensions   38
        b. Turgor Phenomena   38
        c. Physical Factors   39
     2. Dehiscence   40
        a. Drying and Shrinkage   40
        b. Turgor Mechanism   41

3. *Abscission Behavior*   44
   - A. LEAVES   45
   - B. BUDS, BRANCHES, AND STEMS   52
     1. Buds   52
        a. Shoot-tip Abortion and Abscission   52
        b. Abscission of Adventitious Buds   54
     2. Branches   54
        a. Cladoptosis   54
        b. Woody and Herbaceous Branches   56
     3. Stems   61
        a. Main Stems   61
        b. Disarticulation   62
        c. Abscission of Injured Internodes   62
   - C. FLOWERS AND FLOWER PARTS   63
   - D. FRUITS AND SEEDS   68
     1. Young Fruits   68
     2. Mature Fruits   70

3. Seeds   71
4. Disseminules   71

E. BARK   72
  1. General   72
  2. Sheets   74
  3. Strips   74
  4. Scales   74
  5. Prickles   77
  6. Root Bark and Cortex   77

F. ROOTS   77
  1. General   77
  2. *Azolla* Root Abscission   79
  3. Root Cap   80

G. ABSCISSION AND ARCHITECTURE   81
  1. Morphological Factors in Plant Architecture   81
  2. Extent of the Involvement of Abscission   81
     a. Entire Stem   81
     b. All Branches   83
     c. Lower Main-Stem Branches   83
     d. All Branchlets (Cladoptosis)   83
     e. Shoot-tip Abscission   83
     f. Stem Disarticulation   83
     g. Attritional Abscission   83
  3. Examples of Tree Architecture Influenced by Abscission   84
     a. General   84
     b. Umbrella Trees   85
     c. Sympodial Trees   86
  4. Further Reading   89

H. ADVENTITIOUS AND ANOMALOUS ABSCISSION   89
  1. Seasonal Manifestations of Adventitious Abscission   89
  2. Induced Adventitious Abscission   92
  3. Anomalous Abscission   96

4. *Physiology*   97

A. MAJOR RESEARCH METHODS   98
  1. Field and Greenhouse   98
  2. Explant Methods   99
  3. Testing Devices   103
  4. Abscission Indices   104

viii *Contents*

   B. GENERAL PHYSIOLOGY 105
   1. Temperature Response 105
   2. Oxygen Requirement 107
   3. Substrate Requirement 108
   4. Respiration 109
   5. Interpretation 110
   C. NUTRITIONAL FACTORS 110
   1. Carbohydrate Metabolism 110
   2. Nitrogen Metabolism 111
   3. Mineral Nutrition 111
   4. Preabscission Changes in Organs 112
   D. HORMONAL FACTORS 113
   1. Auxin (IAA) 113
      a. Auxin and Abscission 113
      b. Correlative Occurrence 114
      c. Abscission Responses 114
         (1) Retardation 116
         (2) Acceleration 116
         (3) Stages I and II 118
         (4) Response Curves 120
      d. Auxin Physiology in Relation to Abscission 121
      e. Auxin Gradient 122
      f. Mechanism of Auxin Action 123
   2. Abscisic Acid (ABA) 126
      a. Discovery and Hormonal Nature 126
      b. Correlative Occurrence 127
      c. Abscission Responses 128
      d. Mechanism of ABA Action 129
   3. Gibberellin (GA) 130
      a. Correlative Occurrence 130
      b. Lack of Response to General Applications 131
      c. Responses to Localized Applications 131
      d. Applications to Petioles 131
      e. Responses of Explants 132
      f. Mechanism of GA Action 134
   4. Cytokinin (CK) 134
      a. Correlative Occurrence 134
      b. Abscission Responses 134
      c. Mechanism of CK Action 135
   5. Ethylene (ETH) 135
      a. Occurrence 135

    b. Abscission Responses   138
    c. Interrelation with Hormones and Regulators   140
    d. Mechanism of ETH Action   143
  6. Miscellaneous Abscission Substances, Possibly Hormonal   145
  7. Summary of Hormone Action   148
    a. In the Abscission Zone   148
    b. In the Subtended Organ   150
    c. In Source-Sink Relations   151

5. *Biochemistry and Ultrastructure of Abscission*   153
  A. BIOCHEMISTRY   153
    1. Introduction   153
    2. Cellulases   156
    3. Pectinases   158
    4. Lignase and Other Enzymes   160
    5. Are the Cell Walls of the Abscission Zone in a Dynamic State?   161
  B. CYTOLOGY: INVOLVEMENT OF CYTOPLASMIC ORGANELLES   162
  C. CYTOCHEMICAL LOCALIZATION OF ENZYMES   165
  D. CYTOLOGY: CELL-WALL CHANGES   170
    1. Middle Lamella   170
    2. Primary Cell Wall   177
    3. Lignin and Secondary Cell-Wall Separation   177
    4. Cutin and Suberin   184

6. *Ecology*   185
  A. PHYSICAL AND CHEMICAL FACTORS AND THEIR PHYSIOLOGICAL EFFECTS   186
    1. Light   186
    2. Other Radiation   188
    3. Temperature   188
      a. Cold   188
      b. Heat   189
      c. Fire   189
    4. Water   190
      a. Moisture Stress   190
      b. Rain and Mist   193
      c. Flooding   193

5. Wind   194
6. Soil Factors   195
   a. Mineral Deficiencies   195
   b. Toxicities   197
   c. Salinity and Alkalinity   197
7. Atmospheric Pollutants   198
8. Ecological Effects of Abscised Parts   201

B. PHENOLOGY: SEASONAL ABSCISSION   202
1. Autumnal Defoliation   203
2. Vernal Leaf Abscission   203
   a. Abscission of Marcescent Leaves   203
   b. Abscission of a Portion of the Leaves of Evergreen Trees   204
   c. Abscission of All of the Leaves of "Evergreen" Trees   205
   d. Physiology of Vernal Leaf Abscission   205
3. Drought (and Summer) Leaf Abscission   205
4. "Hygrophobic" Leaf Abscission   207

C. BIOTIC FACTORS IN ABSCISSION   207
1. Pathogenic Microorganisms   207
2. Insects and Mites (Acarina)   210
   a. "Defoliator" Insects   210
   b. General Abscission Effects of Insect Feeding   211
   c. *Lygus*   211
   d. *Anthonomus*   211
   e. Mites   213
   f. Insect Galls   214

7. *Abscission in Lower Plants*   217

A. FERNS, MOSSES, AND LIVERWORTS   218
1. Ferns and Fern Allies   218
   a. Vegetative Organs   218
   b. Reproductive Structures   219
   c. Dehiscence   220
2. Mosses and Liverworts   221

B. ALGAE   224
1. Green Algae   224
2. Blue-green Algae   231
3. Brown Algae   232
4. Red Algae   234

C. FUNGI   235
    1. Filamentous Fungi   235
       a. Somatic Fragments   235
       b. Sporangial Abscission and Dehiscence   235
       c. Conidial Abscission   236
       d. Ascus Dehiscence   242
       e. Basidiospore Discharge   244
       f. Abscission and Dispersal in the Nidulariaceae   247
    2. Yeast   250
D. LICHENS   253
E. BACTERIA   254

8. *Genetics of Abscission and Dehiscence*   257

A. EVIDENCE OF GENETIC CONTROL OF ABSCISSION   257
    1. In Cultivated Plants   257
    2. In Taxonomy   259
B. EVIDENCE OF GENETIC CONTROL OF DEHISCENCE   262
    1. In Cultivated Plants   262
    2. In Taxonomy   262
C. IDENTIFIED GENES AFFECTING ABSCISSION AND DEHISCENCE   264

9. *Paleontology and Evolution of Abscission*   266

A. PALEONTOLOGY   266
    1. The Precambrian Era (more than 600 million years ago)   267
    2. The Paleozoic Era (250–600 million years ago)   268
    3. The Mesozoic Era (100–250 million years ago)   277
    4. Abscission in the Cenozoic Era (from 70 million years ago)   279
B. EVOLUTION   281
    1. Evidence of Recent Evolution and "Fine Tuning" of Abscission Behavior   281
    2. Some General Comments on Abscission Evolution   283
    3. A Theory of the Evolution of Abscission   283
       a. Some General Considerations   283
       b. A Brief Statement of the Abscission Evolution Theory   285
    4. Abscission in the Animal Kingdom   285

10. Basics of Agricultural Abscission   287

    A. CULTURAL PRACTICES THAT AFFECT ABSCISSION   288

    B. LEAF ABSCISSION   290
- 1. Acceleration   290
- 2. Retardation   293

    C. FLOWER ABSCISSION   294

    D. YOUNG-FRUIT ABSCISSION   296
- 1. Fruit Thinning (Promotion of Young-Fruit Abscission)   296
- 2. Prevention of Young-Fruit Abscission   299

    E. FRUIT DEHISCENCE AND SEED ABSCISSION   300

    F. MATURE-FRUIT ABSCISSION   301
- 1. Prevention   301
- 2. Promotion   303

    G. BRANCH ABSCISSION   305

    H. BARK REMOVAL   306

EPILOGUE   307

APPENDIX: LITERATURE ON ABSCISIC ACID IN ABSCISSION   311

LITERATURE CITED   315

INDEX   355

Preface

In 1943 I was asked to investigate the anatomical changes associated with leaf removal of guayule (*Parthenium argentatum*) necessary before the shrub could be ground and the rubber extracted. In 1947, after I had joined the University of California, G. A. Livingston and I were persuaded to study the physiology of leaf abscission in *Citrus*. For that work, we developed the explant method. Soon thereafter, I began research on the physiological problems of cotton defoliation and became committed to continuing investigations of abscission. My associates and I concentrated our initial efforts on anatomical and physiological work. The literature available to us in those days was sparse, and it was obvious that we needed all of the ideas we could get if we were to progress in our understanding of abscission. Gradually, we became aware that the literature on horticulture and ecology contained numerous but widely scattered comments on factors influencing abscission: abscission not only of leaves, but of flowers, fruits, and buds as well.

In due course, the materials gleaned from the literature formed the basis of review articles on horticultural and ecological aspects of abscission. Those studies revealed that pathogenic organisms, both plant and animal, could have considerable influence on abscission. During this period, also, we became aware of the many manifestations of abscission in the lower plants. These were often conspicuous in mosses, ferns, and their relatives; but abscission is a far from insignificant part of the life of algae and fungi. Further, as we surveyed the variety of abscission and dehiscence activities in the higher plants we were impressed by the extent of their involvement in both the function and the structure of the plant.

This book, then, is an outcome of thirty-two years of activity in the University of California's Agricultural Experiment Station.

One of the major objectives of this book is to outline, if only in broad, general terms, this involvement of abscission in both the structure and the morphology of plants, as well as in the functioning of individual plants.

Another objective is to draw the reader's attention to the vast literature available on abscission. For various reasons, the early extensive work of the German school has been largely unknown to English-speaking plant scientists, but it contains much that deserves our respect and attention. In addition, many references on abscission are difficult to find, often appearing as casual comments, incidental to the principal thrust of a horticultural or botanical publication. I hope this book will help the interested reader to trace the antecedents of current investigation and to become aware of areas of abscission research that could easily be overlooked.

Some of the subjects treated represent interests that developed more as hobbies than as serious scientific investigations. For example, the role of abscission in the determination of the architecture of a plant has fascinating aspects, but I can claim no real authority on this subject, and the discussion in Chapter 3 should be considered merely an introduction. A similar, almost casual curiosity stimulated my interest in the evolution of abscission and when in the paleontological record abscission actually began. Before the Devonian the record is sparse indeed, but the consideration of the evidence (Chapter 9) led to an interesting, simple, and I hope convincing conclusion.

In writing this book I have resisted the temptation to make it an extended review of the literature, gratifying though it might have been to cite each of the more than 2,500 references assembled in the course of the work. Practical considerations have held me to citing only a fourth of these. However, I have attempted to cite the significant contributions and to mention at least once the name of each researcher who has been active in abscission. An overriding concern has been to bring together a body of knowledge in a consistent and comprehensive way, without overwhelming the reader. Emphasis has been given to important interpretations and relationships, especially where these may not have been previously touched upon in the literature. Comments have been phrased so that the reader should have no difficulty in distinguishing between experimental results and my interpretations of them.

I hope this book will be of interest to many plant scientists, particularly those who for one reason or another would like to get acquainted with the scope of abscission phenomena. I view it as something of a handbook for the plant scientist—not to tell him in detail about the abscission aspects of his own field, but to tell him in broad terms of things in other fields that may be of interest or importance to his own.

Acknowledgments

I gratefully acknowledge my indebtedness to the many colleagues who have contributed to my biological education and who, over the years, have nurtured and expanded my appreciation of the living world. Collectively, they gave added stimulus to my curiosity and broadened and rebroadened my concept of the plant world.

My parents deserve the primary credit for awakening me to the beauty and wonders of nature. These were a part of our family life from its beginning. This early influence seemed to lead directly to the study of biology. My wife, Alice B. Addicott, gave enthusiastic encouragement to my development as a biologist. Her infectious love of nature and of the out-of-doors continually heightened my awareness of the fascinating variety of activities in living things. Her perceptions enriched each summer holiday and each sabbatical leave as we shared adventures in the field.

Professionally, I feel greatly indebted to my fellow graduate students at Stanford University and the California Institute of Technology. From them I learned what biology was all about. From my teachers at those universities—particularly L. R. Blinks, C. B. van Niel, F. W. Went, and James Bonner—I learned what research was all about. The students and associates who joined in our pioneering physiological investigations at UCLA were a tremendous stimulus to imaginative and rigorous thinking about abscission.

I have been fortunate to have worked under enlightened and supportive administrators. I am especially appreciative of R. W. Hodgson, A. G. Norman, P. F. Sharp, J. H. Meyer, and C. E. Hess. Without their support at critical times, my research on abscission would surely have suffered. For the preparation of this book Dean C. E. Hess made possible the assistance essential to the composition of the manuscript and the bibliography.

I acknowledge with gratitude the responses of more than a hundred colleagues who provided reprints, data, illustrations, and personal comments that have been of great value. Information they supplied enabled me to make reasonably meaningful comments about subjects on which my personal knowledge was limited. I am indebted to M. J. Bukovac, H. G. Baker, and G. C. Castor for reading the entire manuscript, and to K. Wells and J. A. Doyle for reading sections of the manuscript; they provided many helpful comments.

The drawings, graphic figures, and uncredited photographs were made by my wife, Alice B. Addicott. Many of the figures are her originals. She gave many hours and days and numerous helpful suggestions at each step in the preparation of this book.

I am grateful to Sheila A. Reed for her excellent assistance in the preparation of the manuscript and bibliography, and particularly for her painstaking library searches for bibliographic materials. We thank Wendy Reed for her superlative typing of the bibliography.

Abbreviations and Chemical Names

ABA	abscisic acid
Alsol	2-chloroethyl-tris-(2-methoxyethoxy)-silane
ATP	adenosine triphosphate
Bromodine	a bromine-iodine formulation
C	degrees Celsius
carbaryl	1-naphthalene-N-methylcarbamate
CHI	cycloheximide
CK	cytokinin
cm	centimeter(s)
cv.	cultivar
daminozide (Alar, B-nine)	butanedioic acid mono(2,2-dimethylhydrazide)
dikegulac	Na 2,3:4,6 bis-0-(1-methylethylidene)=O-L-xylo-2-hexulofuranosonate
dinoseb	2-*sec*-butyl-4,6-dinitrophenol
diquat	6,7-dihydrodipyrido [1,2-*a*:2,1-*c*] pyrazinediium ion
DNA	deoxyribonucleic acid
DPU	diphenyl urea
DPX1840	3,3a-dihydro-2-(*p*-methoxyphenyl)-8*H*-pyrazolo [5,1-*a*] isoindol-8-one
d.w.	dry weight
EM	electron microscopy
endothall	7-oxabicyclo [2.2.1] heptane-2,3-dicarboxylic acid
ER	endoplasmic reticulum
ETH	ethylene
ethephon (Ethrel)	2-chloroethylphosphonic acid
4-CPA	4-chlorphenoxyacetic acid
f.w.	fresh weight

g	gram(s)
GA	gibberellic acid, gibberellin
h	hour(s)
ha	hectare(s)
Harvade	2,3-dihydro-5,6-dimethyl-1,4-dithiin 1,1,4,4 tetraoxide
IAA	indoleacetic acid, indol-3yl-acetic acid
kg	kilogram(s)
m	meter(s)
MCPA	4-chloro-2-methylphenoxyacetic acid
mg	milligram(s)
MH	1,2-dihydro-3,6-pyridazinedione
mm	millimeter(s)
mRNA	messenger ribonucleic acid
NAA	naphthaleneacetic acid, naphth-1yl-acetic acid
NaClO	sodium hypochlorite
naptalam	N-1-naphthylphthalamic acid
NOXA	naphthoxyacetic acid
paraquat	1,1-dimethyl-4,4-bipyridinium ion
p.c.	personal communication
PGU	polygalacturonase
PME	pectin methylesterase
Q_{10}	temperature coefficient
RER	rough endoplasmic reticulum
RNA	ribonucleic acid
2,4-D	2,4-dichlorophenoxyacetic acid
2,4,5-T	2,4,5-trichlorophenoxyacetic acid
2,4,5-TP	2,4,5-trichlorophenoxypropionic acid
TIBA	triiodobenzoic acid

Prologue

Chapter Contents
A. ABSCISSION IN LEGEND AND SONG   1
B. EARLY SCIENTIFIC WRITING ON ABSCISSION   2
C. THE GERMAN SCHOOL 1860–1930   2
D. THE PRESENT IMPETUS   5

The essential element of abscission appeared early in the course of evolution with the development of biochemical processes that enabled cells to separate from each other. As plants became anatomically complex, they evolved the ability to restrict cell separation to precise zones and to control its onset. This ability has been utilized by plants in a great variety of ways, and indeed, it is a fundamental part of the life of innumerable species. In higher plants abscission of weakened organs often enables the survival of the individual, and it is a necessary step in the dissemination of most species. Abscission is truly one of the great phenomena of the living world.

A. ABSCISSION IN LEGEND AND SONG

The abscission of leaves, flowers, and fruits is a striking and thought-provoking event that has long challenged man's imagination. The seasonal coloring and abscission of leaves figures frequently and conspicuously in the legends, literature, and song of many peoples. Similarly, flowering and the drop of flower petals have long been a source of wonder and pleasure. A few of the better-known stories are mentioned below.

In South-West Africa the small *Moringa ovalifolia* trees are leafless most of the year, and the branches are so few that a tree has the appearance of a root system growing skywards. Bushmen legend attributes such trees to the god Thora who, when gardening in paradise, decided that they were too ugly to include. Dissatisfied, he pulled them out, in the process pricking himself on their thorns. Then in a fit of pique, he tossed them over the wall of paradise and they crashed to earth in South-West Africa, upside down with their branches buried in the sand!

2 *Prologue*

One of the most famous legends of Europe is the *Nibelungen-lied* in which Siegfried became almost invulnerable to his enemies by virtue of having bathed in the blood of the dragon Fafnir. Unfortunately for him, an [abscised] linden leaf fell on his back and prevented the blood from completely covering his body. It was only after the location of that spot became known to his enemy, Hagen, that Hagen was able to kill Siegfried.

Perhaps the most famous story involving abscission is that of Isaac Newton and the [abscised and] falling apple. As a young man at Cambridge in 1665, Newton returned to the family home at Woolsthorpe when the university was closed because of the plague. There he continued his work, absorbed with astronomy and particularly with the orbital relationship of the moon and the earth. One day while he was pondering this problem in the orchard, an apple fell nearby in response to the force of gravity [and the process of abscission]. This small incident redirected Newton's thoughts and led shortly to the enunciation of the law of gravitation (Brewster, 1855; Manuel, 1968; Sullivan, 1938). It seems possible that if, at that time, Newton had been thinking about the growth and development of trees rather than the orbit of the moon, he could well have authored the first treatise on abscission!

B. EARLY SCIENTIFIC WRITING ON ABSCISSION

Among the writings of the distinguished Greek philosophers, the work of Theophrastus has been of special interest to botanists. His extensive *Enquiry into Plants* recorded what was known and believed about plants and their responses. The section on the shedding of leaves classified leaf abscission habits of higher plants and cited numerous examples. Further, Theophrastus made perceptive comments on abscission behavior, anticipating substantial aspects of present-day abscission ecology: "It appears also that position and a moist situation conduce to keeping the leaves late; for those which grow in dry places, and in general where the soil is light, shed their leaves earlier, and the older trees earlier than young ones. Some even cast their leaves before the fruit is ripe, as the late kinds of fig and pear" (Theophrastus, 285 BC, translation of A. Hort).

Over the centuries further knowledge of abscission undoubtedly accumulated among farmers and horticulturalists, but records of the development of such information are, indeed, sparse.

C. THE GERMAN SCHOOL 1860–1930

Present-day authors who touch upon the history of abscission research usually begin with von Mohl (1860a) and his important work that drew a

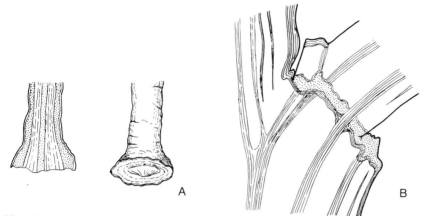

Fig. P:1.
Abscission zone of woody twigs. **A.** Median section and surface view, *Populus nigra*. **B.** Diagram of a section through a node with a leaf scar and a branch in the process of abscission, *Salix fragilis*. (Redrawn from von Höhnel, 1880)

clear distinction between the processes of separation and protection in abscission. His research in turn built on earlier writings and investigations going back into the previous century. It provided a firm point of departure for the anatomists and morphologists of the latter part of the nineteenth century who utilized their improved microscopes with Germanic thoroughness to examine the tissues of higher plants and to explore the diversity of anatomy and morphology throughout the plant kingdom. Among the more notable abscission contributions of the period were those of von Höhnel (1880) on the abscission of woody twigs (Fig. P:1), von Bretfeld (1880) on protective layers in leaf abscission, and Reiche (1885) on the anatomical changes in perianth abscission. The early works were seldom illustrated, although the anatomical changes were usually well described. Illustrations became more common with the turn of the century. Figures from the work of Kubart (1906) and Löwi (1907) are reproduced here (Figs. P:2A,B.). It is interesting that the early investigators regularly observed cell separation ("maceration") in the abscission zone. This appears to be the result of examination of freshly sectioned material. (In contrast, later investigators, especially between 1940 and 1960, who studied sections that had been fixed and embedded in paraffin, rarely detected cell separation, owing at least in part to the artifacts to which paraffin methods are prone.) By the 1920s the literature touching on abscission was extensive, and in 1925 it was concisely summarized by Mühldorf. This was followed shortly by Pfeiffer's 1928 encyclopedic compilation of the anatomical literature on the abscission tissues

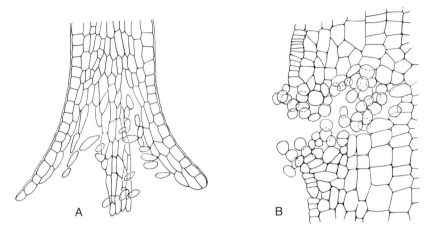

Fig. P:2.
A. Section through the base of a freshly abscised corolla, *Nicotiana tabacum*. Presumably drawn from a fresh, unstained preparation. Note extensive separation of intact cells. (Redrawn from Kubart, 1906). **B.** Section through the abaxial portion of a leaf abscission zone, *Strobilanthes isophylla*. Note separation of intact cells. (Redrawn from Löwi, 1907)

of plants. The number of German publications cited by Pfeiffer is impressive: 444. In contrast, his bibliography lists only 65 publications in English and 55 in French.

Many of the current concepts of ecology, physiology, and biochemistry of abscission appear to have had their origins in the perceptive work of Wiesner (1871–1905) and his students. He drew attention to the role of temperature, water, and light in the timing of leaf abscission. Molisch (1886) reported some of the early physiological experiments establishing important water and temperature relations. He was the first to show the essentiality of oxygen for abscission and to demonstrate that separation involved the action of an enzyme on the walls of cells. Goebel's (1897) extensive writings on correlations in plants laid the groundwork for the present-day concept of abscission as a phenomenon of correlation and homeostasis.

With the advent of the twentieth century, physiological experiments increased. Notable among these is Fitting's (1911) extensive investigations of corolla abscission. Neger and Fuchs (1915) explored the water relations of conifer needle abscission; their work included classical anatomical studies of the needle abscission zones. A short time later, Küster (1916) published his experiments on the role of photosynthesis in abscission in which he used debladed petioles for the first time. This technique became a powerful tool in the study of the physiology of abscission.

It is illuminating and sobering to realize that what we now consider to be the salient features of the process of abscission were, with few exceptions, identified by the early German botanists. Their work and discoveries have been almost unknown and unappreciated by English and American plant scientists. One of the purposes of this book is to draw attention to the significance of their many contributions.

D. THE PRESENT IMPETUS

The German botanists were not entirely alone in the investigation of abscission. In 1848 Inman read a short paper on "The Causes That Determine the Fall of Leaves" before the Literary and Philosophical Society of Liverpool. From observations with his light microscope he described the leaf abscission zone, the development of the protective layer, the pre-abscission deposition of starch, and separation of cells. As often happens to pioneering investigations, Inman's work lay unnoticed for many years, and later researchers continued unknowingly to repeat it. Much later Tison (1900) in France and Lee (1911) in England each published extensive investigations of the anatomy of leaf abscission, particularly of trees (e.g., Fig. P:3). Shortly thereafter, American botanists entered the field. Among these, Lloyd published an important series of contributions on both botanical and agronomic (cotton) aspects of abscission. Several fine theses were published. Noteworthy among them was Sampson's 1918 histochemical study of *Coleus* leaf abscission and Kendall's 1918 thesis on the abscission of flowers and fruits in the Solanaceae. Kendall was the first researcher to use excised abscission zones (Fig. P:4).

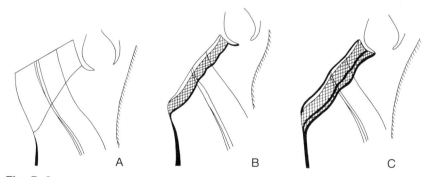

Fig. P:3.
Diagrams of sections through the leaf abscission zone and leaf scar, *Castanea sativa*. **A.** Abscission zone at the time of abscission. **B.** Leaf scar in first winter. **C.** Leaf scar in second winter. (Redrawn from Lee, 1911)

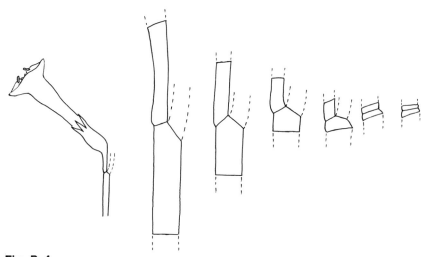

Fig. P:4.
Explants of flower-pedicel abscission zones, *Nicotiana tabacum*. (Redrawn from Kendall, 1918)

In the 1930s Laibach's discovery that auxin could delay abscission led quickly to the successful control of preharvest drop of apples by auxin-regulators. In the same period the first indications appeared that agricultural chemicals could be adapted for fruit thinning and for defoliation. These discoveries greatly stimulated agricultural research on the use of abscission-regulating chemicals. They also justified financial support of the basic research essential to the understanding of the process of abscission. Much of the research on the cytology, physiology, biochemistry, and agricultural applications discussed in this book received its principal impetus from those modest but important discoveries of the 1930s.

1. Introduction

Chapter Contents

A. SCOPE OF ABSCISSION PHENOMENA   7

B. SOME ABSCISSION HABITS OF HIGHER PLANTS   9
   1. Abscission Habits of Leaves   9
   2. Branch Abscission Habits   10
   3. Flowers, Fruits, and Seeds   10

C. ABSCISSION AS A CORRELATION PHENOMENON   11
   1. The Concept of Correlative Plant Behavior   11
   2. Inhibition and Senescence as Factors in Abscission   12
      a. Inhibition of Abscission   12
      b. Abortion and Senescence in Abscission   13
   3. Organ Competition and Abscission   13
   4. Role of Abscission in Plant Homeostasis   15

D. BENEFITS OF ABSCISSION   16

E. THE ABSOLUTE REQUIREMENTS FOR ABSCISSION   19

F. ABSCISSION TERMINOLOGY   20

A. SCOPE OF ABSCISSION PHENOMENA

Most of us in the plant sciences think of abscission as the separation of leaves, flowers, fruits, seeds, and perhaps a few of the other conspicuous organs of higher plants. These are the familiar examples of which some manifestation can be seen almost any time we step into a garden. When we begin to look for further examples, we soon find that many other kinds of plant structures can be abscised, and eventually it becomes apparent that probably every discrete plant organ is capable of being abscised by one or another species (Table 1:1). Further, we find that abscission is relatively common in the lower plants, where there are numerous and widespread examples of abscission, particularly of propagules. Such a propagule can be as

8   Introduction

Table 1:1.   *Plant Parts that can be Abscised*

vegetative buds	trichomes	roots
branches	flower buds	galls
bark	flowers	gemmae
prickles	sepals (calyx)	sporangia
spines	petals (corolla)	sori
glochids	stamens	opercula
cotyledons	styles	hormogonia
leaves	ovaries	filament fragments
leaflets	fruits	yeast buds
leaf stalks	mesocarps	conidia
stipules	seeds	basidiospores

small as a single cell—for example, the bud of a yeast. This leads to consideration of related forms which propagate by fragmentation of filaments or other colonial aggregates of cells. From even a casual survey of the growth habits of higher and lower plants, it is difficult, if not impossible, to enunciate criteria that can set apart the abscission of leaves and fruits from the fragmentation of filaments of cells. The underlying process of cell separation appears essentially the same in all cases. Thus we can view abscission as a process that is very widespread in the plant kingdom, affecting many species and in many ways.

A kind of paradox exists in the manifestations of abscission in various groups of plants. In some, the appearance of abscission is highly variable, both among species and among the cultivars of a species. Some of these will abscise leaves or flowers more easily or at different times than others. In contrast, some manifestations of abscission are so consistent that they become useful taxonomic characters. For example, the abscission of cotyledons (or lack of it) serves to distinguish certain species of *Lupinus*. Abscission of leafy branchlets (cladoptosis) is restricted to certain genera or species of conifers (e.g., *Sequoia*). Another example is the circumscissile dehiscence of the calyx in *Eucalyptus* and of the capsule in *Anagallis*.

The familiar examples of abscission involve physiological activity in specialized tissues within an abscission zone. In these examples the separation layer must be alive and able to utilize the energy of metabolism for the synthesis and secretion of hydrolytic enzymes. As the various examples of abscission are surveyed for evidence of physiological activity, cases can be readily found where the physiological processes are slow or only partly effective. An excellent series of examples of this range of abscission behavior can be found in branch abscission. Some species abscise branches rapidly and cleanly, indicating effective secretion of hydrolytic enzymes. Other species abscise their branches more slowly, and the hydrolytic processes may not go to completion. In such cases, the mechanical assistance of wind or some similar factor must come into play before the branches separate from the tree. In still other species, there is little indication of hydrolytic activity, but the tissues at the base of a branch are weak, less lignified, and hence more susceptible to attack by microorganisms and to separation by mechanical factors. Similar ranges of abscission behavior can be readily found among leaves, flowers, fruits, and other abscising organs. Again, in these considerations it appears impossible to identify a line of demarcation within what appears to be a broad and continuous range of function of the physiological components of abscission.

B. SOME ABSCISSION HABITS OF HIGHER PLANTS

1. Abscission Habits of Leaves

A leaf may live from a few weeks to several years before it becomes senescent and dies. Typically, senescence is followed by abscission. For some plants, abscission of leaves commences within a few weeks of seed germination and progresses more or less uniformly through the growing season. A plant such as cotton that produces a sizable canopy of leaves abscises an appreciable number of its early leaves as they become shaded, or in other ways lose out in competition with later leaves. In woody plants a similar sequence of leaf abscission can occur as growth progresses. In trees growing in regions having a relatively uniform climate, such as a tropical rain forest, leaf abscission can occur steadily in correlation with the growth and vigor of the tree.

A number of plants, particularly trees and shrubs, are deciduous, abscising all of their leaves during a short period. Several patterns of deciduous-leaf abscission are now recognized. Among these, the autumnal deciduous habit is the most conspicuous in the Northern Hemisphere. This habit appears to have evolved in the Cretaceous in response to the development of a strongly seasonal climate. The abscission is stimulated by several environmental factors, primarily shortening photoperiod, and cold.

Another deciduous habit is found in trees and shrubs of tropical and subtropical regions that have a dry season. In this case, the stress of drought is the factor initiating abscission. Typically, this is an annual event; the species will remain leafless until rain comes. Then they leaf out and hold their leaves until the stress of drought returns. Some species can go through the cycle of foliation and defoliation whenever there are rains, up to several times a year. However, other species, including most conifers and species of *Eucalyptus*, are evergreen and rarely, if ever, abscise all of their leaves. These species do abscise their older leaves from time to time, usually after a few days of relatively hot, dry weather.

A third pattern is that of the vernal-leaf-abscission habit that typically

10 *Introduction*

takes place in the spring as buds are expanding (Treiblaubfall of Wiesner). The vernal deciduous habit is common in broad-leaved evergreens of subtropical and temperate regions. Typically, it takes place as the new flush of leaves and flowers appear in the spring and results in the abscission of all of the leaves of the previous year. Trees showing this kind of abscission are technically evergreen, almost always having some expanded foliage; but the behavior is deciduous in that the removal of all the previous year's foliage is accomplished. In addition to the vernal abscission of leaves, many plants will show considerable abscission activity in the spring as bud scales are abscised, tassels and catkins shed their pollen and are abscised, petals function and are abscised, and aborted flowers and fruit are abscised.

2. Branch Abscission Habits

Branches can be abscised as buds, leafy branchlets, small twigs, or branches up to several cm in diameter. Each of the foregoing patterns can have a significant influence on the pattern of growth and general morphology of a woody plant. Shoot-tip abortion and abscission is a regular occurrence in many kinds of sympodial growth habits. Typically, shoot-tip abscission occurs at the end of a period of rapid shoot growth. Abscission of leafy branchlets (cladoptosis) is characteristic of a number of conifers, such as *Sequoia*, *Taxodium*, and *Larix*. For some of the species of these genera, branchlet abscission is the mechanism of an autumnal deciduous habit.

Small branches can be abscised by a number of species in a kind of self-pruning habit. The ease and smoothness with which these branches are detached by many species indicate that active hydrolytic processes occur in the branch abscission zone. In other cases, such activity is incomplete, and the assistance of mechanical factors such as wind is required to bring about twig abscission. For a great many species where evidence of hydrolytic activity at the base of the small branches cannot be found, the bases are weakly lignified, and in due course the dead twigs break away. The distribution and intensity of branch abscission greatly affects the growth habit (architecture) of a tree. For example, if the abscission of branches along the main trunk of the tree occurs early in the life of the tree, the trunk can develop into a massive bole of highly prized timber.

3. Flowers, Fruits, and Seeds

As higher plants evolved the reproductive mechanisms of flowers, fruits, and seeds they also evolved a wide variety of strategies for pollination, seed production, and seed dispersal. Many of these strategies take advantage of plants' inherent capacities for abscission.

One of the strategies to insure reproduction is the production of large numbers of flowers. This is often used by species that normally would develop only a relatively few fruit. The excess flowers are present as a kind of insurance, and after a few fruit have set and begun to develop, the excess flowers are abscised. The abscission often involves disarticulation and abscission of parts of the stalk of the inflorescence and accessory structures. Flower parts, such as calyx and petals, may persist indefinitely, particularly if they are chlorophyllaceous and contributing to the nutrition of the fruit. In many cases, however, their function is served within a very brief period, and the calyx is commonly abscised promptly, and in many species the petals are abscised shortly after pollination.

Even if a number of flowers have been abscised before pollination, many species will set more young fruit than they can develop to maturity, and the excess young fruit are often abscised. Further, adverse weather and pathogenic organisms can lead to the malfunction and abortion of young fruit. Most species abscise defective fruit rather promptly.

Plants have evolved several major strategies to aid in the dispersal of seeds. The first and most ancient of these is the abscission of the seeds themselves, separating from the placenta. The dispersal of the loosened seeds comes about easily if the fruit is one of the dry dehiscent types. Dehiscence itself, which is a form of abscission, can greatly facilitate dispersal if the mechanical factors in the structure of the fruit lead to an explosive opening, such as occurs with a number of legumes. Some dry fruits utilize the strategy of remaining closed until a fire, after which dehiscence can occur. With some species the seeds remain within the fruit, and dispersal is facilitated in a variety of ways. Adaptations exist for dispersal of indehiscent fruits by wind, by water, and by numerous animals. In these cases the progress of abscission is timed to make the fruit readily detachable by the dispersal agent.

Abscission habits are discussed in greater length in Chapters 3 and 7.

C. ABSCISSION AS A CORRELATION PHENOMENON

1. The Concept of Correlative Plant Behavior

In the growth and development of an organism, events in one part of the organism regularly influence events and activities in other parts. Plants show many kinds of such correlative behavior. In the course of his extensive investigations Goebel defined correlation as the reciprocal influence of one organ on another. He described a number of the relationships among buds, leaves, and shoots. The most conspicuous of these is the dominance of the apical bud over the lateral buds. A vigorous apical bud inhibits the development of lateral buds.

Many of the manifestations of abscission involve relationships that typify classical correlative relations. The most conspicuous of these is in leaf abscission where an active, vigorous leaf blade inhibits change at the leaf abscission zone, which is usually some distance away at the base of a petiole. It is only after the vigor of the leaf blade is greatly reduced through injury or senescence that abscission occurs. The change in vigor of the leaf blade is perceived by the abscission zone, and it responds by the physiological activity that culminates in separation. Some such correlative relationship appears to be involved in almost all cases of physiological abscission, and there are many variations.

Flower petal abscission includes two interesting variations. In a number of species petal abscission occurs within a relatively short time after anthesis. For example, the flowers of some species of *Linum*, *Geranium*, and *Erodium* regularly abscise the afternoon of the day of anthesis. Petals of cotton flowers abscise the day after anthesis, and the petals of *Eschscholzia californica* regularly abscise on the fifth day after anthesis. In these cases the petals undergo a kind of programmed senescence that initiates separation in a predictable manner.

In other flowers, such as species of *Clarkia* and *Digitalis*, abscission of the petals occurs promptly after pollination. However, if pollination is delayed, then petal abscission is correspondingly delayed, sometimes for several days (Stead and Moore, 1979).

In plants whose flowering and fruiting covers an extended period, development of early fruits can lead to the abortion and abscission of later developing fruits. If for any reason the early flowers do not develop, later flowers can set, and the fruits develop normally without abscission.

While the early studies of correlations in plant behavior did not provide much in the way of explanations, they did provoke awareness of this important aspect of the coordination of plant growth and development. Curiosity about the mechanisms of correlations led eventually to the discovery of plant hormones and to the enunciation of the other physiological factors in the coordination of plant development.

2. Inhibition and Senescence as Factors in Abscission

a. Inhibition of Abscission

In the correlations of abscission, prolonged inhibitions are usual. For example, a healthy leaf inhibits its abscission. Such inhibitions are the general rule; for most of the life of organs that are abscised abscission is inhibited, typically, by the presence and activity of the organ subtending the abscission zone. After injury or infection, the organ may no longer be able to inhibit abscission, and the process will go forward.

The mechanism of inhibition appears to be largely hormonal, though modified by many other physiological factors. As is discussed in Chapter 4, the secretion of IAA and to a lesser extent GA and CK by leaves, fruits, seeds, and similar organs is the primary mechanism whereby the activities of abscission are inhibited in abscission zones. Changes that lead to a diminution of secretion of these hormones result in a release of the abscission zone from inhibition. This view of abscission is also supported by the anatomical and cytological evidence that shows that the cells of the abscission zone are kept in a state of arrested development relative to adjacent tissues; that is, their full maturation is inhibited. When, as in some cases, the cells of an abscission zone do mature, abscission is no longer possible. This is a regular occurrence in some fruits, such as the fruits of cotton. Also inhibited is the secretion of the enzymes of abscission, which normally may be produced in small amounts; but after abscission is initiated, the secretion increases greatly.

b. Abortion and Senescence in Abscission

When an organ degenerates physiologically, it is no longer able to inhibit its abscission, and the process is initiated. This occurs commonly with abortion of buds, flowers, fruits, and other organs, and with senescence. Plants in their strategies for survival have programmed for abscission zones to be maintained in a state of arrested development with adequate substrates on hand for the cells to function in abscission at any stage in development. Thus, even though aborting buds and young fruits are physiologically weak and low in nutrients, their abscission zones are not so deficient and can function to separate the weakened organs.

Ultimately, most senescent leaves and fruits and other organs are separated by abscission. Abscission remains possible because the cells of the abscission zone are kept functional and supplied with reserve substrates. The zone can remain functional for some time after the subtending organ has come to the end of its cycle of activities. With senescence the chemical signals from the leaf or fruit change. These are detected in the abscission zone, and the activities of separation are initiated.

At one time it was suggested that the process of abscission itself was a kind of senescence. In view of the cytological and metabolic activity in the abscission zone during separation, it is difficult to consider that activity as a real phenomenon of senescence. While the activity is commonly brought on by the senescence (or abortion) of the subtending organ, I consider that the processes of separation and of development of the protective layers should be viewed as *activities* of the main plant body which have the functions of (i) separating an organ no longer useful and (ii) protecting the underlying tissues, which otherwise would be exposed to injury and infection.

3. Organ Competition and Abscission

Plants have the ability to initiate and at least partially develop many more buds, leaves, flowers, and fruits than their stem and root system can possibly supply with adequate nutrients. The result is that the organs of the plant

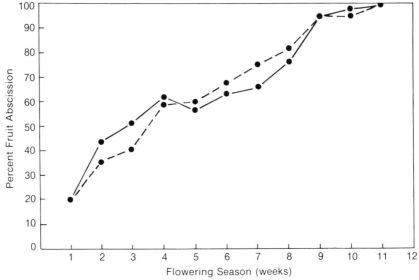

Fig. 1:1.
Changes in young fruit abscission of two varieties of cotton during the flowering season. Data are the averages of three seasons (1977–79) at the West Side Field Station, California. Note the low percentage of abscission early in the season, which rose steadily to 100 percent at the end of the season. (Redrawn from El-Zik et al., 1980)

are in competition for limited nutrients and often for exposure to light. To some degree, every organ is in competition with every other organ. Each leaf is in competition with every other leaf, particularly for nutrients, and with adjacent leaves it is in competition for light. The disadvantaged leaves lose out in the competition and are abscised. Similarly, each flower and fruit is to some degree in competition with the other flowers and fruits on the plant. Often the first-set fruits have a clear advantage and the later-formed fruits are abscised (Fig. 1:1).

There is also competition between the vegetative organs and the reproductive organs. Most species will not produce flowers until there has been a substantial amount of vegetative growth and the foliage is adequate to support at least some reproductive activity. However, some species do have the ability to flower early in their life cycle, but in such plants the flowers and fruits are likely to be abscised if they develop before a substantial degree of vegetative maturity has been attained, or if vegetative development becomes especially vigorous (see Farrington and Pate, 1981a).

Individual plants can often be in competition. In the dense vegetation of the wet tropics or in a densely planted field there can be considerable competition for light, as well as for soil nutrients. The plants that lose out in this competition will at first abscise a number of leaves, and of course, eventually, the weakest plants will not survive.

In the competition for nutrients, some organs serve as *sources* of nutrition, and others are *sinks*. During its life cycle a leaf usually functions in the early stages as a strong sink and is supplied with nutrients from roots, storage organs, and older leaves. When the leaf has fully expanded, it becomes a source supplying carbohydrate nutrients to sinks in other parts of the plant, and at this stage it remains a strong sink, attracting mineral nutrients in particular. As the leaf ages or becomes disadvantaged, its sink strength diminishes, and its nutritional reserves tend to be solubilized and exported to stronger sinks. The final step in the life cycle of a typical leaf is its abscission, the common fate of organs that are no longer either a vigorous sink or a vigorous source.

Flowers and fruits go through a similar life cycle. Often there is a conspicuous abscission of flowers or young fruits because many species produce far more flowers than they can develop into mature fruits. In many species there is continued attrition as the fruits are developing, with the weaker fruits being abscised (see, e.g., van Steveninck, 1957; Yager, 1959; Crane et al., 1973).

In organ competition the abscission zone is where the ultimate decision is made as to whether an organ is to be retained or abscised. In the abscission zone the hormonal influences of the organ meet and interact with the influences of the rest of the plant. The balance of hormonal influences at the abscission zone provides the information that tells the cells whether they should become active and abscise the organ, or remain inhibited and retain the organ. Hormones and their interactions are discussed in Chapter 4.

4. Role of Abscission in Plant Homeostasis

The concept of *homeostasis* originated in animal physiology as a convenient term for the ability of an individual to maintain relatively stable conditions within its body. Plants have a similar ability to develop and maintain a desirable balance in their developmental and physiological activities. Manifestations of homeostasis are readily apparent in plants. Regardless of the size or extent to which they have developed, plants are usually able to maintain an effective balance among the various phases of their activities. They are able to develop and maintain themselves: first, as functional vegetative individuals, and later as reproductive adults in spite of wide fluctuations in the environment and varying degrees of development in size. Homeostasis is well illustrated by those annual plants which under poor conditions produce only a few leaves and but one or two flowers and fruits, but in favorable environments will grow large, producing many leaves and many fruits.

Homeostasis also expresses itself in many less obvious ways. Leaf development is kept in balance with root development and also with available light. Flowers and fruits are kept in reasonable balance with the number of leaves available to provide carbohydrate nutrients for the fruits.

As with most aspects of plant behavior, homeostasis involves participation of a number of physiological and morphological factors. Among these, abscission has an obvious and important part. Abscission can rapidly bring the organs of a plant into balance with one another when that is required.

A corollary to homeostasis is the concept of *compensatory growth*, which often involves a delay of abscission. If a plant loses an unusually large number of buds, leaves, or fruits, the growth of the remaining structures is often stimulated, with the obvious result of at least partial compensation for the loss. A classical example of compensatory growth is the development of lateral buds after the removal of an apical bud of a plant. Other examples come to mind. If an abnormally large number of young fruit are lost or removed from a tree, the remaining fruit compensate by becoming unusually large. The ability is well known to horticulturalists and regularly utilized by fruit farmers to produce larger fruit that will bring a higher price. If most of the buds and young leaves are systematically removed from a plant, the remaining leaves can grow to be very large, and their abscission will be delayed indefinitely until long after the leaves would have fallen from a normal plant. The foregoing are but a few examples of the strong homeostatic abilities of plants.

Homeostasis in plants may not be as rigid as in animals. The homeostatic balance can be modified by nutrition, as well as by a number of other factors. For example, with some species the availability of high levels of nitrogen in the soil will lead to strongly vegetative plants and greatly delayed leaf abscission. Under such conditions cotton and tomatoes, for example, will produce few if any flowers or fruit. In partial contrast, conditions that favor the accumulation of carbohydrate in plants tend to favor the early onset of reproductive activity, and at the same time high carbohydrate favors increased fruit retention and delayed leaf abscission. Further comments on the physiological aspects of nitrogen and carbohydrates in relation to abscission are in Chapter 4.

D. BENEFITS OF ABSCISSION

The processes of abscission are sensitive to a variety of internal and external conditions, as will be discussed in Chapters 4 and 6. It is not surprising, therefore, that plants have found many ways to utilize abscission for the benefit and survival of the species and the individual. In more fashionable terminology, during their evolution plants have repeatedly utilized abscission as they evolved successful strategies of growth, development, and survival.

Table 1:2. *Benefits of Leaf Abscission*

1. Removal of senescent leaves
2. Removal of injured or infected leaves
3. Removal of excess foliage from stressed plants
4. Defoliation of deciduous trees
5. Recycling of mineral nutrients
6. Protection by development of (a) leaf scars (b) spinescent petioles
7. Vegetative propagation
8. Facilitation of pollination and seed dispersal by animal agents
9. Inhibition of seed germination of competitors (allelopathy) through the leachate from fallen leaves

In the development of an individual plant, leaf abscission is frequently utilized to maintain homeostasis among the vegetative parts of the plant. Similarly, if the species is one which typically produces far more flowers than the individual plants can develop into fruits, surplus young fruits abort and commonly are abscised. Further, leaf abscission is a strategy for removing senescent leaves and, also, injured or infected leaves. Where soil nutrients are in limited supply, the abscission of leaves of deciduous species facilitates the recycling of mineral nutrients. Abscission often involves a protective function. The development of leaf scars is an important part of the abscission process. In a few species, abscission of leaf blades with or without some of the petiole will leave behind a spinescent structure with considerable protective value. In a sizable number of species with fleshy leaves, abscised leaves can function in vegetative propagation by the development of buds and roots. Leaf abscission may precede flowering and fruiting and thus facilitate the work of animal pollinators and seed-dispersal agents. Also, abscised leaves can bring growth-inhibiting substances to the soil around the plant. When leached from the leaves, these substances inhibit seed germination of either the same or competing species. Thus, abscised leaves can have a strong effect on the composition of the plant population in the vicinity (Table 1:2).

The benefits of branch abscission are very similar to the benefits of leaf abscission, especially with respect to the removal of weak or senescent branches and to propagation of the species through the rooting of abscised branches. In addition, the pattern of abscission of branches or tips of branches can have considerable influence on the structure of a tree. For example, shoot-tip abortion that occurs in a number of woody species facilitates the termination of growth and results in a sympodial pattern of branch development. The various combinations of bud development and abscission

Table 1:3. *Benefits of Branch Abscission*

1. Termination of growth facilitated by shoot-tip abortion
2. Development of tree architecture
3. Removal of excess branch structure
4. Vegetative propagation
5. Generally—benefits as for leaf abscission

Table 1:4. *Benefits of Flower, Fruit, and Seed Abscission*

1. Removal of excess flowers
2. Removal of flower parts after function
3. Removal of aborted, diseased, or excess fruits
4. Separation of mature fruit
5. Dispersal of seeds

Table 1:5. *Benefits of Bark Abscission*

1. Hindrance or prevention of the establishment of epiphytes and parasites
2. Recycling of nutrients

of less-favored branches contributes substantially to the kinds of architecture found in trees (Table 1:3).

In the evolution of flowers and fruits abscission has been utilized in a number of ways. One way is a process adapted to the rapid removal of excess flowers. This usually occurs only after earlier flowers have set fruit. Abscission removes flower parts after they have functioned. For example, in many species the petals are abscised promptly after pollination. As with leaves, abscission is a method for the removal of aborted or diseased fruits. A very important function centers on the dispersal of seeds. With many fruits, the seeds themselves are abscised from the placenta, and their dispersal is facilitated by the dehiscence of dry-fruit walls. Other species utilize fleshy edible fruits as a device to facilitate dispersal by animals. With these, the fruit is often abscised or at least loosened to facilitate its removal and transport of the seeds (Table 1:4).

The abscission of bark occurs in a number of species. It has the evident advantage of helping to prevent the establishment of epiphytes and parasites when the abscission of layers comes at sufficiently frequent intervals. Bark abscission also contributes to the recycling of nutrients (Table 1:5).

E. THE ABSOLUTE REQUIREMENTS FOR ABSCISSION

Over the years there has been repeated speculation as to which physiological factor was actually required to bring about abscission. There has been a tendency to consider that a single change, such as heat, drought, or frost, was a universal causative factor in abscission. The early speculations were often narrowly based on limited observations, and now seem largely of historical interest. For the purposes of this introduction, it will be more valuable to consider which of the various physiological factors affecting abscission are actually essential to the process. These and other factors, which are important but secondary, are treated and their interrelations are discussed in Chapters 4–6.

So many factors can influence abscission that it has been difficult to identify those physiological factors that are essential to the process. At present, however, three physiological factors are identified that do, indeed, appear to be essential.

The first of these essential factors is a lowering of auxin levels in the abscission zone. Conversely, abscission can be delayed, usually indefinitely, by maintaining a high level of auxin at the abscission zone.

A second essential factor is oxygen. Substantial evidence shows that abscission can be accomplished only in the presence of oxygen. The evidence indicates that oxygen is required to provide energy (through ATP or its equivalent) for the synthesis and secretion of hydrolytic enzymes.

Thus, the third essential factor includes a source of energy, usually carbohydrate, and the constituents for the synthesis of hydrolytic enzymes, primarily amino acids.

Further, the hormones ABA and ETH are very important in abscission; however, their essentiality has not been established. ABA appears to be present in every species of higher plant, and it has promoted abscission in every species on which it has been tested. However, in some experiments the acceleration induced by ABA could be counteracted by other hormones and physiological treatments. Although it is a sound assumption that ABA is present whenever abscission is occurring, it cannot yet be said with certainty that it is one of the absolute requirements.

For ETH there is now considerable evidence that it is not essential to abscission even though it is a strong promotor of the process, as well as of numerous other respiration-dependent processes. Abscission still goes forward under hypobaric conditions when concentrations of ETH are infinitesimal, far below the concentrations that can accelerate abscission. Further, while ETH often increases at the time of abscission, there are situations in which ethylene synthesis is not correlated with abscission, and indeed, in which it increases only after abscission has been completed.

The physiology of abscission is the subject of Chapter 4.

F. ABSCISSION TERMINOLOGY

The terms in most frequent use are *abscission* and *abscise*. In some ways these are anachronisms whose original Latin meanings are now inappropriate for our concept of abscission processes. The original meaning of these words was in reference to the cutting off of a structure by an external agent, such as a surgeon's knife. Today we understand abscission as a process of internal physiological activity. For such a physiological process the English words *separate*, *shed*, or *cast off* would have been more appropriate. However, *abscission* and *abscise* are now in wide use in English scientific literature and are well understood. Thus, it is of practical importance to maintain the current usage and to reduce and, if possible, eliminate the usage of less well recognized synonyms in order to minimize confusion and foster clear communication.

Abscission—The separation of cells, tissues, or organs from the remainder of the plant body. The more conspicuous examples of abscission are brought about by physiological changes that weaken cell walls; but these intergrade with mechanical factors, and some abscission appears to be the result of purely mechanical forces. Abscission includes the disarticulation of stems and the fragmentation of filaments in lower plants. The absence of abscission is retention. In horticulture, the retention of leaves, fruits, or flowers is often of far greater interest than their abscission. *Abscise* is the verb for separation by abscission. This verb is now by far the most widely used in scientific writing. For clarity, one should avoid the use of related verbs, such as *shed, absciss, abscind, throw off*, and *cast off*.

Abscission Zone—Most discrete organs have at their base a region where, if abscission occurs, the cellular changes of separation will take place. This region is called the *abscission zone*. Commonly, abscission zones can be recognized because of external differences in color or shape. Internally, their anatomy and cytology are quite distinctive, as is discussed in Chapter 2.

Adventitious Abscission—Abscission that occurs in tissues away from a recognizable abscission zone in positions that are variable. Most commonly, adventitious abscission occurs across or within the blade of a leaf, or across a pedicel or internode.

Anomalous Abscission—Separation of cells or tissues that normally do not separate.

Cladoptosis—The abscission of leafy branchlets, such as those that are abscised by *Sequoia, Taxodium, Larix,* and *Pinus*. Typically, species showing cladoptosis abscise all of their leaves in that manner.

Deciduous—This term is used to describe the periodic defoliation of plants: for example, a drought-deciduous tree. In this treatise we will limit the use of *deciduous* to more or less synchronous, complete defoliation.

Other terms that are sometimes used to describe briefly retained structures include *caducous, ephemeral,* and *fugacious.* Such words should be used with care and with attention to their precise botanical meanings.

Dehiscence—The opening of a structure, typically in a predictable way. It involves the separation of adjacent parts and hence may be considered a form of abscission. In the case of fruit dehiscence, separation usually occurs longitudinally along the sutures or midrib of the carpels of a fruit. The segments that have split usually remain attached to one another.

Hormone—A substance active in very small amounts which moves from a site of production to a site of function. The net effect of a hormone is to coordinate some aspect of the growth and development of an organism. Auxin, gibberellins, abscisic acid, cytokinin, and ethylene are the major hormones of higher plants, and several dozen others have been identified.

Protective Layers—After separation of an organ, the cells remaining on the surface of the scar modify somewhat and become the *primary protective layer*. Development of the layer may involve cell divisions either before or after separation. The layer usually becomes suberized, and other substances may be deposited as well. *Secondary protective layer(s)* can develop within or beneath the primary protective layer. These usually develop much as a periderm and usually become continuous with the adjacent periderm of the stem.

Senescence—The terminal, irreversible phase in the functioning of a living entity, such as a cell, an organ, or a plant.

Separation Layer—A layer of cells within an abscission zone which becomes physiologically active and secretes the hydrolytic enzymes that weaken the cell walls and permit separation. Avoid use of the synonyms *absciss layer, abscission layer*.

Other Botanical Terms—I shall not attempt to define here the many other botanical terms that are used in this book. My usage follows the standard botanical dictionaries and Webster's Third New International Dictionary (1971). The spelling of botanical names follows Willis (1973) and Ainsworth (1971).

2. Anatomy of Abscission

Chapter Contents

A. THE ABSCISSION ZONE 23
B. SEPARATION LAYER 26
C. PATTERNS OF SEPARATION 27
D. PROTECTIVE LAYERS 28
E. DEHISCENCE 31
   1. Fruit Dehiscence 32
      a. General 32
      b. Ontogeny of Dehiscence Zones 32
      c. Circumscissile Dehiscence 34
      d. Anther Dehiscence 34
F. MECHANICAL FACTORS IN ABSCISSION AND DEHISCENCE 38
   1. Abscission 38
      a. Tissue Tensions 38
      b. Turgor Phenomena 38
      c. Physical Factors 39
   2. Dehiscence 40
      a. Drying and Shrinkage 40
      b. Turgor Mechanism 41

Von Mohl (1860a, b, c) in his pioneering examination of the abscission process pointed out two significant parts, *separation* and *protection*. In leaf abscission, the two processes go on almost simultaneously. The development of protective layers may precede or follow the development and activity of the separation layer. As will be brought out in subsequent discussion, the two processes are not necessarily linked; either can occur without the other. The cell divisions that characterize the development of the protective layer may be completely absent in some cases and the protective function accomplished by the suberization of exposed tissues. In other species the protective layers may develop across the base of a leaf and become continuous with the periderm in the complete absence of any separatory activity. It may be of interest to note in passing that the latter pattern is common in desert shrubs (e.g., *Parthenium* and *Larrea*), which grow in such arid environments that few microorganisms gain entry to the plant through the leaves, hence there is little need for abscission of moribund leaves.

A. THE ABSCISSION ZONE

At the base of most discrete organs of plants is a region called the abscission zone, within which the changes of abscission occur. It has distinguishing external and internal features. The following comments, although taken largely from descriptions of leaves, leaflets, pedicels, and internodes, are pertinent to abscission zones generally.

In leaves the abscission zone is at the base of the petiole where it joins the stem (Fig. 2:1). In internodes it is at the base adjacent to the next lower node. In compound leaves that abscise their leaflets, a leaflet abscission zone lies at the point of attachment of the leaflet to the leaf stalk (rachis). When a pulvinus is present, the abscission zone lies at the junction of the pulvinus and the supporting tissues.

The abscission zone may be swollen or constricted relative to the diameter of the subtended organ. The abscission zones of many leaves are constricted dorsi-ventrally and expanded laterally. Such expansion reaches an extreme manifestation in monocots and some dicots where the base of the leaf, including the abscission zone, completely encircles the stem. A noteworthy example is the coconut palm (*Cocus nucifera*), whose step-like leaf scars on the trunk show the unusual size of the leaf abscission zone. A

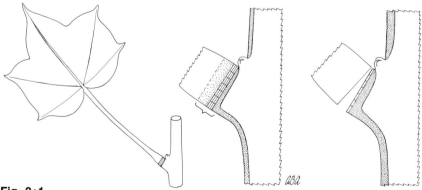

Fig. 2:1.
Diagrams of the leaf abscission zone showing typical locations and arrangements of tissues.

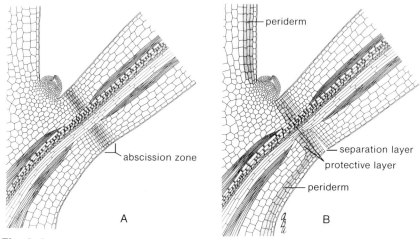

Fig. 2:2.
Diagrams of the tissues in a typical leaf abscission zone. **A.** Shortly before separation. Note that abscission zone cells are smaller and less differentiated than cells in adjacent regions. **B.** At time of abscission. Note compression of vascular tissues by the developing protective layer.

groove, or constriction, is a common feature of abscission zones. When present, it usually encircles the zone. The active processes of separation frequently start at or near the groove, sometimes a few cells distant from the most constricted part.

Cells of the abscission zone are smaller and more densely filled with cytoplasm than cells in adjacent regions (Fig. 2:2). Tracheary elements are less well developed, with a lesser deposition of lignin. Cells and structures that may exist in the tissues distal to the abscission zone (e.g., fibers, laticifers, resin canals) may be absent or much less developed in the abscission zone. For example, the mucilage canals in the pedicel of the *Hibiscus* flower do not pass through the pedicel abscission zone. In *Parthenium argentatum* the resin canals on the adaxial (upper) side of the petiole stop when they reach the abscission zone, while those on the abaxial (lower) side pass through the zone into the cortex of the stem.

Vascular bundles usually pass directly through the abscission zone. However, it is common to find vascular anastomosis in the vicinity of the abscission zone, usually immediately proximal to the zone. For example, the vascular bundles in the petiole of *Phaseolus vulgaris* are arranged as an interrupted ring, but at the leaflet abscission zone they fuse into a solid core, resembling a protostele. In this species the separation layer forms across the crotch of the fusing vascular bundles (Brown and Addicott, 1950). In other situations a kind of vascular plexus exists at or near the abscission zone. For example, the abscission zone of an internode lies immediately above the vascular anastomosis of the node immediately below. Similar evidence of vascular anastomosis has been observed in the leaflet abscission zone of *Citrus* (Livingston, 1948). Another example is in cotton where the vascular tissues in the stem adjacent to the leaf abscission zone form a kind of plexus with connections to the stipules, to the leaf proper, and to the leaf trace.

Differentiation does not proceed to full maturity of the cells and tissues in the abscission zone. The cells are shorter and less vacuolated, and deposition of cell-wall materials is much less than in tissues adjacent to the zone. Thus, it is clearly a region of arrested development. In abscission zones that have an extended life span, maturation may proceed slowly. For example, in the leaves of *Citrus*, which normally live for several years, there is slow maturation of the abscission zone on its distal and proximal periphery, so that as time progresses, the number of relatively undifferentiated cells diminishes (Livingston, 1948).

The limited cell-wall development of cells in the abscission zone gives the impression that it is a zone of weakness. However, that is often not the case during most of the life of an organ. For example, the petioles of many fully developed leaves can be broken, but the break will not necessarily come most easily at the abscission zone. On the other hand, in the mechanical abscission of dry, persistent leaves and of dead, dry branchlets, the break of separation occurs most commonly across the thin-walled cells of the abscission zone.

The early literature (e.g., Lee, 1911) makes frequent mention of lignification of tissues in and near the abscission zone. In the leaves of a number of species of deciduous trees considerable lignin may be deposited at the base of the petiole adjacent to the abscission zone (Fig. 2:3A). Also, it is common in such species for there to be some development of a protective layer(s) well in advance of separation. The protective layer also is usually lignified. Consequently, by the time the leaves of such species are mature and ready to abscise, most of the cells on each side of the abscission zone show some degree of lignification. This does not occur to an appreciable extent in herbaceous species. Indeed, the factors both of limited time and of limited lignification generally in herbaceous plants mitigate against such deposition. Although positive reactions to phloroglucinol-HCl reagent are common in the parenchyma tissues at the base of petioles of most species as abscission approaches, the reaction is no longer considered specific for lignin.

Examination of abscission zones leaves unanswered the important question as to how abscission events are programmed. It is clear that in the normal course of events chemical changes in the leaf blade or fruit initiate the processes of abscission in the abscission zone. The nature of the chemical stimuli, the mechanism by which they are perceived at the abscission zone, and the mechanism whereby the response is limited to a narrow band of

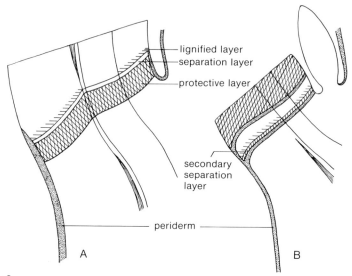

Fig. 2:3.
A. Diagram of abscission zone tissues of *Betula verrucosa* shortly before separation. Note the lignified layer distal to the separation layer. **B.** Diagram of leaf scar tissues of *Cornus sanguinea* in second spring. Note secondary separation layers that will cut off outer layers of stump. (Redrawn from Lee, 1911)

cells across the abscission zone (the separation layer) are crucial questions, to which we have as yet only fragments of information for answers. Nevertheless, these fragments are the leads from which satisfactory answers and adequate explanations seem likely to be developed.

B. SEPARATION LAYER

Within the abscission zone the physiological processes of separation are almost always restricted to a narrow band of cells, the separation layer, often only a single layer of cells thick. These secrete the enzymes necessary for the cell-wall hydrolysis and separation. In internodes and pedicels the layer approximates a plane perpendicular to the axis. The separation layers of leaves and lateral structures are usually parallel to the surface of the supporting stem, but there are significant variations, some of which are described below. The development of the separation layer may be preceded by cell divisions, or the layer may function in the complete absence of cell divisions. Commonly, as abscission approaches, cells in the abscission zone begin to divide and can give rise to several tiers of cells, of which the distal tier usually becomes the functional separation layer. Careful study of the patterns of separation-layer development in a variety of species led to the conclusion that cell divisions are not an essential aspect of abscission (Gawadi and Avery, 1950). Cell divisions, when they occur, appear to be an early manifestation of the development of protective layers.

The position of the separation layer can be predicted fairly accurately if the abscission zone is very short and the transition from the organ to its supporting structure is abrupt. Closely delimited separation layers are found between the leaflets and the rachis of compound leaves, as in legumes (Brown and Addicott, 1950) and in the leaves of *Sambucus* (Osborne and Sargent, 1976a). In species where the abscission zone is not confined morphologically, the position of the separation layer cannot be predicted with accuracy, and it appears possible to modify the position by treatments which would have the effect, among others, of modifying the length of time before abscission during which the cell divisions can occur. In cotton the position of the pedicel separation layer is variable. If restricted wholly to the base of the pedicel, the abscission layer will be approximately circular; but it often extends into the adjacent stem and extends for some millimeters below the pedicel attachment. In other instances the separation layer includes tissue both above and below the pedicel attachment and leaves an oval scar after detachment. Occasionally, the separation layer will pass from the base of the pedicel down through the whole length of the next internode and cut off a long segment of cortex that remains attached to the pedicel (Lloyd, 1920b).

Adventitious separation layers can develop in positions outside of the recognized abscission zones. The precise locations vary with the nature and intensity of the initiating factors. Adventitious abscission is discussed in Chapter 3.

C. PATTERNS OF SEPARATION

Considerable variation exists among species as to the part of the separation layer in which active changes commence. In leaf abscission the usual pattern is for separation to commence at the abaxial (lower) surface and progress upward across the petiole (Lee, 1911). It is noteworthy that in leaf abscission of cotton the normal pattern can be reversed by application of GA and other chemical regulators which induce abscission to commence at the adaxial surface (Bornman et al., 1967; V. T. Walhood, p.c.). Separation activity of axial organs such as internodes and pedicels appears to commence at the periphery and move inward (Gilliland et al., 1976; Jensen and Valdovinos, 1967). In other situations, such as leaflet abscission of the primary leaf of beans, separation often commences centrally in the crotch of vascular bundles at the abscission zone and moves outward. Lloyd's (1916) careful observations of internode abscission in *Mirabilis jalapa* disclosed that the first steps in abscission occurred in the innermost cells of the cortex at two points in a plane normal to the plane of the opposed leaves. From these points, abscission changes developed outwardly, inwardly, and around the stem, developing more rapidly toward the center of the pith than toward the epidermis. Separation in the pith was usually completed before the epidermis ruptured. Internode abscission of *Impatiens sultani* commences immediately above the axillary bud and progresses to the opposite side of the stem (Lloyd, 1914).

In a detailed analysis of the leaf abscission pattern of *Impatiens sultani*, Sexton (1979) found separation first detectable at 13 h (after deblading) in a small group of cortical cells just beneath the adaxial surface of the abscission zone. From that location, middle-lamella breakdown spread downward (abaxially) and laterally. The collenchyma and epidermis along the sides and abaxial surface of the abscission zone and the central vascular tissues were the last to separate. To determine if separation was a contagious phenomenon, requiring cell-to-cell contact in order to spread, Sexton dissected the abscission zone into as many as thirty-six sections. He found that each section was independently capable of separation, which demonstrated that the pattern of abscission is not a contagious phenomenon.

D. PROTECTIVE LAYERS

The primary protective layer develops immediately proximal to the separation layer. In herbaceous species it often consists of little more than the tissues exposed by separation which deposit suberin and lignin in the outermost cells, and there may be little or no cell division (Gawadi and Avery, 1950). However, in most species, cell divisions occur proximal to the separation layer, usually commencing well before abscission. The result of such activity is the development of a kind of periderm that usually becomes continuous with the periderm of the stem. Sometimes the outermost cells of the protective layer collapse shortly after exposure to the air, but still form what appears to be an effective protective layer. If water relations and other conditions are favorable, the exposed cells of the primary protective layer expand in response to the turgor pressures within them. But even under these conditions, cells of the primary protective layer eventually become suberized and a part of the periderm.

One or more secondary protective layers can develop beneath the primary protective layer. Secondary protective layers are characteristic of species of *Cornus, Tilia, Gleditsia*, and many other genera (Fig. 2:3B). In some cases a separation layer also develops distal to the secondary protective layer by means of which the primary protective layer and some contiguous tissues are abscised. With few exceptions, the secondary protective layers across leaf scars become continuous with the periderm of the stem (Lee, 1911).

The development of protective layers often involves some expansion of parenchyma tissues, which tends to crush and close off the functional phloem.

One of the regular features of abscission is the development of tyloses in the xylem vessels near the protective tissues. They develop in the usual way by the expansion and intrusion of the protoplasts of xylem parenchyma. The intensity of tylosis varies with the species, being less pronounced in herbaceous species, and tyloses may appear to only a limited extent in the xylem distal to the abscission zone (Bornman, 1967b).

In addition to xylem vessels, resin canals can also become plugged at the abscission zone. The epithelial parenchyma of the abaxial resin canals of the leaf of *Parthenium argentatum* hypertrophies in the abscission zone as leaf abscission approaches. The new tissue plugs the resin canal with a mass of parenchyma cells that deposit suberin and lignin. The plug becomes continuous with the protective layer and the periderm of the stem (Addicott, 1945).

Many, possibly all, laticifers also are closed off during abscission. Lee (1911) describes cross walls formed in the laticifers of the abscission zone just before the leaf fall in *Morus alba*, *M. nigra*, *Ficus carica*, and *Broussonetia papyrifera*. Parkin (1900) found that laticifers in the petioles of *Hevea brasiliensis* and *Plumeria acutifolia* are open through the abscission zone in young leaves, but become blocked in the abscission zone as the leaves reach full maturity. The blockage in these instances appears to result from pressure of expanding parenchyma cells in the abscission zones.

Leaf scars have an appearance that is often characteristic of the species with a distinctive outline and pattern of bundle scars (Figs. 2:4–2:6). If separation has involved a rapid and complete hydrolysis of middle lamellae, the scar will present a smooth appearance. On the other hand, if hydrolytic activity is minimal or absent, the subtending organ may break away and leave a relatively rough surface. Thus, something of the physiological nature of the separatory processes can be estimated from examination of the surface of a scar. Many tree ferns show little evidence of physiological involvement in the separation of their leaves. In a number of species the dead leaves droop and hang about the trunk in a rather massive skirt. In other tree fern species the dead leaves break away, leaving the base of the petiole attached to the trunk. A few species show what appears to be a physiological separation and the activity of a separation layer at the very base of the petiole. A rather unusual manifestation appears in *Cyathea medullaris*, the black tree fern of New Zealand. The dead leaves droop and hang for awhile along the trunk, but eventually they break away, leaving a ragged scar to which stumps of vascular bundles 1–3 cm long are still attached. The protective

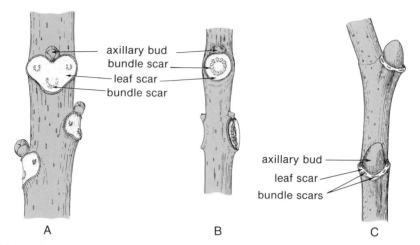

Fig. 2:4.
Leaf scars of autumnal deciduous species. Note characteristic scar shapes and arrangement of vascular bundles. An axillary bud is immediately above each scar. **A.** *Juglans regia.* **B.** *Catalpa bignonioides.* **C.** *Platanus acerifolia.* (Courtesy of Weier et al., 1974)

Fig. 2:5.
Some distinctive leaf scars. **A.** *Carica papaya.* (Photograph courtesy of Jean A. McKinnon) **B.** An aroid in the Botanical Garden, Bogor, Indonesia. Note that the leaf bases completely encircle the axillary buds. (Photograph courtesy of W. R. Philipson)

Fig. 2:6.
Leaf scars of palms. **A.** *Cocos nucifera*. (Photograph courtesy of Jean A. McKinnon) **B.** *Roystonea regia*. Note the separation of leaf sheath and abscission in progress at the leaf base.

layers in the outermost surface of the stem become heavily lignified. The projecting stumps of vascular bundles remain attached until polished off by some mechanical force. Commonly, this occurs six to eight feet below the crown of the tree where the hanging dead leaves rub against the stem. In this case the smooth leaf scars are the result of mechanical action rather than of physiological activity, as might first be imagined. When leaf scars develop rapidly and vigorously, they become effective barriers to the invasion of pathogenic organisms. Species in which the protective layer(s) develop slowly or weakly are often susceptible to invasion (see Chapter 6). In rare cases, as in *Theobroma cacao* infected with *Phytophthora*, the leaf scar itself may become a site of development and dissemination of the pathogen (Gregory, 1974).

E. DEHISCENCE

The changes in dehiscence zones are so similar to those that take place in abscission zones that the phenomena of abscission and dehiscence must be considered two aspects of the same basic process. The similarity is strongest in the nature of separation. Typically, dehiscence involves breakdown and separation of parenchyma after middle-lamella and cell-wall degradation.

However, there are numerous differences. In the higher plants the morphology of dehiscent sutures differs in many respects from that of the typical leaf or fruit abscission zone. Mechanical factors have an important role in facilitating dehiscence, but they usually have only a minor part in abscission. Also, in dehiscence protective layers do not develop, nor are they needed.

1. Fruit Dehiscence

a. General

Dehiscence of capsules and other dry fruits shows a surprising variety of patterns. For most, separation occurs either at the sutures (septicidal dehiscence) where the carpels have fused, or along a line of weakness in the midrib of the carpel (loculicidal dehiscence). This dehiscence can be restricted so that only a relatively small pore opens. More commonly, the separation is extensive and the entire suture separates, usually remaining attached only at the base.

A number of genera show circumscissile dehiscence. Such abscission resembles leaf and stem abscission in that the separation layer comes at right angles to vascular tissue of the fruit wall and cuts through it.

Dehiscence patterns are often uniform within a taxon and, consequently, are valuable characters for taxonomic identification.

b. Ontogeny of Dehiscence Zones

The fruits of higher plants consist of one or more carpels, each of which is homologous with a leaf. In the early development of many fruits the carpels are distinct, but the edges touch and fuse in the course of fruit development. The degree of fusion varies greatly from one that is loose, with the appressed epidermis readily detectable, to one in which the margins of the carpels can hardly be detected (Puri, 1951). Stages in such a fusion are shown in Figs. 2:7, 2:8. Consequently, when septicidal abscission takes place, it may involve a simple loosening of the appressed cell walls, or more extensive breakdown in other cases. In the example of *Firmiana* shown in Fig. 2:9, it appears that separation has been largely the result of middle-lamella dissolution with some small amount of breakage of cell walls (Yen, 1932). General observations of dehiscing fruits, particularly fruits that dehisce before they are completely dry, indicates that separation is usually facilitated by middle-lamella dissolution. As yet, however, there is little cytological evidence on this separation in the literature.

There has been even less attention given in the literature to loculicidal dehiscence. Here the anatomical changes of separation take place within what is essentially a midrib of a highly modified leaf. The smooth scars that are characteristically exposed in loculicidal separation indicate weakening and at least partial dissolution of middle lamellae in a highly localized zone of dehiscence.

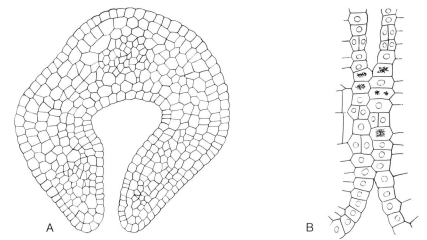

Fig. 2:7.
Ontogeny of the dehiscence zone of the follicle of *Firmiana simplex*. **A.** Before the fusion of the carpel edges. **B.** During fusion of the carpel edges. (Redrawn from Yen, 1932)

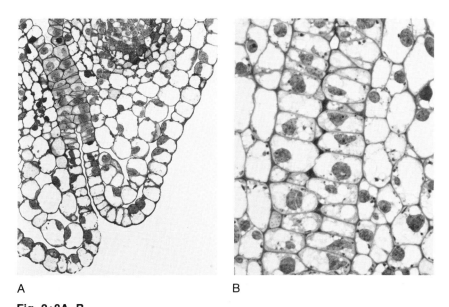

Fig. 2:8A, B.
Carpel-fusion region of *Caltha palustris* stained with toluidene blue O. (Photomicrographs courtesy of R. L. Peterson)

34 *Anatomy of Abscission*

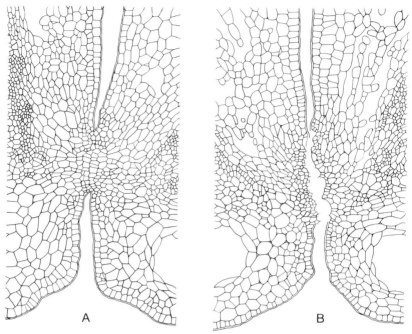

Fig. 2:9.
Suture of the mature follicle of *Firmiana simplex*. **A.** Before separation. **B.** After separation. Note that most cell walls at the fracture are intact, indicating that separation was mainly through the middle lamellae. (Redrawn from Yen, 1932)

c. Circumscissile Dehiscence

This dehiscence, while accomplishing the same end as septicidal and loculicidal dehiscence, is morphologically distinct. An abscission zone may be apparent early in the ontogeny of the fruit, as in *Hyoscyamus*, or it may not be apparent until much later. For example, in *Anagallis* and *Plantago* the abscission zone does not become evident until the fruit approaches full size (Fig. 2:10). In the few examples that have been studied, separation may involve considerable dissolution, such as Rethke (1946) observed in *Plantago pusilla*, where the separation layer consists of small, recently divided cells whose separation involves both cell-wall dissolution and some hypertrophy (Fig. 2:11). In other cases separation appears to involve a limited amount of pectic dissolution and some breakage of cells, much as is illustrated for *Firmiana* (Fig. 2:9).

The fruits of the Brazilian sapucaia (*Lecythis ollaria*, *L. usitata*, and related species) produce edible nuts that are also a source of oil. The fruits are

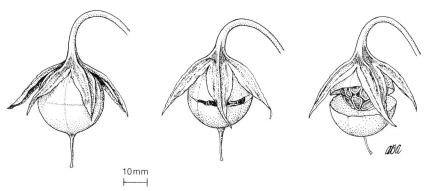

Fig. 2:10.
Circumscissle dehiscence of the fruit of *Anagallis arvensis*.

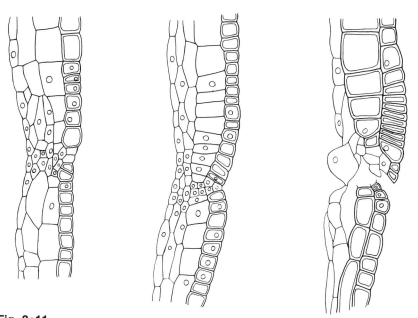

Fig. 2:11.
Anatomy of circumscissle dehiscence in the fruit of *Plantago pusilla* at three developmental stages. Note that the small cells of the dehiscence zone appear to enlarge and separate to accomplish dehiscence. (Redrawn from Rethke, 1946)

Fig. 2:12.
Circumscissle dehiscence of the fruit of *Lecythis usitata*. (Photograph courtesy of Chicago Natural History Museum)

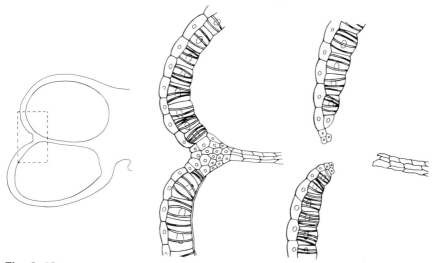

Fig. 2:13.
Anatomy of anther dehiscence in *Memecylon*. Note that many of the small cells in the dehiscence zone have disappeared by the time of dehiscence; cf. Fig. 2:11. (Redrawn from Venkatesh, 1955)

several inches in diameter and more or less spherical, as are other fruits in the Lecythidaceae. Incidentally, the size and weight of these fruits make it hazardous to be beneath the trees when the fruits are being abscised. What is remarkable about sapucaia is its circumscissile dehiscence, which abscises a woody operculum two or three inches in diameter at the distal end of the fruit (Fig. 2:12). This abscission permits ready access to the seeds that otherwise would be unavailable to animals other than man (Prance and Mori, 1977).

d. Anther Dehiscence

Anthers are morphologically very different from fruits, and it is not surprising that the ontogeny of the dehiscence zone is different. Basically, an anther is a highly modified leaf with its own peculiar tissues. The anatomy of dehiscence has been studied in a few cases. *Memecylon* (Fig. 2:13) appears fairly typical. Cells in the dehiscence zone remain small, and by the time of opening, some of the cells have completely disappeared, indicating considerable hydrolytic activity. Opening of the anther walls is facilitated by dehydration and the special thickenings of wall cells. Anther dehiscence by pores is common in the Ericaceae, where the plate of cells covering the pore appears to shrink and fall inward, permitting the pollen grains to escape (R. Sexton, unpubl.) (Fig. 2:14). As with fruits, there are many variations both

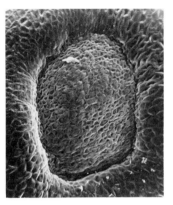 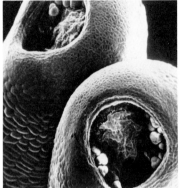 

Fig. 2:14.
Scanning electron micrographs of anther-pore dehiscence in *Rhododendron* sp. (Ericaceae). Note that the plate of cells that covers the pore collapses inward. (Courtesy of R. Sexton)

in the gross morphology of anther dehiscence and in the anatomical details (see Stanley and Kirby, 1973).

F. MECHANICAL FACTORS IN ABSCISSION AND DEHISCENCE

1. Abscission

a. Tissue Tensions

As plants develop, cells of the various tissues tend to develop at somewhat different rates, and the osmotic relations of tissues can vary considerably. In our early research with both *Citrus* and beans we observed that if the abscission zone was cut cleanly with a razor at right angles to the axis, the exposed surfaces promptly became concave, a result of the relative tensions and pressures at the leaflet abscission zone. These tissue tensions were particularly conspicuous in the bean-leaflet abscission zone where the anatomical observations showed that abscission commenced in the central tissues and frequently resulted in a lens-shaped separation (Brown and Addicott, 1950). Such tissue tensions are common in the abscission of herbaceous material that is relatively young. In woody plants or those that have had an opportunity for the tissues to mature, and at the same time to deposit heavy cell walls, tissue tensions are much less noticeable.

b. Turgor Phenomena

Shrinkage due to loss of turgor is a significant mechanical factor in abscission, particularly of leaves and floral parts. One of the effects of abortion and senescence is membrane breakdown and relatively rapid water loss.

As the distal tissues shrink and contract, separation at the abscission zone is facilitated.

When water relations are favorable in the tissues proximal to the abscission zone, the cells tend to maintain their turgor and size. Further, parenchyma cells in the proximal portion of the abscission zone (primary protective layer) may elongate or expand as the abscising tissues fall away, the expanding cells often taking on a clavate shape. At one time it was believed that this "turgor mechanism" was a major causal factor in abscission (Pfeiffer, 1928). This view has been in disfavor for some years with the increasing emphasis on the role of cell-wall dissolution as the central factor in abscission. However, recent detailed studies indicate that in favorable circumstances at least, turgor-induced changes in cell shape at the time of abscission can, indeed, facilitate separation (Sexton and Redshaw, 1981).

The explosive fruit abscission of *Ecballium elaterium* is an impressive example of the force of turgor. An abscission zone is present between the edge of the fruit and the peduncle (Figs. 2:15, 2:16) (Jackson et al., 1972). When the fruit is ripe, osmotic pressures build up to very high levels, and in due course the abscission zone gives way, permitting the explosive discharge of the rather fluid fruit contents and the seeds (von Guttenberg, 1926).

Other kinds of explosive opening of fruits involve dehiscence rather than abscission and are discussed in a following section.

c. Physical Factors

Abscission is facilitated and often accelerated by physical factors such as wind, accumulation of ice, and the like. These factors can bring about earlier abscission of leaves, fruits, and branches that would normally fall unassisted.

In cases where the physiological processes of separation move slowly or

Fig. 2:15.
The squirting cucumber, *Ecballium elaterium*. Fruit and peduncle before and after abscission. (Courtesy of Jackson et al., 1972)

40 *Anatomy of Abscission*

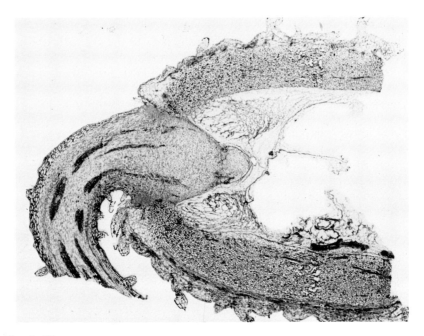

Fig. 2:16.
Section of the fruit base of *Ecballium elaterium* shortly before abscission. (Courtesy of Jackson et al., 1972)

are incomplete, wind is the usual agent that removes the organs. Similarly, where there is little or no physiological activity, but where the abscission zone is structurally weaker than the subtending parts, the physical agents are the effective ones in abscission. For many tree species, this is the usual way in which branch abscission takes place.

Brodie (1955) has described a number of interesting examples in which raindrops activate a springboard dispersal mechanism. Such mechanisms are frequent in seed dispersal of the Labiatae, including species of *Salvia*, *Ocimum*, and *Prunella*. In *Kalanchoë tubiflora* the plantlets of vegetative propagation are borne on small, strap-like extensions of the leaves. Raindrops striking the plantlets depress the strap-springboards, which, as they return to their normal position, throw the plantlets for distances up to 1.5 m.

2. Dehiscence

a. Drying and Shrinkage

For most dehiscent fruits, opening of the sutures is achieved by shrinkage of tissues between the sutures. The events in the opening of a cotton fruit are typical (Fig. 2:17). When the fruit is fully mature, the loculicidal sutures open slightly and remain that way until atmospheric moisture falls to the point that the cells of the ovary wall begin to dry. This loss of moisture is accompanied by shrinkage and some curling back of the segments of ovary wall. This continues until the surface of the wall has been reduced to less than half its original area, exposing the seeds and their fiber. Among other species, the many variations in extent and pattern of opening are related to the anatomical arrangement of the various tissues in the fruit wall (von Guttenberg, 1926).

In the dry explosive fruits, such as legumes, the drying of tissues proceeds to the point of high tensions within the fruit wall. At the same time, there is relatively little activity in the dehiscence zones, which resist opening for a considerable time. Consequently, when they do break apart from the strong tensions of the fruit wall, the curling back is extremely rapid and forceful (Fig. 2:18).

An interesting variation occurs in the dry fruits of many Acanthaceae which require some re-exposure to moisture to dehisce explosively. When the fruit matures, the middle lamellae holding the valves together do not weaken sufficiently to permit dehiscence. However, upon brief exposure to moisture the middle lamellae soften and weaken sufficiently to permit the fruit to open explosively (Bremekamp, 1926).

b. Turgor Mechanism

A number of fruits dehisce explosively as a result of turgor pressures within the fruit. *Impatiens* is the classical example. The dehiscence zones of

Fig. 2:17.
Stages in loculicidal dehiscence of a capsule: cotton, *Gossypium hirsutum*.

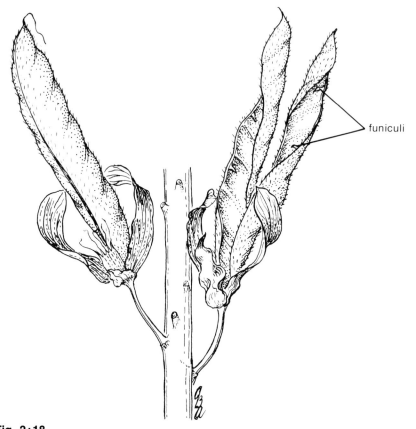

Fig. 2:18.
Explosive dehiscence of a legume, *Lupinus polyphyllus* (before and after). Note funiculi, at edge of the open carpel, from which the seeds have been thrown.

the capsule are very weak, and as the fruit matures turgor builds within the fruit wall. The cell structure of the tissues of the fruit wall is such that when the dehiscence zones finally give way, the capsule wall curls back with great rapidity (Fig. 2:19), expelling the seeds (Overbeck, 1924; von Guttenberg, 1926).

Another impressive example of explosive dehiscence is that of *Arceuthobium* and related dwarf mistletoes, where the fruit is one-seeded with a dehiscence zone near its base. Here, also, turgor pressures build within the ripe fruit. The explosion projects the seeds at a rate allegedly too fast to be seen by the human eye. Study of the phenomenon with photographic instruments disclosed that several species project their seeds with initial velocities of 24 m/sec (Kuijt, 1969).

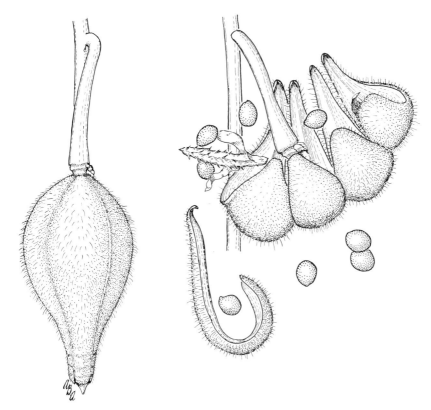

Fig. 2:19.
Fruit of *Impatiens balfouri* before and after dehiscence.

3. Abscission Behavior

Chapter Contents

A. LEAVES   45
B. BUDS, BRANCHES, AND STEMS   52
    1. Buds   52
        a. Shoot-tip Abortion and Abscission   52
        b. Abscission of Adventitious Buds   54
    2. Branches   54
        a. Cladoptosis   54
        b. Woody and Herbaceous Branches   56
    3. Stems   61
        a. Main Stems   61
        b. Disarticulation   62
        c. Abscission of Injured Internodes   62
C. FLOWERS AND FLOWER PARTS   63
D. FRUITS AND SEEDS   68
    1. Young Fruits   68
    2. Mature Fruits   70
    3. Seeds   71
    4. Disseminules   71
E. BARK   72
    1. General   72
    2. Sheets   74
    3. Strips   74
    4. Scales   74
    5. Prickles   77
    6. Root Bark and Cortex   77
F. ROOTS   77
    1. General   77
    2. *Azolla* Root Abscission   79
    3. Root Cap   80

G. ABSCISSION AND ARCHITECTURE   81
  1. Morphological Factors in Plant Architecture   81
  2. Extent of the Involvement of Abscission   81
     a. Entire Stem   81
     b. All Branches   83
     c. Lower Main-Stem Branches   83
     d. All Branchlets (Cladoptosis)   83
     e. Shoot-tip Abscission   83
     f. Stem Disarticulation   83
     g. Attritional Abscission   83
  3. Examples of Tree Architecture Influenced by Abscission   84
     a. General   84
     b. Umbrella Trees   85
     c. Sympodial Trees   86
  4. Further Reading   89

H. ADVENTITIOUS AND ANOMALOUS ABSCISSION   89
  1. Seasonal Manifestations of Adventitious Abscission   89
  2. Induced Adventitious Abscission   92
  3. Anomalous Abscission   96

A. LEAVES

Leaf abscission appeared in the Devonian and was a well-established phenomenon in the Carboniferous, as the fossil record indicates. It was the rule among the more immediate ancestors of present-day higher plants. Those plants that do not abscise their leaves appear to have lost the ability in the course of recent evolution, and other deviations from the usual pattern would likewise appear to be of relatively recent origin (see Chapter 9).

The length of time that a leaf remains attached before it is abscised varies greatly from a few weeks to a few years. Leaves that appear early in the growing season may likewise be abscised early; for example, a cotton plant growing under agricultural conditions abscises its cotyledons in about four weeks. As development progresses through the growing season, the main-stem leaves are abscised at fairly frequent intervals correlated with the development of the leafy branches. By the end of the flowering season, four months from germination, the main stem and branches will have abscised all but their uppermost leaves. In contrast, an evergreen tree such as *Citrus* may hold its leaves in functional condition for three or four years, and some of the pines for much longer. Trees of *Pinus longaeva* commonly retain their needles for 25 to 35 years, and in at least one individual, for over 40 years (D. K. Bailey, 1970 and p.c.).

46 *Abscission Behavior*

Fig. 3:1.
Ephemeral leaves (and leaf scars) on a succulent *Euphorbia* sp.

In the normal course of events, older leaves fall first and younger leaves last. Indeed, for most deciduous trees, leaf abscission commences during the summer in a limited way with the abscission of the leaves that developed first in the season as they become disadvantaged in competition with later-developing foliage. Similarly, in evergreen trees the normal course of events is for the oldest leaves to fall before the younger ones. There are a few exceptions to this rule. One is *Fagus sylvatica*, which in its autumnal defoliation first abscises the oldest and youngest leaves on a shoot, and later abscises the middle leaves.

Leaf abscission is no respecter of size. The great leaves of palms (Fig. 2:6) abscise almost as effectively as the ephemeral leaves of cacti and euphorbias (Fig. 3:1). In compound leaves the abscission of leaflets commonly commences before the abscission of stalks of pinnae or the main petiole. This tendency is shown even when the compound leaf consists of one leaflet, such as the primary leaf of *Phaseolus vulgaris* and the leaves of *Citrus sinensis* and *Parthenocissus tricuspidata* (Brennan, 1966). In a highly divided leaf a prodigious number of abscission zones can be functioning almost simultaneously. From the descriptions in Bailey (1935), the leaf of *Acacia decurrens* var. *mollis* can have as many as 4,800 (leaflets and) abscission zones, and the leaf of *Acacia filicina* can have over 6,000!

Cotyledons are abscised in much the same manner as leaves. Cotyledonary abscission zones have been used extensively in physiological investigations, and their behavior appears similar to that of young foliage leaves. Cotyledons show several patterns of morphological and functional behavior, and correlated with this are differences in abscission behavior. In many species the cotyledons remain beneath the ground and do not emerge (hypogeal germination). For many other species, the elongation of the hypocotyl raises the cotyledons above the surface of the ground (epigeal germination). In some cases the cotyledons expand and function as foliage leaves (e.g., cotton). In other cases they remain compact and shrivel as reserve substances are translocated away (e.g., beans). In both those kinds of epigeal germination the cotyledons eventually abscise, usually persisting for only a few weeks, but sometimes much longer. An interesting modification of epigeal germination occurs in several tropical genera, including *Strombosia*, *Durio*, *Rhizophora*, and *Dipterocarpus*. In these cases the cotyledons are raised above the surface of the ground by the hypocotyl, but are abscised well before their reserves have been fully utilized (durian germination) (Fig. 3:2) (Ng, 1978).

The spines on the long shoots of the Fouquieriaceae consist of the petiole and a slicing portion of the midrib that remains after the blade has been abscised away (Henrickson, 1972). This is a remarkable exception to the usual rule that abscission zones lie at approximately right angles to the axis of the organ (Fig. 3:3).

In other genera the petiole may persist as a spinescent structure after the leaf blade dries and falls away—e.g., *Sarcocaulon*, *Combretum*, and *Astragalus* spp. In some palms the bases of the petioles remain as spinescent structures that serve as deterrents to climbing herbivores.

A number of species have developed mechanisms for the abscission of only a portion of the blade of a leaf. In species of *Narcissus*, *Galanthus*, and *Leucojum*, and undoubtedly other genera in the Amaryllidaceae, the bases of leaves form a bulb. As the leaves senesce at the end of a growing season, the distal portions of the leaves above the swollen bulbous portions abscise by a separation layer, and protective layers are developed (Parkin, 1898). Similar abscission has been described in *Ammocharis*, *Crinum*, and *Cybistetes* (Milne-Redhead and Schweickerdt, 1939).

48 *Abscission Behavior*

Fig. 3:2.
Cotyledon abscission in the durian pattern of germination that shows seemingly premature cotyledon abscission, *Strombosia javanica*. (Courtesy of F. S. P. Ng)

In the leaf abscission of the arborescent Cycadaceae the senescent leaves droop and eventually die and decay to a point a few cm from the cortex where a separation layer develops and abscises the remains of the petiole, leaving a relatively smooth scar over a protective layer. The closely packed stumps of the petioles form an armor about the trunk. Beneath the first-formed protective layer new separation layers develop in succession and cut off thin sheets of petiole tissue, with the result that the older portions of the trunk have a smaller diameter than the younger portion (Chamberlain, 1935).

In some instances the development of divided and compound leaves involves abscission processes. The best-known example is that of *Monstera* and related genera whose leaves develop lobes and holes early in the development. This is brought about by necrosis of patches of cells (Pfeiffer, 1928). A very different kind of separation takes place in the development of the pinnae of palm leaves. In these the blade splits very early in its ontogeny in either a pinnate or a palmate pattern (Pfeiffer, 1928). Current cytological investigation is disclosing that the separation of the pinnae involves pectic dissolution in a manner that appears essentially identical to abscission (D. R. Kaplan, p.c.).

Leaf abscission for the apparent purpose of vegetative propagation is common in a number of genera, particularly in the Crassulaceae. In these plants the abscission zone is relatively weak and the leaves are usually fleshy. Thus, abscission is easily accomplished, and the water and reserve materials in the leaves are often sufficient to support the initiation and development of a new plantlet.

In genera that are commonly deciduous in the autumn a few species retain an appreciable number of marcescent (apparently dead) leaves through the winter and abscise them in the spring (Fig. 3:4). The blade and most of

Fig. 3:3.
Abscission of the leaves of the long shoots of *Fouquieria*. Note that lower portions of the petiole and midvein remain as a spine.

Fig. 3:4.
Marcescent leaves on *Quercus* sp. Photograph taken in March 1979. Note autumnal deciduous *Quercus* sp. in the right background, and an evergreen *Quercus* in the left background.

the petiole die in the autumn, but the tissues at the very base of the petiole, including the abscission zone, remain alive and hold the marcescent leaf on the trees through winter. Tison (1900) described the anatomy of such abscission in *Carpinus betulus*, *Fagus sylvatica*, *Quercus hispanica*, and *Q. pedunculata*. Abscission of marcescent leaves of the American species *Quercus coccinea*, *Q. velutina*, *Q. marilandica*, *Q. rubra*, *Fagus grandifolia*, *Ostrya virginiana*, and *Acer saccharum* has been studied by Berkley (1931), and that of *Quercus palustris* and *Q. coccinea* by Hoshaw and Guard (1949). Examination of the abscission zones showed that in some cases abscission commenced in the autumn and then was arrested, presumably by the advent of cold weather. In any event, the abscission processes of marcescent leaves follow an anatomical pattern essentially identical with that of leaves that are abscised in the autumn (Berkley, 1931). The retention of marcescent leaves through the winter appears to be a juvenile character, at least in some species. Young trees of *Fagus sylvatica* in Denmark retain their leaves through the winter, but the retention disappears with age (Schaffalitzky de Muckadell, 1961). A similar tendency is shown by oaks in the United States to the extent that young trees retain a larger percent of their leaves through the winter than do older trees.

The abscission of marcescent leaves is interpreted as a kind of delayed autumnal abscission, delayed in part because the species concerned are slow to initiate autumnal abscission, and in part because cold weather slows metabolic activities. With the higher temperature of spring, the process of abscission is resumed and goes to completion. It is possible that hormones from the developing buds may contribute to abscission of marcescent leaves, as they appear to do in the case of vernal leaf abscission. Hoshaw and Guard (1949) observed in *Q. coccinea* that, although the majority of the leaves had fallen by the time the buds began to swell, some abscission did not occur until after that.

At the end of its life the typical marcescent leaf withers, dies, and remains attached to the parent plant indefinitely. This habit is common in ferns, grasses, and palms, and in herbaceous annual plants generally. In woody species the marcescent leaves may eventually break off, but the separation is due to mechanical forces and does not involve physiological (biochemical) weakening of cell walls in the abscission zone. For example, on the desert shrub *Parthenium argentatum*, marcescent leaves do not wither appreciably but become dry and brittle. In due course, the dry leaves are broken off by such mechanical factors as wind. Commonly, the leaf breaks from the stem at the abscission zone; this is the weakest part of the leaf because of generally thin cell walls and absence of fibers (Addicott, 1945). Some marcescent leaves of *Quercus* spp. may also be broken off during winter. In this case the relatively brittle petioles snap at their narrowest point, and the petiole stumps are abscised later by physiological action in the abscission zone (Hoshaw and Guard, 1949).

Trichomes (epidermal hairs) of many types are widely distributed among the families of higher plants. Often, these are easily detached, particularly as the leaf matures. Two well-known examples are *Quercus margaretta* and *Lithocarpus densiflora*, whose leaves are conspicuously pubescent when young, but abscise the trichomes by the time the leaves are mature (J. M. Tucker, p.c.). The species of *Quercus* vary greatly in the kinds of trichomes on their leaves and in the rates at which the different trichomes abscise (disappear) as the leaves mature (Hardin, 1979). In *Ledum groenlandicum* the two kinds of trichomes on the upper surface are abscised shortly before the leaves are fully expanded. Separation is achieved "by disintegration of the pectic portion of the walls of the unvacuolated cells at the base of the hair . . . cellulose and cuticle remained intact, and the break invariably came at the middle lamella" (Sifton, 1963).

B. BUDS, BRANCHES AND STEMS
1. Buds
a. Shoot-tip Abortion and Abscission

Growth of shoots is an intermittent process for most woody plants; periods of active growth alternate with periods of rest or dormancy. In temperate climates this alternation tends to be annual, correlated with alternations in temperature or rainfall or both. In tropical and subtropical climates there may be more flushes. For example, *Xanthophyllum curtisii* of Malaya regularly has two flushes of growth each year; *Citrus* may have three or four. As development slows at the end of a period of growth, any of several kinds of change may take place at the shoot-tip according to the species. In temperate regions the most familiar change is the development of bud scales that enclose and protect the shoot-tip during dormancy. At the end of the dormant period the shoot-tip resumes its growth. This pattern of shoot growth is called monopodial. The usual alternative to that pattern of growth involves the abortion of the shoot-tip, followed shortly by its abscission. In *Citrus*, for example, growth of the terminal few millimeters ceases, but adjacent leaves continue to expand. An abscission layer develops in the stem immediately distal to the node of the last vigorous leaf, and the aborted shoot-tip is abscised (Fig. 3:5). In *Ulmus americana* the separation layer develops and functions without previous cell divisions. A protective layer develops with cell divisions shortly after separation has been completed (Millington, 1963). When growth is resumed, it is the axillary bud of the terminal leaf that develops and becomes the apical bud of the resulting shoot. This pattern of growth is called sympodial. It occurs in *Salix, Betula, Corylus, Castanea, Ulmus, Platanus, Robinia, Rhamnus, Tilia, Catalpa,*

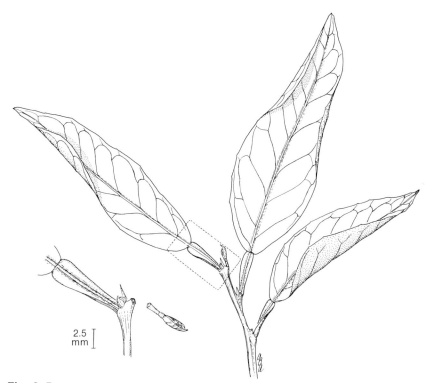

Fig. 3:5.
Abscission of aborted shoot-tip, *Citrus sinensis*.

Xanthophyllum, and *Crataeva*, among many others (Koriba, 1958; Romberger, 1963; and Millington and Chaney, 1973). The amount of shoot growth prior to tip abortion and abscission varies with growing conditions, particularly temperature, moisture, and nutritional factors, so that growth will be less and shoot-tip abscission will come earlier when conditions are adverse and will be delayed when growing conditions are favorable. For example, in experiments with *Salix pentandra*, shoot growth was arrested and tip abscission initiated by short photoperiods and by growth-retarding chemicals. In contrast, growth continued and tip abscission was greatly delayed after application of GA (Junttila, 1976). The length of the aborted bud varies from 2 or 3 mm in *Citrus* to about 15 mm in *Ulmus*, which can include seven to eight young leaves (Millington, 1963). In *Tilia americana*, shoot-tips up to 4 cm long are aborted and abscised (Romberger, 1963).

Sympodial growth habits can also result from the development of a flower at the apex of a shoot followed by the abscission of the flower or fruit and the resumption of shoot growth from the activity of an axillary bud. A variety of growth patterns can arise from variations in the pattern of growth following such tip abortion (Hallé et al., 1978). Trees of striking form can result, such as the "pagoda" form of *Alstonia scholaris* and the seemingly dichotomously branched crown of *Sapium discolor* (Koriba, 1958, and see sec. G.).

b. Abscission of Adventitious Buds

A number of genera produce adventitious buds that function as vegetative propagules. Commonly, these are easily abscised. *Bryophyllum* is well known for the buds produced on its leaves at the tips of veins. Other genera, including *Agave*, *Polygonum*, *Saxifraga*, *Juncus*, and *Cardamine*, also produce buds on various parts of the plant (Pfeiffer, 1928).

In *Kalanchoë tubiflora* the adventitious buds are cup-like and develop at the end of narrow outgrowths from the edge of the parent leaf. When a raindrop strikes the cuplet squarely, it depresses the outgrowth, which then acts as a springboard. In its rapid return to its normal position the springboard throws the bud for a distance of up to 1.5 m (Brodie, 1955).

2. Branches

a. Cladoptosis

Cladoptosis is the abscission of discrete leafy branchlets, the leaves and stem falling as a unit. Typically, cladoptosis is characteristic of the deciduous, coniferous genera *Taxodium*, *Metasequoia*, and *Larix*. The deciduous species of these genera abscise all of their small branchlets annually. Evergreen species also show cladoptosis, but abscise only a portion of their branchlets each year. These include *Sequoia* (Fig. 3:6A), *Sequoiadendron*, *Libocedrus*, *Cupressus*, *Thuja*, *Juniperus*, and *Araucaria*. A number of variations in the pattern of cladoptosis occur among these genera; for example, the branchlets shed by *Sequoia* are often compound and commonly include the flushes of growth for a period of two or three years (Stark, 1876). Other genera similarly shed leafy branches having one to several years' growth.

At the abscission zone the cortical tissues at the base of the branchlet are usually enlarged. As abscission approaches, the separation layer and protective layers develop. In *Metasequoia* the development of the protective layer precedes that of the separation layer and forms across the proximal portion of the branchlet abscission zone. Eventually, the protective layer joins with the periderm in the adjacent stem (Böcher, 1964). In *Taxodium distichum* the separation layer develops toward the end of the growing season. Active cell division leads to a layer seven to twelve cells thick. Eventually, the middle lamellae of the most proximal cells of the separation layer are dissolved, enabling the cortical cells to separate. The xylem is relatively weak, and the branchlet can break away rather easily at this stage. No evidence was observed of weakening of the vascular tissue. Apparently, it became brittle as

Fig. 3:6.
A. Cladoptosis, abscission of leafy branchlets: (1) *Sequoia sempervirens*, (2) *Taxodium distichum*, (3) *Metasequoia glyptostroboides*. **B.** Abscised twigs of (1) *Quercus suber* and (2) *Zelkova serrata*. Note the swollen base and smooth abscission scar of the *Quercus*, whereas the *Zelkova* twigs break away irregularly at the base.

the branchlet desiccated (C. Barnard, 1926). In general, the vascular tissues show little evidence of having been weakened, but because of the separation of the cortical tissues, abscission of the branchlets takes place readily and at an accelerated rate if mechanically agitated by factors such as wind (Stark, 1876). In *Pinus* the needles occur in fascicles of one to five (according to the species) on very short branchlets. When the needles are abscised, the fascicle falls as a unit. Some of the needles of *Cedrus* occur on short shoots which can be abscised as units. The foregoing are the most common examples of cladoptosis. With the evergreen trees, cladoptosis is a method of removing the weaker and disadvantaged foliage and commonly takes place in the autumn or after a period of dry weather.

b. Woody and Herbaceous Branches

Many tree species abscise their small weakened branches and thereby accomplish a kind of self-pruning (Fig. 3:6B). The anatomy of the abscission varies considerably. Often the base of each small branch is swollen (Fig. P:1), and the tissues are very low in lignin.

A detailed description of the anatomy of the branch base of *Agathis australis* was made by Licitis-Lindbergs (1956). The conspicuous swelling at the base of each small branch is largely due to the expansion of the cortex by meristematic activity of elongated islets of dividing cells. The pith is also expanded within the swollen region. The xylem tapers abruptly to less than half its diameter as it approaches the abscission zone (Fig. 3:7). Separation of the cortical tissues and development of a secondary protective layer (periderm) take place before actual abscission of the branch. Although Licitis-Lindbergs did not study the histology of xylem separation, she did describe the xylem stumps that project from the exposed abscission surface of a fallen branch. These are tapered and rounded and typically covered with adhering portions of the periderm. These observations indicate that considerable softening and separation of the xylem occurred before the branch actually fell from the tree. A stub of the branch trace remains in the cortex of the tree trunk after abscission of the small branch. Much later it is abscised with a flake of bark (Licitus-Lindbergs, 1956). This abscission of a sizable segment of xylem vascular tissue could only have been accomplished by the hydrolytic activity of living cells in a separation layer. Such observations are interpreted as further evidence of physiological separation of lignified tissues. In a similar study of the anatomy of branch abscission in *Perebea mollis* and *Naucleopsis guianensis*, Koek-Noorman and ter Welle (1976) examined the histology of the abscission zone and the regions on each side. In the abscission zone the xylem differed from that on either side by the presence of shorter, unlignified fibers with small, simple pits. Similar observations have been made on the abscission zones of small branches of *Populus* (Höster et al., 1968). In common with all other abscission zones, the branch ab-

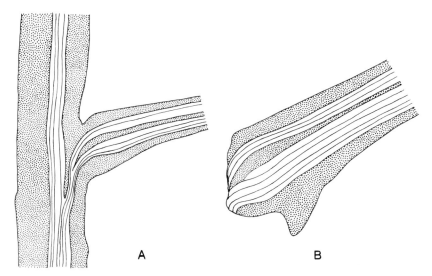

Fig. 3:7.
Branch base of *Agathis australis*. **A**. Intact. **B**. After abscission. (Redrawn from Licitis-Lindbergs, 1956)

scission zones of *Quercus robur* have unusually high levels of acid phosphatase (Böhlmann, 1970).

The size of abscising branches varies widely among the higher plants. Perhaps the smallest branches (daughter fronds) are those of the Lemnaceae (*Lemna*, *Spirodela*, *Wolfia*). Because of the herbaceous nature of the tissues and the ease with which the growth and development can be studied, considerable attention has been given to the group. *Spirodela*, for example, abscises the daughter fronds by activity in an abscission zone lying where the daughter frond joins its stalk (Fig. 3:8) (Witztum, 1974; Newton et al., 1978). Another kind of abscission that can be considered abscission of a small branch is that of the axillary buds of the avocado (*Persea americana*). These buds abscise readily when subjected to stress, a characteristic that makes successful grafting very difficult for the horticulturalists (Chandler, 1950). Axillary buds may also abscise from the intact plant, thereby affecting the growth habit.

For the most part, branch abscission affects smaller branches, 1 to 2, or at most a few, cm in diameter. There is very little information on the maximum size of branches that can be abscised; possibly this is because most of such manifestations involve rain-forest trees that occur in regions little frequented by trained botanists. The largest branch abscised of which I am aware was observed by a colleague (J. Dransfield, p.c.). While standing beneath a large tree of *Anthocephalus chinensis* in a North Sumatra forest, he

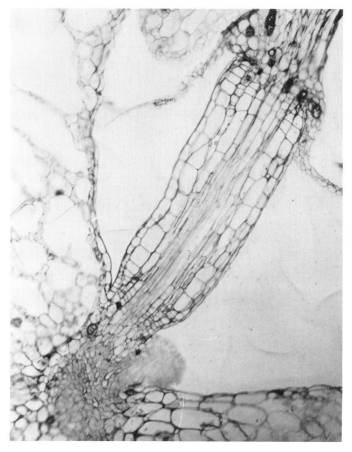

Fig. 3:8.
Branch abscission zones in *Spirodela polyrhiza*. (Courtesy of Newton et al., 1978)

narrowly missed being struck by a branch falling from the high umbrella-like crown. The branch was approximately 10 cm in diameter and 5 m long. From the growth habit of *A. chinensis*, such branch abscission must be a regular occurrence. Better evidence of the abscission of large branches is found in the paleobotanical literature. Various Lycopods were already abscising sizable branches in the Carboniferous period. Leaf scars 9 to 12 cm in diameter occur on specimens of *Bothrodendron*. Similar scars and specimens, which can only be interpreted as abscised branches, have been described for several other genera as well (C. Barnard, 1926; Jonker, 1976).

In branch abscission there appears to be a variable amount of weakening in the abscission zone through the activity of enzymes. However, enzymatic activity is not always complete, and the assistance of a mechanical factor such as wind is usually required to abscise branches of such genera as *Populus* and *Quercus*. In some species the small branches are sometimes abscised with a few green leaves attached. This is the case with *Agathis australis*, *A. dammar*, and *Podocarpus falcatus*, for example, and see van der Pijl (1953). In other species separation does not take place until after all of the leaves have fallen and the branch is dead. In these cases fairly strong wind is required to dislodge the branch. Some species, of which *Zelkova serrata* is a good example, lack an active abscission zone, but still abscise many small dead branches (Fig. 3:6B2), and an occasional live one, because the tissues at the base of the branches are weaker than the distal portions. Almost every good wind will break off some such branches. This type of self-pruning can be very efficient and can keep a tree almost completely free of dead branches. Such abscission leaves a ragged stump, which offers ready access for microorganisms of decay. Unless the species is one that produces resinous or other antiseptic agents, the tree may suffer severely and eventually die from the activity of microorganisms invading the wounds.

Interesting examples of abscission of dead branches occur in *Eucalyptus*. Four phases of branch abscission have been recognized by M. R. Jacobs (1955). These occur in differing degrees among the *Eucalyptus* species, and in some other genera as well. In the first phase the wood at the base of the branch becomes brittle. This brittleness is correlated with desiccation of the branch, and possibly with some activity of fungi. In the second phase of dead-branch abscission the branch breaks at the brittle zone and falls away, leaving a stub projecting from the main stem. (A tree with many such stubs is shown in Fig. 3:30.) Over a period of two or more years, the portion of the branch that is within the stem (branch trace) erodes and is replaced by a soft layer of gum containing no xylem or organized structural elements (Ewart, 1935). This erosion of the xylem elements of the branch stub weakens it. At the same time, pressure of the bark against the outer portions of the stub eventually break it away from its connections to the xylem in the main stem. In the fourth phase the branch stub becomes so loose that it falls away from the main stem (Fig. 3:9). Ewart concluded that the erosion of the lignified branch trace is brought about by the activity of lignase(s). His evidence strongly supports the involvement of lignases in abscission.

A less well developed system of abscission of small living branches occurs in some species of *Salix*, particularly *S. fragilis*. The base of the small branches is thickened by enlargement of the cortex. Both the cortex and the phloem lack fibers, which are abundant elsewhere in the plant, and the xylem is less lignified in the region than elsewhere. These structural weaknesses facilitate the rupture of the branch when it is subjected to mechanical stress (Eames and MacDaniels, 1947).

Two methods whereby the branch stumps or scars become healed have been described in the preceding paragraphs. Additionally, in *Salix* the stub

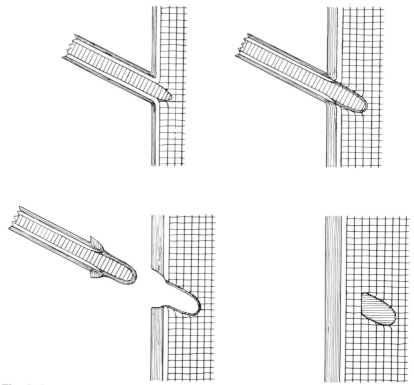

Fig. 3:9.
Diagrammatic representation of branch disarticulation (abscission and scar healing) in *Eucalyptus* sp. at 0, 2, 3, and 5 years. (Redrawn from Ewart, 1935)

of the abscised branch persists, and is eventually buried by secondary growth of the stem. If the small branches of *Salix* or *Populus* escape abscission, additional increments of xylem over a period of a few years strengthen the zone so that it is no longer fragile (Eames and MacDaniels, 1947).

From time to time, large branches ranging up to 50 cm in diameter break away from trees. This phenomenon has been observed repeatedly among ornamental trees in regions of California having a desert climate. Typically, the branches break a short distance from the main trunk, but no evidence of disease, decay, or other possible causal factors has been detected (R. W. Harris, p.c.). Such branch abscission has occurred in *Platanus acerifolia, Cedrus deodara, Ficus retusa,* and more than one species of *Ulmus, Eucalyptus, Pinus,* and *Quercus*. Although no cause for this abscission has been determined, it has always affected trees growing in well-watered areas, in or near lawns, and usually occurs late on a hot (35–40°C), still summer afternoon. Large-branch abscission has also been observed in southern England, Australia, and South Africa, but the observations have not yet been published (R. W. Harris, p.c.).

3. Stems

a. Main Stems

The entire above-ground portions of a plant can abscise when mature. This habit is characteristic of the herbaceous plants known as "tumbleweeds," and appears to have evolved independently in several families, including Chenopodiaceae, Amaranthaceae, Asteraceae, Boraginaceae, Brassicaceae, and Fabaceae. Typically, there is an abscission zone at the base of the plant which permits separation and enables the entire plant to function as a disseminule. In *Psoralea argophylla* (Fig. 3:10), one or more separation layers form in the base of the stem, and abscission takes place after the dissolution of pectins in the middle lamellae (Becker, 1968). In other species, such as *Kochia scoparia* (and presumably *Salsola iberica*, Fig. 3:11), cell walls in the abscission zone are largely unlignified, while other parts of the stem become normally lignified. After the plant dies and desiccates, the abscission zone becomes brittle, but the rest of the stem remains tough because of its lignification. The consequence is that the entire plant can be broken off by the wind to form a tumbleweed. In such species there is no evidence in the abscission zone of physiological hydrolysis of pectins or celluloses. However, abscission is facilitated when a soil fungus (*Rhizoctonia* sp.) attacks the abscission zone (Becker, 1978).

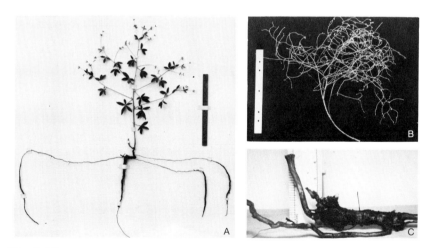

Fig. 3:10.
Psoralea argophylla. **A.** An intact plant. **B.** A mature shoot abscised from its underground stem. The leaves abscised earlier. **C.** An underground stem showing a branch stump from which the shoot has abscised. (Courtesy of D. A. Becker)

Fig. 3:11.
A mature tumbleweed, *Salsola iberica*, broken from its root by the wind. This species does not have a separation layer.

b. Disarticulation

Stems of a number of plants can disarticulate through the action of an abscission zone that is typically located immediately distal to a node. Such abscission is common in *Ephedra*, *Euphorbia*, *Crassula*, and other genera. A similar disarticulation occurs in the Cactaceae, but in this family each disarticulated "joint" consists of many nodes.

c. Abscission of Injured Internodes

It has been known for many years that a number of species will abscise internodes that have been injured. Such internode abscission was studied by Lloyd in 1914 and 1916 in *Mirabilis jalapa* and *Impatiens sultani*. After injury to an internode, an active separation layer develops at its base adjacent to the subtending node. Abscission involves the hydrolysis of middle lamellae of one or more cell layers. Lloyd found no evidence of cell-wall breakage in *Impatiens*; separated individual cells would float away from the sections.

The physiological control of internode abscission appears to be similar to, if not identical with, that of leaf abscission. The abscission can be prevented by application of IAA (Beal and Whiting, 1945). An interesting related observation was made by Carns et al. (1961), who found that GA applied to the stem stump (first internode) of the excised cotyledonary node of cotton would often induce abscission of the stump. The observation is noteworthy because internode abscission does not occur in field-grown cotton plants. The anatomical changes of the stem abscission are similar to those in the cotyledonary abscission zone (Bornman et al., 1968).

C. FLOWERS AND FLOWER PARTS

The abscission of flowers and flower parts varies as greatly as does the abscission of other organs. Flowers of some species open on the day of anthesis, and the corolla withers or abscises or both a few hours later. In other species flower parts may persist for many days before withering or abscission. In still other species flower parts are never abscised.

The flower of *Magnolia grandiflora* is typical of those that abscise all of their flower parts. The bracts which tightly enclose the flower bud split and abscise before the flower opens. Shortly thereafter, the stamens are abscised, and after about two days the petals abscise (Fig. 3:12). When the seeds are mature, the follicles dehisce, the seeds abscise, and the dry fruit is abscised at the base of the peduncle. Most flowers go through some combination of abscission similar to that of *Magnolia*.

The calyx may abscise very early or may persist until the fruit is mature. In *Eschscholzia* the calyx is abscised as a calyptra as the petals begin to open (Fig. 3:13). In *Eucalyptus* the woody calyptra consists of fused sepals or petals or both (Fig. 3:14); its morphology varies with the species (Penfold and Willis, 1961; Pryor and Knox, 1971). When the fruit of *Citrus* is mature, it abscises from the receptacle, leaving behind the persistent five-pointed calyx button (Fig. 3:15). In *Cryptantha circumcissa* the calyx persists until the fruit is mature, and then the outermost portions (approximately 60 percent) are abscised in a circumscissile manner.

As was mentioned above, petals of some species are abscised early in the life of the flower. *Vitis* presents an extreme example of this. In this genus the calyx is a small torus, and the corolla, which encloses the anthers and pistil in the bud, abscises as the flower opens. The petals separate from the receptacle and from each other at their bases, but remain attached at their tips, the corolla falling as a star-shaped calyptra (Fig. 3:16). Species of *Linum*, *Geranium*, and *Erodium* normally abscise their petals the afternoon of the day the flower opens. Others, such as *Digitalis* and *Gossypium*, do not nor-

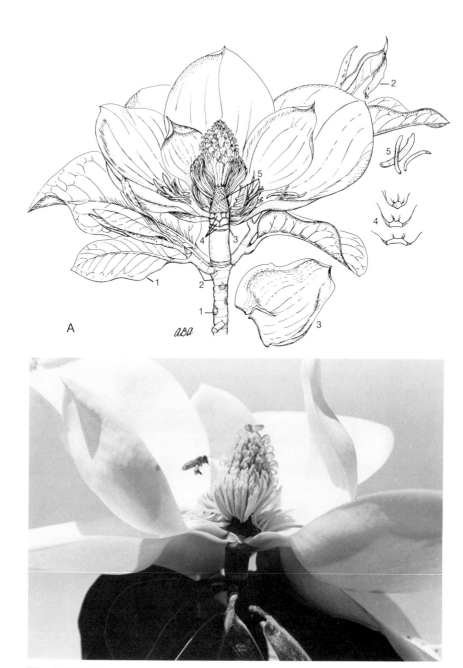

Fig. 3:12A, B.
Flower and branch of *Magnolia grandiflora* showing (1) leaf and leaf scar, (2) stipule and stipule scar, (3) floral bract and scar, (4) petals and petal scars, (5) stamens and stamen scars. (Photograph by Sheila A. Reed)

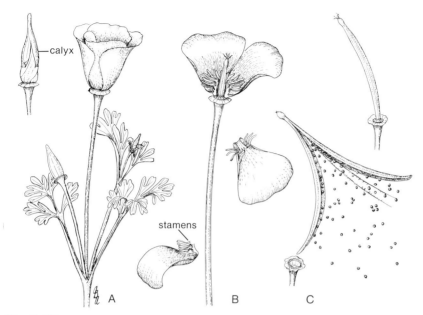

Fig. 3:13.
Abscission and dehiscence in the flower and fruit of *Eschscholzia californica*. **A.** The calyx is abscised as a calyptra. **B.** A portion of the stamens is abscised with each petal. **C.** The mature fruit abscises at about the time of dehiscence.

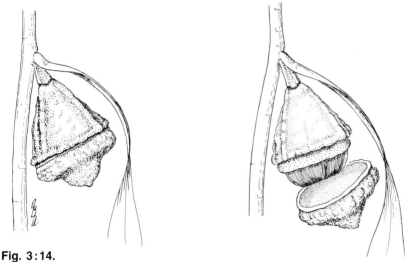

Fig. 3:14.
Calyptra abscission of *Eucalyptus globulus*. (In various species of *Eucalyptus*, the calyptra consists of calyx or corolla or both.)

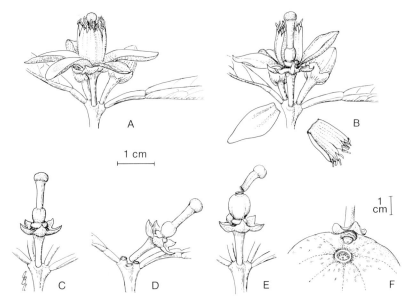

Fig. 3:15.
Abscission of the flower and fruit of *Citrus sinensis* cv. Washington Navel. **A.** Flower at anthesis. **B, C.** During and after petal and stamen abscission. **D.** Young fruit abscission with scar of pedicel abscission. **E.** Style abscission of a young fruit that has set. **F.** Abscission of a mature fruit, between fruit and receptacle.

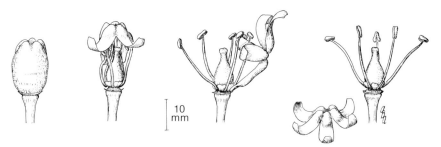

Fig. 3:16.
Corolla abscission of *Vitis labrusca*. Note that the petals abscise at their base but remain united at their tip, forming a star-shaped "calyptra."

mally abscise the corolla until the day after anthesis. The petals of *Eschscholzia californica* persist for several days, opening in response to light and closing by late afternoon. Typically, on the fifth day the flowers open and abscise the petals during the day. Some corollas do not abscise at all, they simply wither in place. The petal abscission zone may be a fairly substantial region, as in the large petals of *Magnolia grandiflora* (Fig. 3:12); or more commonly, the abscission zone lies in a narrow, constricted base of the petal, as in many spp. of the Linaceae, Papaveraceae, and Geraniaceae (Fitting, 1911; Wiatr, 1978). The cytological changes in petal abscission may proceed in the larger petals much as they do in leaf abscission—e.g., *Magnolia grandiflora* (Griesel, 1954). In the more ephemeral petals, cell separation can actually begin before the flower bud is fully opened and the petal is fully expanded—e.g., *Linum lewisii* (Wiatr, 1978). The attachment of some corollas is so tenuous that jarring or shaking the plant will cause the corolla to be abscised. Darwin (1888) noted that "many species of *Verbascum*, when the stem is jarred or struck by a stick, cast off their flowers." Flowers of species of *Veronica* and *Cistus* are similarly sensitive. The mechanism of this rapid abscission has not been studied, but it likely involves the early onset of cell separation, probably well started before the corolla is fully open, as in *Linum lewisii*. Possibly also, the physical stimuli induce some kind of rapid turgor change.

Some corollas abscise poorly or not at all. For example, the corolla of an *Aloë* species is pushed off by the developing fruit (Fig. 3:17).

Stamens are commonly abscised after the pollen has shed. Larger sta-

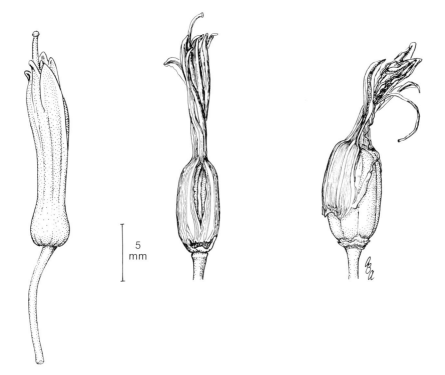

Fig. 3:17.
Corolla abscission in *Aloë* sp. Note that separation is largely mechanical by the developing fruit.

mens will leave a conspicuous scar on the receptacle—e.g., *Magnolia grandiflora* (Fig. 3:12). Others, such as the anthers of *Cecropia*, are abscised in two stages. The first involves breakdown of cortical tissues of the filament, after which the pendulous anthers remain loosely attached by a bundle of stretched spiral walls of the tracheary elements (Berg, 1977). Stamens also may remain attached and wither in place.

The calyx, corolla, and stamens may be abscised as a unit, as in *Anagallis*, Labiatae, and Scrophulariaceae, or they may be abscised in various adherent combinations. Floral-cup abscission in the *Prunus cerasus* is shown in Figure 3:18.

The pistil usually remains intact, with the style and stigma withering in place. However, some abscission does occur; for example, in *Geum hirsutum* the stigma and upper portion of the style abscise, leaving the lignified basal portion, which is curved and functions in seed dispersal (von Guttenberg, 1926). In *Citrus* the style usually abscises (Fig. 3:15) (Goldschmidt and Leshem, 1971), and when it persists, it can become the object of unusual attention. For example, in Jewish tradition a citron of good color and shape with a persistent style is considered a symbol of perfection.

D. FRUITS AND SEEDS

1. Young Fruits

Many species produce far more flowers than they can possibly develop into fruits. The excess fruits or potential fruits may be abscised at any time during development, but abscission takes place most commonly (i) in the early flower-bud stage, (ii) before flower opening, (iii) as the young fruit are developing, and (iv) at fruit maturity. In cotton, toward the end of the growing season or under adverse conditions, there is considerable abscission of young flower buds. Usually there is relatively little abscission at the time of anthesis, but during the period five to ten days after anthesis many varieties abscise approximately two-thirds of the young fruit. This response is largely a reflection of the nutritional status of the plant. Early in the season there is very little abscission of young fruit, but as flowering progresses the percentage of abscising young fruit increases (see Fig. 1:1), and the average abscission under most conditions is about 65 percent of the flowers. In cotton, no further abscission of fruit occurs; fruit that are retained after ten days remain firmly attached.

The ratio of flowers abscised to fruits retained and matured can be very high. For example, the avocado can have as many as 1,000 flowers in an inflorescence from which rarely more than 2 will develop into mature fruit. The aborted flowers and their supporting stalks disarticulate (abscise) shortly after anthesis. The mango has from 300 to 3,000 flowers in an inflorescence, depending on the variety, and develops only a few of these to mature fruit. The remainder are abscised (Chandler, 1950).

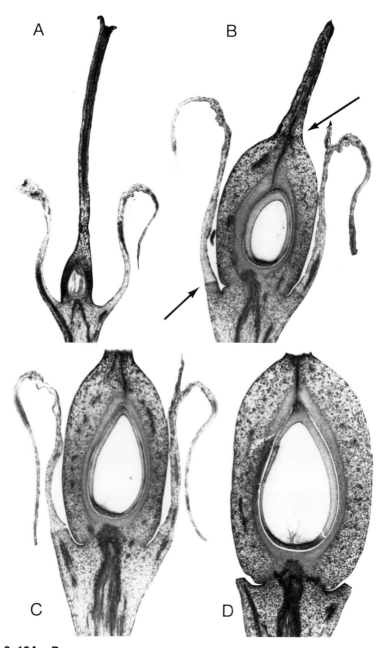

Fig. 3:18A – D.
Style and floral-cup abscission in the cherry, *Prunus cerasus* cv. Montmorency.
(Courtesy of Lott and Simons, 1966)

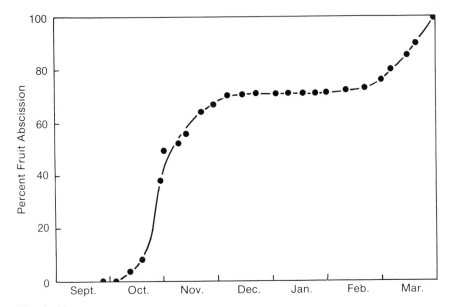

Fig. 3:19.
Time-course of fruit abscission in the peach, *Prunus persica* cv. Golden Queen, at Tatura, Australia. (Redrawn from Chalmers and van den Ende, 1975)

Cultivated tree-fruit varieties often abscise appreciable numbers of young fruit early in development (Fig. 3:19). In the Northern Hemisphere, this is referred to as June drop. The drop can be sufficiently extensive in some varieties of apple and pear to greatly reduce the cost of hand thinning. June drop appears to be the result of competition for nutrients among the fruits, there simply being insufficient nutrients available in most trees to support all the fruits that are retained after flowering. Under some conditions, water stress also appears to be a factor that can promote abscission of young fruit (Lloyd, 1920a). Varieties that retain high numbers of young fruits must be thinned chemically or by hand so that the remaining fruits will reach a desirable size (Chandler, 1951; and see Chapter 10).

2. Mature Fruits

Usually at maturity fleshy fruits are abscised, but dry fruits are retained until after dehiscence and release of seeds. Abscission is utilized in many ways to facilitate seed dispersal. Abscission of fleshy fruits may take place between the ovary and the receptacle—as in *Citrus*, peaches, plums, olives, mango, and some cherries (Fig. 3:20)—or at the base of the pedicel, as in apples, pears, and figs. Most nuts and acorns abscise between the nut and the receptacle.

3. Seeds

Seeds abscise from the funiculus in most dry fruits. The abscission zone is typically composed of thin-walled cells that permit the seed to break away with a minimum of stress. Studies of the anatomy of seed separation have disclosed only limited evidence of physiological activity in the seed abscission zones (Hintringer, 1927; Fahn and Werker, 1972). After seeds have separated from the funiculus in dry dehiscent fruits, they are commonly dispersed with the help of various agents, such as wind and animals.

4. Disseminules

There are a number of interesting cases where seeds are not abscised from the fruit, but where, nevertheless, abscission separates a "disseminule," which effects the dispersal of the seeds. Examples of such cases include beets (*Beta vulgaris*), in which the dried fruit and the attached base of the perianth abscise to form a disseminule. Another is the loment, a dry leguminous fruit which at maturity abscises transversely into segments, each containing one seed. In the case of the mulberry the fleshy flower parts remain attached to the ovary and are abscised as a unit, which is very attractive to birds. In other species the disseminule includes a dry spinescent calyx which facilitates dispersal—e.g., various Lamiaceae (Labiatae). In the foregoing cases, abscission takes place either immediately below the receptacle or at the base of the pedicel. Similar abscission is common in the Asteraceae, where often the compact inflorescence is abscised. In the grasses (e.g.,

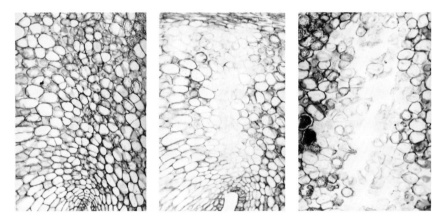

Fig. 3:20.
Anatomical changes in the fruit abscission zone of a sour cherry, *Prunus cerasus* cv. Montmorency. Note reduced affinity for hematoxylin in the separation layer as abscission progresses. (Courtesy of Stösser et al., 1969)

Aegilops), either the fruiting spikes or spikelets may be abscised (Gould, 1968). In *Panicum capillare* the large, spreading panicles usually break away when mature, forming a wind-carried disseminule. In *Allium cratericola* the entire fruiting umbel is abscised from the peduncle and can be moved considerable distances by the wind. In *Salicornia pusilla* the stems disarticulate at maturity. Each segment (composed of a node, an internode, and fruit) forms a disseminule. A great many other examples could be cited where appreciable portions of the inflorescence and sometimes even of the vegetative portions of the plant bearing fruits are abscised and function as disseminules.

Perhaps the most spectacular example of abscission of vegetative structures to assist in the dissemination of seeds is found in the tumbleweeds, in which the entire above-ground portion of the plant is abscised at maturity (see sec. B.3.a.).

Another remarkable disseminule is the seedling of the mangrove. In *Rhizophora* and related genera the seed regularly germinates in the fruit, and the seedling grows to be several inches or more long before the disseminule (fruit with seedling) is abscised at the fruit pedicel abscission zone. The disseminule shown in Figure 3:21 is adapted to penetrate the mud where it falls; other mangrove disseminules are adapted to float on the tidal waters.

E. BARK

1. General

The abscission of bark is a conspicuous characteristic of many species, although it is far from

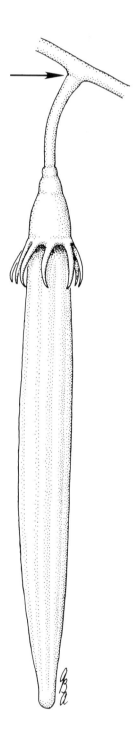

Fig. 3:21.
Disseminule of the mangrove, *Brugiera gymnorhiza*, shortly before abscission. It is about 16 cm long. At this stage, the seedling is largely hypocotyl; the cotyledons remain embedded in the ovary. Separation from the parent plant takes place at the pedicel abscission zone (arrow).

universal. Of the species that do not abscise bark, some will build up rather massive layers of accumulated periderm, as in *Quercus suber*, and the even more massive layers of *Sequoia* and *Sequoiadendron*. Others (e.g., *Citrus* spp.) may keep their bark very thin, sometimes only a millimeter or two thick, by the crushing and resorption of secondary phloem as it presses against the thin periderm.

The abscission of bark takes place in several ways. It may involve the shedding of only small amounts of thin, papery tissue, or it may include sizable volumes. For example, at the base of older trees of *Pinus ponderosa*, heaps of abscised scale bark may accumulate to heights approaching two feet (Fig. 3:23). In Australia the accumulated bark on the ground under mature trees of *Eucalyptus viminalis* has been estimated to amount to 110 tons per hectare (M. R. Jacobs, 1955). Such accumulation as these are, of course, only possible in relatively dry climates where decay and disappearance of plant materials proceeds slowly.

The anatomy of bark, including its abscission, has received considerable attention from plant anatomists. Pfeiffer (1928) summarized much of the early literature, and Eames and MacDaniels (1947) have an excellent introduction to the anatomy and abscission of bark. Chattaway (e.g., 1953, 1955) has published an extensive series of studies on the many kinds of bark in *Eucalyptus*.

The anatomical investigations have disclosed no evidence of physiological activity in connection with bark abscission. Rather, the separation appears to be entirely mechanical and a consequence of the tissues of the bark containing areas of thick-walled cells alternating with areas of thin-walled cells. Bark abscission, when it occurs, results from the rupture of the thin-walled cells. The various patterns of bark abscission are a consequence of the pattern in which the thick-walled tissues are laid down relative to the thin-walled ones. In some cases bark can properly be described as deciduous, essentially all of it being abscised annually. In other cases only a portion of the bark will be abscised in any one year. The abscission is usually seasonal, related to growth and weather. Typically, it occurs in the summer or early autumn. In *Eucalyptus*, for example, "shedding occurs commonly in the early summer when growth is resumed and enormously after any particularly hot, dry spell of summer weather" (Chattaway, p.c.). The two mechanical factors that appear most important in bark abscission are the expansion of the trunk during a period of growth, and the shrinkage of dead bark tissues from desiccation.

The patterns of bark abscission vary greatly among the many species of trees and shrubs. The patterns have been classified variously, but the arrangements published to date have been largely a matter of convenience. For the purpose of this book, we will consider three general patterns: abscission of sheets, abscission of strips, and abscission of scales and plates.

2. Sheets

A number of species abscise very thin sheets of bark. In some genera, such as *Arbutus* and *Arctostaphylos*, the entire bark is shed annually by the peeling of thin, papery sheets (Fig. 3:24A). In other genera, such as *Commiphora* and *Bursera*, sizable sheets will peel away; but the abscission is patchy, affecting only parts of the stem at any one time. *Arbutus* and *Arctostaphylos* are two prime examples of what are commonly termed smooth-bark species.

3. Strips

Many species abscise strips and sheets of bark that are more complex anatomically than the thin, papery bark mentioned above. The strips may be several millimeters thick and contain considerable fibrous tissue. With some species strips of bark are abscised annually, and the species also may be described as smooth barks—for example, the smooth-bark species of *Eucalyptus* (Fig. 3:22). With other species the outer cylinder of bark loosens more or less completely each year. These are sometimes called ring-bark species, and they are found in *Vitis*, *Clematis*, and commonly in the Cupressaceae. A number of other species abscise strips of bark in irregular patterns, and any portion of the bark may remain for several years before abscising. *Fuchsia excorticata* is an excellent example.

4. Scales

Scales and plates are abscised in a wide variety of sizes and shapes. A notable example of scale bark is that of *Pinus ponderosa* (Fig. 3:23) in which the layers show irregular patterns reflecting the patterns of thick and thin-walled cells in the bark. Other species of *Pinus* produce somewhat different scales. *Carya ovata* abscises elongated scales. Various species of *Platanus* abscise large, irregular plates of bark (Fig. 3:24B).

The ability to abscise bark tends to decline with age, so that on older trees of *Arbutus* and the smooth-bark *Eucalyptus* species, bark tends to build up on the lower, older parts of the trunk. Borger (1973) gives a more detailed description of bark development and abscission patterns. M. R. Jacobs (1955) and Penfold and Willis (1961) describe the various kinds of bark and bark abscission in *Eucalyptus*.

Botanists have had difficulty in determining what advantage bark abscission could have for a tree, since a great many species survive successfully without it. It is clear that in some circumstances at least, bark abscission serves to prevent the establishment of epiphytes and parasites. This is particularly evident in the smooth-bark trees where the annual abscission of bark keeps the trunk remarkably free of other organisms. I am not aware of published studies on this subject, but it may be of interest to record a correlation

Fig. 3:22.
Bark abscission of *Eucalyptus grandis*. **A.** On 27 June 1979, at the end of spring growth. **B.** Two weeks later.

Fig. 3:23.
Bark abscission in *Pinus ponderosa*. **A.** Accumulation of abscised bark and needles at the foot of a mature tree. **B.** Detail of abscised bark. Note also abscised cone and needle clusters. (Courtesy of J. F. Addicott)

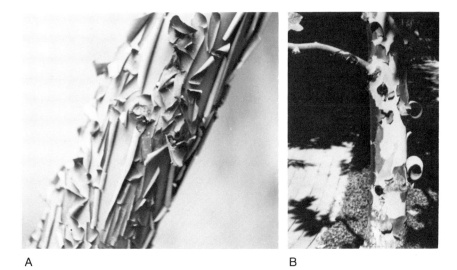

Fig. 3:24.
A. Bark abscission from the trunk (ca. 10 cm diameter) of a young tree of *Arbutus menziezii*. (Photograph by Sheila A. Reed) **B.** Bark abscission in a young tree of *Platanus racemosa*.

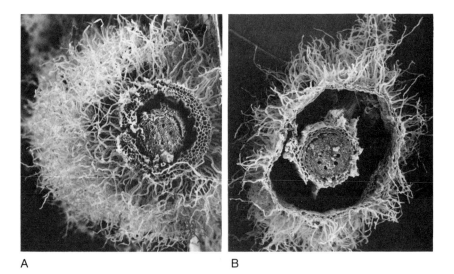

Fig. 3:25A, B.
Two stages in the separation of the rhizosheath of *Oryzopsis hymenoides*. The sand grains had been removed from the rhizosheaths with forceps. Note the almost complete lysis of the cortical tissue. (Courtesy of Wullstein and Pratt, 1981)

observed in California and confirmed by several colleagues. The mistletoe *Phoradendron flavescens* infects a large number of the tree species in western America. However, it has never been reported on *Arbutus menziesii*, one of the species that annually abscises papery bark from all of its branches.

5. Prickles

Prickles are woody emergences on the stems of *Rosa*, *Rubus*, and other genera (and are often incorrectly called thorns). Commonly, they are abscised, particularly when the stem increases in diameter. A protective layer forms in the stem beneath the prickle and becomes continuous with the periderm of the stem. There is an abscission zone of small, thin-walled cells at the base of the prickle adjacent to the protective layer. In due course, these cells break, abscising the prickle in what appears to be a wholly mechanical manner (Mühldorf, 1925).

6. Root Bark and Cortex

Roots also abscise bark, especially roots of woody perennial species. Many of these produce successive layers of bark, which rarely, if ever, build up to a thick covering. The amount of tissue produced is usually small, and the outer layers are subject to decay. Roots that do not develop much in the way of secondary thickenings or periderm may still abscise the epidermis and outer layers of the cortex as a result of cortical lysis (Fig. 3:25A, B; Price, 1911; Henrici, 1929). On such roots, a rhizosheath can form as sandy particles from the soil become cemented to the root hairs of the epidermis (and outer cortex) (Fig. 3:26; Wullstein et al., 1979). It has a protective function, helping to prevent water loss (Blydenstein, 1966; Verboom, 1966).

An interesting example of abscission of secondary phloem has been observed in the root of *Taraxacum kok-saghyz*, the rubber-producing dandelion. As the root ages the outer layers of the secondary phloem are cut off by cambial activity, and when the root is pulled from the ground the older tissues remain behind. The anatomy of this separation has not been studied, but from the observations reported (Metcalfe, 1967), the active tissue functions as a separation layer. The net result is a considerable loss of rubber yield as the rubber-bearing laticifers are concentrated in the phloem of the root.

F. ROOTS

1. General

There are very few species that show an active separation of roots from the parent plant. On the other hand, losses of materials from roots and losses of portions of the root system from various factors have been known

Fig. 3:26.
Rhizosheaths of the roots of *Agropyron dasystachyum*. (Courtesy of Wullstein et al., 1979, reproduced with permission of the publisher.)

for some time. Some of these losses may seem small, but they can be of considerable significance in the life of the plant. For example, the root cap steadily sloughs off cells and secretes mucilage as it grows through the soil. Roots that persist for any period of time shed the epidermis with its root hairs and cortex, often whether or not secondary thickenings develop. The abscission of cortex and bark has been discussed in the preceding section. Finally, there is a significant exudation of organic chemicals by many roots, which has important effects on soil and root ecology (Head, 1973).

In forests and orchards a considerable portion of small roots die each year. It is calculated that up to 1,100 kg/ha become nonfunctional and die each year in a dense forest. The quantity of roots dying each year can approach one-tenth of the quantity of leaves, branches, and bark falling from the trees of a forest (Head, 1973). The dying back of the rootlets may be related to any of a number of factors, including seasonal variation in moisture, such as drought or flooding, the changes in vigor of the roots of the plant, as well as activity of soil organisms and diseases.

Active physiological abscission of roots is extremely rare. Pfeiffer (1928) makes no mention of such abscission, nor does Head (1973). I have consulted several colleagues having authoritative knowledge of roots and root morphology; none of them knows of active abscission of roots, except for *Azolla*, described in the next section. However, separation of dead or dying roots from the living portions of the root system eventually takes place, and protective layers of periderm develop at the base of the moribund roots (Reynolds, 1975). Also, after the contractile roots of *Oxalis esculenta* have contracted, they wither, and a corky protective layer develops in the main root (Duncan, 1925), essentially cutting off the dead root from the remainder of the plant. In a similar way, root nodules of many legumes growing in Western Australia cease functioning and are cut off at the commencement of the dry, hot summer (J. S. Pate, p.c.). In other situations, differing patterns of nodule development, decline, and separation have been recognized (Pate, 1977). Such changes can be considered a kind of abscission similar to that which occurs in leaf abscission of *Parthenium* and other shrubs of arid regions. In that genus the leaf dies, and while it is not actively abscised, a protective layer forms in the stem across the leaf base, and the dry leaf eventually breaks away mechanically (Addicott, 1945).

2. Azolla Root Abscission

Azolla appears to be the only living genus in which roots abscise in a manner involving the physiological and cytological changes typical of above-ground portions of the plant. At the base of each root is an abscission zone of appropriately small cells in which separation takes place (Fig. 3:27) (Konar and Kapoor, 1972). As far as I am aware, the ecology and physiology of *Azolla* root abscission has not been investigated.

80 *Abscission Behavior*

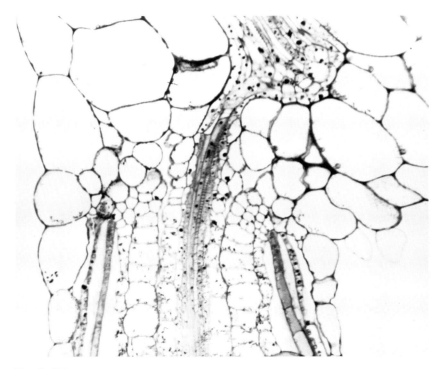

Fig. 3:27.
Root abscission zone of *Azolla* sp. (Courtesy of B. E. S. Gunning)

Root abscission was apparently much more common in earlier geological periods. In the Carboniferous, *Lepidodendron* abscised its root-like stigmarian appendages. Also, the axillary tubercles of *Crenaticaulis* and *Gosslingia* of the Devonian are considered scars of abscised rhizophores (see Chapter 9).

3. Root Cap

The major functions of the root cap include protection of the underlying meristem and facilitation of the growth of the root tip into the soil. As it penetrates the soil the root cap is constantly sloughing off cells. The manner in which these cells separate is surprisingly similar to separation in the above-ground abscission zones. In the case of the root cap the sloughing off of cells is a continuous process that can be conveniently described as "steady-state abscission." The separation of cells results from the activity of enzymes secreted by the cells. The enzymes are synthesized in the membrane systems of the endoplasmic reticulum and dictyosomes, and sequestered into dictyosomal vesicles. The vesicles migrate to the plasmolemma, fuse with it, and discharge their contents into the paramural region and the cell wall (Berjak and Villiers, 1970; Juniper, 1972). The enzymes bring about the separation of the outer cells of the root cap and convert much of the cell-wall material into mucilage. This functions to lubricate the penetration of the root tip into the soil. The younger cells of the root cap contain large amounts of starch, some of which appears to serve as a substrate providing energy for the synthesis of enzymes. It is also possible that some of the carbohydrate of the starch is converted to mucilage and secreted in the dictyosomal vesicles along with the hydrolytic enzymes. The consequence of the enzyme secretion is the separation of the individual cells and the disintegration of the older portions of the root cap as a tissue. Interpretation of root-cap activity as demonstrating a steady state of abscission is strongly supported by the close similarity of the ultrastructure of enzyme secretion and cell separation in the root cap to that in abscission zones.

G. ABSCISSION AND ARCHITECTURE

1. Morphological Factors in Plant Architecture

The structural form of a plant is the result of the interaction of a number of morphogenetic factors which may be expressed in a variety of ways and combinations. These factors include the activity and distribution of terminal and lateral buds, and the extent to which any particular bud may develop. The arrangement of the buds is an expression of the phyllotaxy of the species and may be either compact or diffuse. Branches may grow vigorously or remain as short shoots. They may grow orthotropically (erectly) or plagiotropically (more or less horizontally). The extent and degree of bud, leaf, and branch abscission can have a profound effect on the form of a plant. The position of the flowers may have little effect, or it can strongly influence the pattern of architecture.

Interplay of the above factors results in a variety of plant architectures. These are described by Hallé, Oldeman, and Tomlinson (1978) in their authoritative analysis of tree architecture. It is a remarkably thorough and informative book, which the interested reader will find to be a valuable addition to his library. The following discussion is based in part on the book, but includes additional materials and emphasis on the role of abscission in plant architecture.

2. Extent of the Involvement of Abscission

a. Entire Stem

A few plants, such as tumbleweeds (Figs. 3:10, 3:11), abscise all of their above-ground portions annually. These are the most extreme examples of stem abscission.

Fig. 3:28.
Branch abscission in *Cyphostemma juttae*. **A.** At end of summer showing full development of branches and inflorescences. **B.** Mid-autumn, branch abscission complete.

b. All Branches

Some plants will abscise practically all of their major branches each year. These include species of *Cyphostemma* and *Moringa* (Fig. 3:28). Such plants add to their main stem slowly, as the abscission actually includes all but the basal few nodes of the current year's branches. These basal nodes will add a small net amount to the mass of the trunk each year.

c. Lower Main-Stem Branches

Many trees abscise their lower branches progressively so that foliage is concentrated near the tips of the main stem and main branches. This habit is especially common among forest trees (Fig. 3:29), but it is not restricted to them (Fig. 3:30).

d. All Branchlets (Cladoptosis)

A number of conifers abscise their leaves in branchlets. If all the branchlets are abscised annually, the tree is deciduous, such as *Taxodium* and *Metasequoia*. If only a portion of the branchlets are abscised each year, the tree remains evergreen, as do *Sequoia* and *Cupressus* (see sec. B.1.a.).

e. Shoot-Tip Abscission

In a number of genera, at the end of a flush of growth, the shoot-tip aborts and abscises. This results in a sympodial pattern of growth for the shoots, as resumption of shoot growth requires the development of a lateral bud into a new shoot. This is the usual pattern of growth in *Citrus*, *Syringa*, and a number of other genera (see sec. B.2.a.).

f. Stem Disarticulation

This kind of abscission takes place readily in various species of *Ephedra*, *Euphorbia*, and Cactaceae. Depending on the amount of the plant involved, the structure of the remainder can be greatly affected.

g. Attritional Abscission

The mature structure of many trees is affected by the death and eventual shedding of branches of many sorts. This kind of shedding usually does not involve active abscission, but is the result of the breakage of branches at the region of weakness near the base of the branch. For some genera, such as *Pinus*, it is largely the main-stem branches that are broken away from the trunk (and see Fig. 3:30). In others, such as the broad-leaved deciduous *Zelkova*, breakage is limited almost entirely to small twigs and branchlets. In both situations the final architecture of the tree is strongly influenced by which of the branches are susceptible to attritional abscission.

84 *Abscission Behavior*

3. Examples of Tree Architecture Influenced by Abscission
a. General

Abscission has two principal effects on tree architecture. One is the removal of lower foliage so that the trees assume shapes suggestive of umbrellas. The other is the removal of terminal foliage, leading in most cases to one or another pattern of sympodial growth.

Fig. 3:29.
Tree of *Triplochiton scleroxylon* on cleared land in Nigeria. Shading by pre-existing mid-story trees led to abscission of all but the uppermost branches. (Photograph by F. T. Last, courtesy of J. A. Longman)

Fig. 3:30.
Eucalyptus grandis in an exposed location at Davis, Calif. Lower branches have been broken away by wind.

b. Umbrella Trees

The simplest of the umbrella trees include the tree ferns (e.g., *Cyathea* and *Dicksonia*) that abscise their lower leaves with varying degrees of efficiency and display a flattish crown of active leaves (Fig. 3:31A). Many of the palms and the tree species of *Agave* show a similar growth habit. Among cultivated plants, *Carica papaya* (paw-paw, papaya) is also in this category. Umbrella trees may be either monoaxial (single trunked) or polyaxial (several trunks, each with a crown of leaves).

The foregoing examples were all concerned with leaf abscission. The umbrella pattern of architecture is also achieved by the abscission of lower branches. Notable examples are found in *Canthium*, *Castilla*, and *Cordia* (Fig. 3:31B). Among forest trees, the Araucaceae and the Podocarpaceae are conspicuous for the ease with which lower disadvantaged branches are abscised. This habit is, indeed, characteristic of upper-story forest trees, al-

86 *Abscission Behavior*

though only a few of the many other species involved abscise their lower branches as readily as the Araucaceae. In *Euphorbia* a number of the larger species abscise their lower branches and become candelabra trees (Fig. 3:32).

c. Sympodial Trees

In this discussion I shall be using the term *sympodial* rather loosely to indicate architecture that is influenced by the abscission of buds, branches, and reproductive structures.

The role of shoot-tip abortion in sympodial growth of shoots has already been discussed (sec. B.1.a.). However, the variations can be very impressive, particularly when the subterminal growth consists of well-developed plagiotropic branches (Fig. 3:33A).

Striking dichotomously branched trees can also result from shoot-tip abortion and abscission. Notable examples are found in *Aloë*, *Dracaena*, *Cornus*, and *Rhus*.

When inflorescences are terminal on branches, the fruit is usually ab-

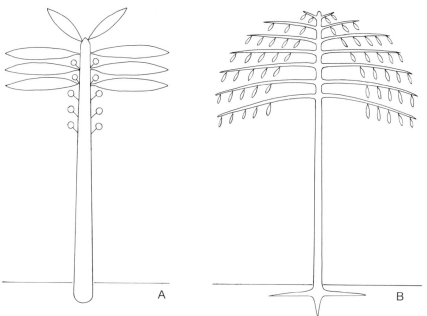

Fig. 3:31.
A. Architectural diagram of an umbrella tree in which the crown is composed of leaves—e.g., *Cocos nucifera*, *Carica papaya*. **B.** Diagram of an umbrella tree in which the crown is composed of leafy branches—e.g., *Castilla elastica*. (Redrawn from Hallé et al., 1978)

Fig. 3:32.
A candelabrum tree, *Euphorbia* sp.

scised, and vegetative growth continues from an axillary bud in various patterns of sympodial growth. This is diagrammed in Figure 3:33B. One of the common patterns is shown by *Magnolia grandiflora*. A flower develops at the end of each branch. Usually, two lateral buds develop just below the peduncle, and at maturity the fruit abscises at the base of the peduncle. Typically, the more horizontal branch becomes dominant, resulting in a distinctive branch system (Fig. 3:34). There are innumerable variations of such patterns (Hallé et al., 1978).

With many species there is a general thinning of the crown of the tree by abscission, usually attritional, of smaller and weaker branches. The net effect of this abscission is a considerable reduction in the number of small branches. It is difficult to imagine how shrubby a mature tree might become

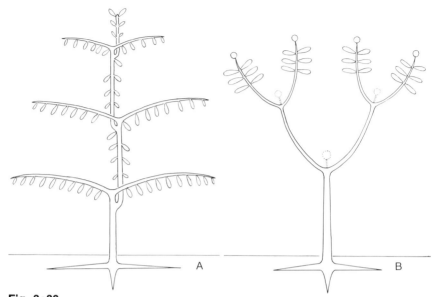

Fig. 3:33.
A. Architectural diagram of a tree with a sympodial main trunk—e.g., *Theobroma cacao*. **B.** Architectural diagram of another sympodial branching pattern—e.g., *Ricinus communis*. (Redrawn from Hallé et al., 1978)

Fig. 3:34.
Sympodial branch of *Magnolia grandiflora*. Note fruit scars (white arrows). One to three (usually two) lateral branches develop about the time the fruit abscises; cf. Fig. 3:12.

without such thinning. Evidence of this activity can be found when one examines the crown of most deciduous temperate-zone trees or notes the numbers of small branches on the ground after a strong wind. Attritional thinning of larger branches contributes to the impressive architecture of the great baobab trees (*Adansonia digitata*) of Africa (Figs. 3:35, 3:36). Branchlet abscission, if any, is inconspicuous in the baobab (K. Coates-Palgrave, p.c.).

4. Further Reading

Anyone interested in tropical trees and their architecture will find a fascinating store of information in Hallé, Oldeman, and Tomlinson (1978), with excellent illustrations and diagrams. Van der Pijl's (1952, 1953) descriptions and analyses of branch abscission in tropical trees are perceptive and enlightening; they are highly recommended to the temperate-zone plant scientist.

H. ADVENTITIOUS AND ANOMALOUS ABSCISSION

Adventitious abscission results from the activity of a separation layer functioning in a position that is not predetermined by the morphology of the plant. The location of adventitious separation layers can vary widely; apparently, the precise location is determined by interaction of various physiological factors. In contrast, the regular abscission zones occur in a limited number of morphological locations. (Adventitious abscission has sometimes been called secondary abscission [Pierik, 1971], but that is somewhat ambiguous, whereas the well-known adjective *adventitious* is fully appropriate for the phenomenon.)

Anomalous abscission results from activities of tissues or other structures that do not normally participate in the process of separation. Although, conceivably, there could be numerous kinds of anomalous abscission, there is as yet but one example known which appears appropriate for the category, and that is the abscission within the vascular cylinder described at the end of this section.

The fact that adventitious abscission can take place, and does so in a wide variety of species, is important and significant, and emphasizes the totipotency of plant cells. It appears that when they are properly stimulated, most cells, possibly all cells, can secrete enzymes necessary to bring about the separation of the cell from its neighbors.

1. Seasonal Manifestations of Adventitious Abscission

An outstanding example of seasonal adventitious abscission occurs in the monophyllous species of *Streptocarpus* and in a few of the species having more than one leaf. The leaves of the monophyllous species are actually

Figs. 3:35, 3:36.
The baobab tree, *Adansonia digitata*. (Courtesy of the South African Tourist Corporation)

phyllomorphs, as they are perennial and develop and grow more or less continuously from a basal meristem. At the end of a growing season, up to two-thirds of the distal portion of the leaf senesces and is cut off by the activity of a protective layer that develops across the axis of the leaf. There is no evidence of a separation layer. The amount of tissue cut off depends on the condition of the plant. Less is abscised if the plant is young or growing vigorously (Fig. 3:37) (Hilliard and Burtt, 1971; Noel and van Staden, 1975).

A number of succulent plants will abscise a part of their leaves as the leaf ages or growing conditions become adverse. For example, in *Aloë asperifolia* the tips of the older leaves have a tendency to die back for several cm, and then to be cut off by a separation layer (which usually functions imperfectly, leaving the dead portion still attached). Proximal to the separation layer, a thick protective layer develops. This process can be repeated several times, and because the activity of the separation layer is seldom complete, sometimes two or three abscission zones can be found on the same leaf (Fig. 3:38). Similar abscission also occurs on older portions of leaves of *Kalanchoë*, on the scale leaves of *Hyacinthus* (Kamerbeek and de Munk, 1976), and on the persistent leaf bases of some cycads (Stopes, 1910).

 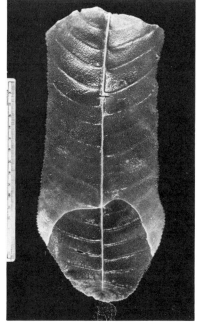

Fig. 3:37.
Autumnal adventitious abscission of the distal portion of the phyllomorph of *Streptocarpus molweniensis*. (Courtesy of Noel and van Staden, 1975)

Fig. 3:38.
Adventitious abscission (arrows) of the distal portion(s) of the leaf of *Aloë asperifolia*.

2. Induced Adventitious Abscission

Lloyd (1914) made a number of observations of the abscission responses in the stem of *Impatiens* following injury. While some of the responses occurred in the usual abscission zone just above the nodes, most of the responses involved the development of separation layers in patterns determined in part by the position of the wound, and in part by the proximity of nodes, buds, and leaves. He induced separation layers which in cross section were V-shaped, C-shaped, and S-shaped.

In *Bryophyllum pinnata* an irregular abscission zone develops in the lower part of an internode after excision of the distal node and shoot. Development can be accelerated by ethephon and inhibited by auxin (Horton, 1976).

Another interesting kind of adventitious abscission results from infections caused by the "shot-hole" diseases. These are caused by various microorganisms that attack the leaves of plums, peaches, and cherries, among other species in the genus *Prunus*. Samuel (1927) published a careful investigation of the shot-hole disease caused by *Clastosporium carpophilum* on the almond (Fig. 3:39A). When infection occurs, the cuticle is penetrated and the microorganism injures and brings about the death of some of the leaf cells. A zone of healthy mesophyll cells, forming a circle about the injured cells, becomes active. The inner portion of the active cells becomes impregnated with lignin, tending to wall off the infected tissues. Immediately outside the lignified zone, cells separate by dissolution of the middle lamella, and the infected disk of tissue drops away (Fig. 3:39B). Cell divisions in the outer portions of the abscission zone lead to the development of a protective layer, which becomes suberized and sometimes lignified, with the result that the margin of the shot-hole is completely protected. If the leaf is old or the tree is deficient in water, separation may not occur. It is of interest to note that the shot-hole abscission response on leaves of species of *Prunus* can be induced by application of drops of a number of toxic materials, such as $CuSO_4$, CH_2O, $HgCl_2$, HCl, as well as by injury with a hot needle (see Samuel, 1927). Further examples of infection-induced adventitious abscission are descibed by Whitney (1977).

In a number of other situations adventitious abscission zones develop beneath injured tissues, particularly of succulent and herbaceous plants. For example, after injury to the stem of *Opuntia basilaris*, protective and separation layers develop beneath the wound. The injured tissue is abscised, and the underlying stem tissues are covered by the adventitious periderm of the protective layer. In most species, however, when protective layers develop beneath a wound, the damaged tissues are not abscised.

Mid-petiole abscission has been induced in *Ipomoea batatas* by bud grafting. In that species, a small shoot can be successfully grafted into the split petiole of a young leaf (Fig. 3:40A). The scion shoot develops vigorously, and in about twelve weeks the portion of the petiole distal to the graft union together with the blade is abscised (Fig. 3:40B–D). In similar grafts, mid-petiole abscission was found occasionally with *Lycopersicon esculen-*

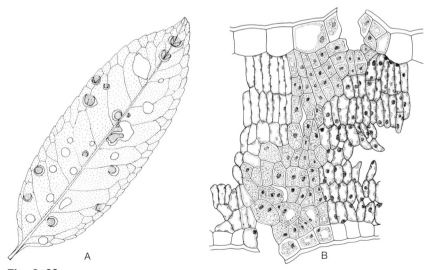

Fig. 3:39.
A. "Shot-holes" in a leaf of almond, *Prunus amygdalus*, infected with *Clasterosporium carpophilum*. **B.** Separation layer of a "shot-hole." (Redrawn from Samuel, 1927)

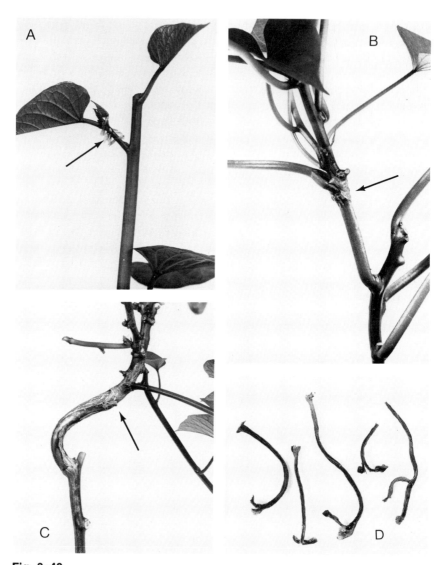

Fig. 3:40.
Adventitious abscission following mid-petiole grafts in *Ipomoea batatas*. **A.** Freshly made bud graft to the mid-petiole. **B.** Four weeks after grafting. Arrow points to the graft union. The distal portion of the petiole is to the left, and the proximal portion is below the arrow. **C.** Sixteen weeks after grafting. Arrow points to the graft union and the scar left after abscission of the distal portion of the petiole. **D.** Distal portions of petioles after abscission. Note the splits made at the time of graft insertion. Withered blades have been removed. (Courtesy of P. M. Warren Wilson)

tum, but did not occur in *Datura stramonium, Phaseolus multiflorus, Helianthus annus*, or *H. tuberosus* (P. M. Warren Wilson, p.c.).

An interesting series of adventitious abscission responses to injurious chemicals have been described in the literature. One of the first of these to be described was the response of young bean plants (*Phaseolus vulgaris*) to applications of TIBA. The chemical is persistent in the plant, and induces abscission in all of the usual abscission zones, such as the base of leaflets and petioles, and induces as well a number of abnormalities of growth. In addition to those effects, TIBA induces adventitious abscission in young internodes. Usually the abscission zone develops about midway in the internode, in contrast to the usual location of internode abscission zones at the very base of the internode, just above the subtending node. Under appropriate circumstances, up to three abscission zones are induced in adjacent internodes of the main stem of the plant (Whiting and Murray, 1948).

A similar kind of adventitious abscission was induced by ETH when excised portions of the main stem of *Phaseolus vulgaris* were exposed to it. The explants consisted of the first node and approximately 1 cm each of the internodes above and below the node. Abscission zones were induced in the segment of the upper internode in positions varying from midway in the internode to the very base. Sometimes more than one abscission zone was induced, additional ones forming below the first, each initiated after the zone above had become delimited (Webster and Leopold, 1972).

Adventitious abscission zones have also been induced in the leaflet pulvini of *Phaseolus vulgaris*. These were observed in tissues that had been excised and treated with NaClO as a surface sterilant. The result was the development of one or two adventitious abscission zones in each pulvinus (Morré, 1968). It is interesting to note that a common factor of disturbed auxin metabolism is likely in the foregoing examples. Both TIBA and ETH are well known for their effects on auxin metabolism. TIBA is a strong inhibitor of auxin transport, and ETH both reduces transport and increases inactivation (see Chapter 4). Although I am unaware of information on the effects of NaClO on auxin, its strong oxidative action could well promote inactivation of auxin.

In contrast to the foregoing examples, the hormones auxin and gibberellin themselves can induce adventitious abscission in susceptible tissues. Carns et al. (1961) observed that high concentrations of GA, when applied to the stem stump of excised cotyledonary nodes of cotton, induced abscission of the stem, something that has never been observed in nature. Bornman et al. (1968) studied the anatomy of this abscission and found that the anatomical changes in the stem were identical to those occurring in the nearby cotyledonary abscission zones. This finding is not too surprising, as the excised tissues were taken from 14-day-old seedlings in which both petiole and stem tissues are very little lignified and quite herbaceous. In experiments with excised young flowers and pedicels of apple and pear, Pierik (1980) found that IAA would induce adventitious abscission in the apple pedicels and that both IAA and CK could induce adventitious abscission in the pear pedicels.

3. Anomalous Abscission

A remarkable longitudinal splitting and separation of segments of the vascular cylinder occurs in *Eriogonum ovalifolium* var. *nivale*. The mature specimen is a cushion plant and is actually a colony of distinct individuals derived from a single parent plant by abscission of segments of the vascular cylinder. Separation is achieved by the development of anomalous periderms from secondary xylary parenchyma around and between segments of the original vascular cylinder, followed by apparent dissolution of middle lamellae between the outer cells of adjoining periderms (Fenwick and Moseley, 1979). The activity of these anomalous periderms has a threefold result: (i) the cutting out of a central core of primary xylem and pith, (ii) a radial splitting of the axis accompanied by (iii) a sloughing off of successive portions of secondary xylem toward the center of the axis (Fenwick and Moseley, unpubl.). The resulting colony consists of several intertwining axes which join independent leafy branches, each with a root system. While this is the only example of anomalous periderm abscission of which I am aware, it seems probable that it will be found to occur in other species, at least to some degree. The active secretion of middle-lamella-degrading enzymes shown by the periderm cells of *Eriogonum ovalifolium* var. *nivale* is unusual and quite in contrast with the abscission of bark, which, as far as we now know, involves only mechanical separation of dead tissues.

4. Physiology

Chapter Contents

A. MAJOR RESEARCH METHODS   98
  1. Field and Greenhouse   98
  2. Explant Methods   99
  3. Testing Devices   103
  4. Abscission Indices   104

B. GENERAL PHYSIOLOGY   105
  1. Temperature Response   105
  2. Oxygen Requirement   107
  3. Substrate Requirement   108
  4. Respiration   109
  5. Interpretation   110

C. NUTRITIONAL FACTORS   110
  1. Carbohydrate Metabolism   110
  2. Nitrogen Metabolism   111
  3. Mineral Nutrition   111
  4. Preabscission Changes in Organs   112

D. HORMONAL FACTORS   113
  1. Auxin (IAA)   113
     a. Auxin and Abscission   113
     b. Correlative Occurrence   114
     c. Abscission Responses   114
        (1) Retardation   116
        (2) Acceleration   116
        (3) Stages I and II   118
        (4) Response Curves   120
     d. Auxin Physiology in Relation to Abscission   121
     e. Auxin Gradient   122
     f. Mechanism of Auxin Action   123
  2. Abscisic Acid (ABA)   126
     a. Discovery and Hormonal Nature   126
     b. Correlative Occurrence   127

          c. Abscission Responses   128
          d. Mechanism of ABA Action   129
       3. Gibberellin (GA)   130
          a. Correlative Occurrence   130
          b. Lack of Response to General Applications   131
          c. Responses to Localized Applications   131
          d. Applications to Petioles   131
          e. Responses of Explants   132
          f. Mechanism of GA Action   134
       4. Cytokinin (CK)   134
          a. Correlative Occurrence   134
          b. Abscission Responses   134
          c. Mechanism of CK Action   135
       5. Ethylene (ETH)   135
          a. Occurrence   135
          b. Abscission Responses   138
          c. Interrelation with Hormones and Regulators   140
          d. Mechanism of ETH Action   143
       6. Miscellaneous Abscission Substances, Possibly Hormonal   145
       7. Summary of Hormone Action   148
          a. In the Abscission Zone   148
          b. In the Subtended Organ   150
          c. In Source-Sink Relations   151

A. MAJOR RESEARCH METHODS

Methods effective in the study of abscission range from observation of intact plants in the field to excised abscission zones in rigorously controlled environments. Each method has advantages that can be exploited, but also limitations as to how much physiology can be learned from it. It is essential to be aware of the limitations, as well as the advantages, of a selected research method.

1. Field and Greenhouse

Field observations introduced botanists to many of the important factors in the physiology of abscission. Theophrastus (285 B.C.) noted various conditions and circumstances that influenced abscission of leaves, for example. Among other things, he pointed out the involvement of water availability in the control of leaf abscission. More recent botanists such as Wiesner (1871, 1904a, b, c), who was a keen observer of plant behavior, learned a great deal from repeated observations of the same plants in field and garden. Such observations enabled the identification of a number of factors, such as light intensity, heat, and frost, that are correlated with the initiation of leaf abscission. Further, experiments with intact plants are essential to research designed to understand or modify the abscission behavior of plants. In many such situations, information learned in the greenhouse or the laboratory must be tested in the field to determine its validity under field conditions. With few exceptions, the only valid test of a new chemical or agricultural practice is in the field with plants being grown under agricultural conditions. Many kinds of experiments actually can be conducted on plants in the field, but such research is subject to hazards of weather and numerous other factors beyond the control of the researcher.

There are substantial advantages to abscission research in the greenhouse or controlled-environment chamber, where growing conditions can be kept fairly uniform and experiments can be repeated with little variation in growing conditions. Treatment and manipulation of plants is usually much easier in the greenhouse, as is also the keeping of records. In a similar way, it is sometimes convenient to study branches that are brought indoors where they can be treated and manipulated. By the use of controlled environments, large numbers of uniform plants can be produced. Even without such indoor growing conditions, valuable physiological information can be obtained from experiments with plants or cuttings in which, for example, the leaves are in different stages of development. W. P. Jacobs's (1955, 1958, 1962) elegant experiments with leaf abscission of *Coleus* exploited the fact that each pair of leaves on a shoot is in a physiological state different from its neighbors.

2. Explant Methods

For intensive study of the basic physiology of abscission, explants (excised abscission zones) have been an invaluable tool. Kendall (1918) made the first abscission-zone explants during his investigation of flower abscission in *Nicotiana*. The methods now in use stem from early work with explants of *Citrus* and *Phaseolus vulgaris* leaflet abscission zones (Addicott et al., 1949). Explants of the leaflet abscission zone of the primary leaf of *Phaseolus vulgaris* have become the most widely used material. Mature *Citrus* leaves proved unsatisfactory, as they were infected with mold spores from many months on the tree, and disinfection without damage to the leaf tissue could not be accomplished. On the other hand, when explants are taken from young seedlings grown in a reasonably clean environment, surface microorganisms are rarely a problem. Very satisfactory explants come from *Phaseolus vulgaris* that are about three weeks old and from cotton, *Gossypium* spp., that are two weeks old (Addicott et al., 1964). Other explants that have been widely used have been cut from *Coleus*, and a variety of other species have been used from time to time. The choice of material is largely a matter of convenience and scientific objective of the research. Some explants are more versatile than others. The cotton cotyledonary node offers the opportunity of treatment of one or both petiole stumps, in positions distal to the abscission zone; and of either the stem stump or the hypocotyl stump, positions that are proximal to the abscission zones. Explants consisting of a node of Coleus or Phaseolus vulgaris with two petiole stumps offer similar opportunities. In contrast, the leaflet abscission zones of Phaseolus have only two ends for treatment: the pulvinus end, distal to the abscission zone, and the petiole end, proximal to the abscission zone.

The amount of tissue included in an explant is somewhat a matter of convenience, but the length of the petiole has a strong influence on abscission and on the response to physiological treatments (see, e.g., Prakash, 1976). Abscission will take place when the explant includes only 1 or 2 mm of tissue distal to the abscission zone. However, such small stumps are difficult to test and observe. With the cotton cotyledonary node explant, a 5 mm petiole stump has proved a very satisfactory length. Lyon (1964) studied the effect of petiole length and found long petiole stumps abscised at a much slower rate than short stumps. However, retardation from length declined as the age of the seedlings increased, and when explants were cut from 27-day-old seedlings (rather than the standard 14-day seedlings), the abscission-retarding effect of long stumps was gone. Carns (1951) obtained satisfactory results with bean petiole explants that were 3 mm long overall.

The abscission retardation of long petioles can also be modified by application of various chemicals (see Pathak and Chatterjee, 1976a). What appears to happen is that toxic materials injure the petioles and reduce their capacity to retard abscission, so that the treated petioles abscise more rapidly than controls. In contrast, when such chemicals are applied to short petiole stumps, the toxic material can reach the abscission zone quickly and inhibit cell function. Consequently, abscission of the short petiole stumps is greatly retarded or completely inhibited relative to the controls. A similar difference in response has been obtained with bean leaflet explants in which the distal portions were of different lengths. Gaur and Leopold (1955) used explants in which the distal portions were 3 cm long. When the tips of the distal ends were placed in agar containing NAA at low concentrations, abscission was accelerated. In similar experiments that differed primarily in the fact that the distal portion of the explant was only 5 mm long, Biggs and Leopold (1958) found that immersion of these distal portions in agar containing NAA gave no acceleration at low concentrations and inhibited abscission at the higher concentrations of NAA. Therefore, one cannot assume that the response to a chemical applied to an explant is due solely to reactions in the abscission zone. The tissues at the site of application and between the site and the abscission zone can have a critical influence on the response.

Davenport and Marinos (1971) made the remarkable observation that abscission proceeded normally in longitudinal slices (0.25 mm thick) of Coleus leaf explants kept on agar in a small chamber. Following this lead,

COTTON EXPLANT TECHNIQUE

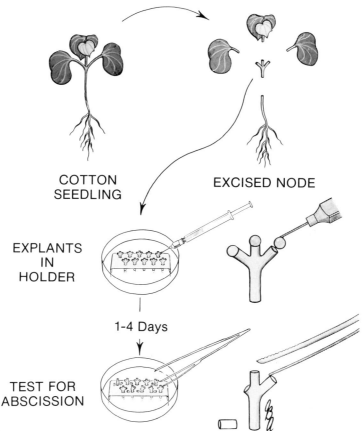

Fig. 4:1.
Diagram of the cotton explant method for the investigation of abscission.

Sexton (1979) observed normal abscission in longitudinal blocks (approximately 0.5 x 0.5 mm in cross section) of *Impatiens* abscission zones. Such micro-explants permit simultaneous observation of cytological details and physiological responses.

A major advantage of explants is that they can be held under accurately controlled conditions. Since the explants require relatively little space, appropriate incubators are not a costly item relative to much present-day research equipment. It is necessary to support the explants in some sort of rack or holder. These can be made of glass, stainless steel, or inert plastic materials (Fig. 4:1). Usually, it is convenient to place the holders with explants in petri dishes. For some experiments, gas-tight vessels must be used.

Careful attention must be given to the water relations of the explants.

Obviously, the explants must be protected from excessive water loss. On the other hand, in our earliest experiments we found that abscission was completely inhibited if the explants were floated on aqueous solutions. Eventually we learned that abscission could proceed actively if the explants were supported on agar gel, on moist filter paper, or on racks above a layer of water. Even a film of water will influence abscission. Immersion of explants in water for as short a period as one minute at the start of an experiment results in an appreciable retardation of abscission (Livingston, 1950); and similarly, a droplet of water placed daily on the distal end of an explant produces significant retardation (Carns et al., 1951). A very convenient way to apply test solutions to explants is to add about 0.75 percent agar to the solution. The resulting gel is soft and can be easily dispensed. And of equal importance, the gel holds the applied solution to the precise site of application. Without the agar, solutions tend to spread from the point of application.

The rate of abscission in an explant is in part also a function of the age of the tissue, particularly the tissue being abscised. In general, the older the tissue the more rapidly it will abscise after the stimulus of excision (but see Trippi and Boninsegna, 1966). With explants of beans and cotton, for example, the life span of the leaves involved in the explant is usually only a few weeks. Consequently, care must be exercised to select material of uniform age if results of different experiments are to be comparable (and see Pathak and Chatterjee, 1976b). This change in basal abscission rate gives the experimenter an opportunity to select material that will abscise at a convenient rate. If the experimenter wishes to study abscission retardation, he can select rapidly abscising material; and conversely, if he wishes to study acceleration, he can select material abscising at a slow rate; and with special care, he can select material that will abscise at an intermediate rate and will enable him to detect either acceleration or retardation from a basal control rate.

Response to physiological treatments also vary considerably with age of explant material. One of the most important of these changes is the response to applied IAA. Low amounts of IAA accelerate abscission of young cotton explants but retard abscission of older explants (Lyon, 1964). Somewhat similar differences have been observed in bean explants of different ages when treated with NAA (Biggs and Leopold, 1957). In contrast to IAA, the accelerated abscission response of cotton explants to GA was relatively little affected by the age of the material (Lyon, 1964). The foregoing examples are but indications of the many kinds of physiological changes that take place progressively during the development of an organ such as a leaf, changes that can strongly affect abscission responses.

When experimental materials are grown in the field or in the greenhouse, seasonal differences in rate of abscission and responsiveness to hormones are likely. For example, abscission of debladed twigs of *Ervatamia divaricata* and *Coleus blumei*, grown in the field at Burdwan, India, was faster by 10–20 percent in the winter months than in the summer (Choudhuri and Chatterjee, 1970). However, the summer material was more responsive to IAA. Retarding treatments were slower in the summer, and acceleration treatments were faster. Similarly, response to ETH was faster in the summer months (Mukherjee and Chatterjee, 1969). In contrast, experiments with debladed petioles of *Catharanthus roseus* at Meerut, India, found that in the summer months abscission was faster by 20–30 percent than in the winter months. As before, retardation by IAA was greater in the summer than in the winter (Prakash, 1974). The difference between the response of *Catharanthus* at Meerut and *Coleus* and *Ervatamia* at Burdwan is probably related to differences in climate. The more rapid abscission at Meerut may be correlated with higher summer temperatures, and the more rapid winter abscission at Burdwan may be a result of low light intensity. Unfortunately, the publications cited above do not give sufficient details of either growing conditions or experimental conditions to enable an appraisal of the differences.

The growth status of the plants from which explants are taken can also affect both the rate of abscission and the response to physiological treatment. Particular attention must be given to insure that light and temperature are appropriate to the test species. For example, if the light regime is inadequate, abscission material can be so deficient in carbohydrate that it will abscise slowly and will respond with accelerated abscission when provided with sucrose (Biggs and Leopold, 1957). On the other hand, with most materials additional sucrose *retards* abscission relative to untreated controls. Further, the amount of nitrogen available to the plants from which explants are taken will have a profound effect upon abscission behavior. A high-nitrogen plant is a high-auxin plant (Avery and Pottorf, 1945; Carns, unpubl.). High-auxin plants show slower abscission than plants that are low in auxin (Pathak and Chatterjee, 1976b). Levels of other mineral nutrients in the plant can have similar, although usually lesser, effects upon abscission. However, if the explants are being taken from relatively young seedlings, the nutrient reserves in the seed will be adequate for the development of the seedling. Additional mineral nutrients in the soil medium of such seedlings have the effect of slowing the abscission rate of explants.

3. Testing Devices

It is desirable to determine the time of abscission of an explant or other experimental material by the use of a test instrument. While abscission can be determined visually without touching the material and by waiting until the distal portion falls, under such circumstances small variations can have large effects. Also, abscission can be tested mechanically in a number of subjective ways, such as touching the explant with a needle. Such tests, where the amount of pressure applied is at the discretion of the experimenter, are always open to question, particularly when the experimenter knows the treatments that were applied. To eliminate the possibility of bias in testing, various instruments have been devised and have been called, for want of a better name, *abscissors*. Most of these employ the principle of a spring balance in which the calibrated displacement of a spring determines the amount of pressure applied to the explant. Several versions of the original abscissor (Addicott et al., 1949) have been used from time to time. One of the simplest is illustrated in Figure 4:1. It consists merely of a round wire spring attached to a pair of forceps. More sophisticated versions have been developed, including an electrified model in which a light is turned on when the test pressure is reached (Mitchell and Livingston, 1968; Craker and Abeles, 1969). Most abscissors are calibrated to deliver pressure from two to five times the weight of the leaf or other organ supported by the abscission zone. While separation can be achieved with the application of even larger pressures, it is wise to use a test that is reasonably close to what the tissues would be subjected to in nature. Interesting and valuable information about the course of abscission has been obtained by use of the Instron linear stress-strain analyzer (Morré, 1968). The preparation of a single explant for test in the instrument is laborious, and this method has, therefore, been used infrequently.

4. Abscission Indices

Results of experiments on the physiology of abscission commonly record the time-course of the incidence of abscission in a population of abscission zones. Such data are usually presented graphically as the cumulative percentage of abscission with time. Such time-course curves are identical with those obtained in studies of seed germination. For many purposes, visual examination of time-course curves gives adequate information about the abscission experiment. When it is desired to compare such curves, various measures can be used. The most common is "hours to 50% abscission." If two time-course curves differ at all, they usually differ with respect to this measure. A similar measure would be to determine the percentage of abscission of different treatments at a fixed time. The use of this method is especially hazardous if the time is not carefully selected. If the experimenter waits too long, he may find that all the material has abscised and no differences are apparent (Brian et al., 1955), when it is possible that significant differences existed at an earlier time in the experiment.

Comparison of time-course curves by use of a single measure is sometimes unsatisfactory because the curves vary in three major attributes: (i) time of onset; (ii) slope of the curve (rate of abscission); (iii) total amount of abscission achieved during the experiment. A measure of a time-course curve which is influenced by all three factors is the mathematical integral—i.e., the area under the curve. Lyon and Coffelt (1966) developed a mechanical device (a modified planimeter) to integrate time-course curves. The resulting number representing the area under the curve is called the "abscission index," and has proved to be valuable for summarizing the results shown on time-course curves (see Figs. 4:12, 4:14). (The reader is referred to the literature on seed germination for other methods of recording and analyzing time-course curves.)

Which method is selected for the study of the physiology of abscission depends primarily on the objective of the investigation and also, of course, to some extent on the facilities available. If the object is the study of the extrinsic control mechanisms—i.e., mechanisms outside the abscission zone—then one must use plants in a reasonably intact condition in order to be able to treat distal organs such as leaves and fruits, and probably also to treat proximal organs and tissues. On the other hand, if the primary object is to study the mechanism of separation, then explants enable such investigations with a minimum of influence from the rest of the plant. This is feasible and entirely practical because the cells of the abscission zone contain in their nuclei and cytoplasm all of the machinery and material substrates to accomplish the changes of separation. In view of this, it is not surprising that abscission can take place in very young abscission zones and in those of mature vigorous organs, as well as in the abscission zones of senescent organs. The principal precaution in planning experiments with explants is to remember that the abscission zone proper, the distal tissues, and the proximal tissues are three different physiological entities. If this fact is overlooked, valid interpretations of the results can seldom be made.

Physiological experiments in which the influence of various treatments upon abscission is studied have resulted in a great deal of valuable information, but such experiments have a serious limitation in that the physiological and biochemical events that lie between the application of a treatment and the abscission response are still obscure. Usually, the experimenter can only speculate as to what lies in the gap between the treatment and the end point. The great need in abscission research, as well as in most other research in plant physiology, is to narrow this gap. In this and subsequent chapters, recent progress will be outlined. The progress has been especially gratifying in recent years. Some of the steps immediately preceding separation are now identified, steps that involve the synthesis and release of the hydrolytic enzymes of separation.

B. GENERAL PHYSIOLOGY

1. Temperature Response

Early workers sometimes had difficulty in distinguishing the two aspects of temperature relations of abscission. On the one hand, extremes of tem-

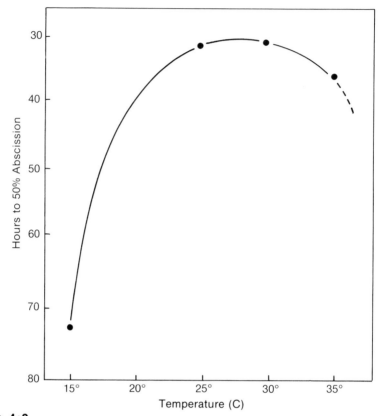

Fig. 4:2.
Effect of temperature on rate of abscission in bean leaflet explants. (Redrawn from Yamaguchi, 1954)

perature impose stresses that serve as initiating factors of abscission (see Chapter 6); on the other hand, temperature determines the rate of the processes of abscission. It is the latter aspect that will be discussed here. In his extensive experiments with petal abscission, Fitting (1911) recorded increasing rates of abscission as the temperature increased, but did not carry the matter further. More recent work, with the object of defining the temperature response, has produced curves typical for physiological reactions generally. Such curves have been determined for explants of bean (Fig. 4:2) (Yamaguchi, 1954) and cotton (Marynick, 1976), and for petal abscission of *Linum lewisii* (Wiatr, 1978). The curves show maxima characteristic of the species; 25°C for beans, 30°C for cotton, and at least 35°C for *L. lewisii*. In appropriate regions the curves indicate a Q_{10} of about 2, characteristic of chemical reactions. Clearly, the temperature data indicate that the process of abscission is not a passive one, or one in some way involving mere diffusion of substances, but they show that chemical reactions are an important and limiting step in the process.

2. Oxygen Requirement

Our first experiments with explants were failures because the explants were floated on aqueous solutions. Much earlier, Molisch (1886), in his pioneering experiments with abscission, observed that immersion of branches in water retarded leaf abscission. He concluded, correctly, that oxygen must be essential to abscission. Later, Sampson (1918) showed that lowered concentrations (below 20%) of oxygen in the atmosphere would strongly retard the abscission of *Coleus* leaves. When our explants were floated in water through which air or O_2 was actively bubbled, abscission proceeded normally. In fact, when the O_2 concentration in the gas mixture bubbled through the water was sufficiently high, abscission was actually accelerated over the rate in air (Fig. 4:3) (Carns et al., 1951).

Recently, a more detailed investigation of the O_2 relations of abscission was conducted with cotton explants (Marynick and Addicott, 1976). The data of the abscission response to O_2 took the form of a double sigmoid curve (Fig. 4:4). The rate of abscission rose sharply as O_2 increased to 10%. There was a plateau between 10 and 20%, followed by a sharp rise at about 25% O_2 to the maximum rate at about 30%, above which concentration there

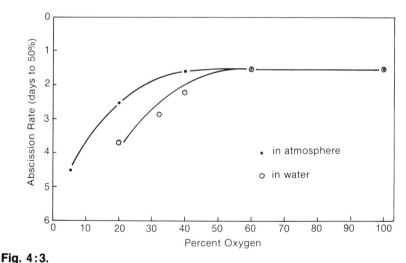

Fig. 4:3.
Effect of oxygen concentration on abscission in bean leaflet explants. Note accelerated abscission with oxygen concentrations above 20 percent. (Redrawn from Carns et al., 1951)

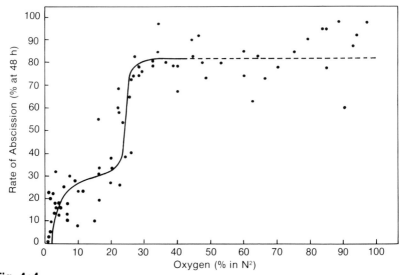

Fig. 4:4.
Double sigmoid abscission response to oxygen concentrations, cotton explants. (Redrawn from Marynick and Addicott, 1976)

was no further increase in rate. The rapid rise in rate at low levels of O_2 (below 10%) is undoubtedly correlated with respiration, in common with most physiological processes that are respiration limited. The acceleration of abscission by O_2 levels above 20% has as yet no certain explanation. It can only be said that the higher levels of O_2 facilitate a reaction(s) that is stimulatory to abscission. A possible candidate for this reaction is the oxidative inactivation of IAA, catalyzed by the enzyme complex known as "IAA-oxidase." Increased IAA-oxidase activity and decreased concentrations of IAA are both promotive of abscission. Possibly also, high O_2 may stimulate ETH production, as one of the final steps in ETH synthesis is O_2 dependent (Lieberman, 1979). However, these possibilities, and others that could conceivably be involved, have not yet been investigated.

3. Substrate Requirement

For reasons that are not entirely clear, little attention has been given to the need for energy-yielding substrates in abscission. Possibly, this is because experimental materials are almost always grown under conditions favorable to substrate accumulation. Further, with very few exceptions, anatomical studies show that the abscission zone, and especially the cells of the future separation layer, contain conspicuous deposits of starch. This is in contrast to adjacent regions that usually contain low levels of starch. However, it is possible that experimental plants would occasionally be deficient in carbohydrate substrates. Such was the case when explants from greenhouse-grown beans were placed in the dark (Biggs and Leopold, 1957). In this particular treatment the rate of abscission was extremely low, but could be brought to a satisfactory level by illumination or by supplying the explants with sucrose. It now seems apparent that there is, indeed, a substrate requirement for abscission that is, in all but rare circumstances, adequately met by the carbohydrates translocated to the abscission zone by the plant.

4. Respiration

While the involvement of respiration in abscission was strongly indicated by the O_2 requirement, that evidence alone was hardly conclusive. Carns (1951) went on to test the effect of a selected series of inhibitors of respiratory enzymes. The list included azide, cyanide, iodoacetate, phenyl mercuric chloride, arsenite, maleic hydrazide, and fluoride. Except for phenyl mercuric chloride, each respiratory inhibitor also inhibited abscission. While phenyl mercuric chloride is a very potent inhibitor of respiration, in the explant system it promotes abscission, and it also promotes abscission when applied in the field. Using microrespirometers, Carns measured O_2 uptake of individual leaflet explants of beans. He found rates in the order of 1–2 mm^3/10 mg·hr. He followed respiration for twenty-four hours and found that the explants that abscised showed a "climacteric" rise in respiration (Fig. 4:5), while those that did not abscise showed a slow decline in respiration. In cotton explants, Marynick (1977) measured CO_2 released in respiration, and obtained rates of from 0.1 to 0.5 mg/g·hr.

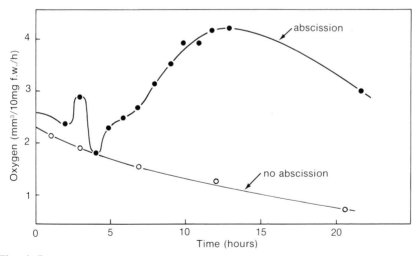

Fig. 4:5.
Climacteric respiration of a bean leaflet explant. (Redrawn from Carns, 1951)

5. Interpretation

The evidence outlined in the preceding sections established that oxidative respiration is essential to abscission. After that, there was little doubt but that abscission involved active physiological processes requiring energy and the functioning of the full machinery of oxidative respiration.

C. NUTRITIONAL FACTORS

1. Carbohydrate Metabolism

Horticulturalists have known for many years that plants that are high in carbohydrate do not abscise their leaves or fruits as readily as plants that are low in carbohydrate. In the field this behavior is related to photosynthesis. Abscission is delayed when conditions for photosynthesis are good. Abscission tends to come sooner or be accelerated if conditions for photosynthesis are poor. Shaded leaves are abscised, and leaves in full sun are retained. Much of this abscission behavior is related to the accumulation of carbohydrate reserves (Mason, 1922; Chandler, 1951). In the case of deciduous fruits, fruit retention is often dependent on the accumulation of carbohydrate reserves in the previous season. If these reserves are low, there will be greater abscission of young fruit. With some varieties, ringing the stem will lead to the accumulation of carbohydrates above the ring and to greater fruit retention.

Application of sucrose to explants has a strong retarding effect on abscission (Brown and Addicott, 1950; Biggs and Leopold, 1957). The sucrose is readily interconverted to other carbohydrates; for example, application of sucrose to bean explants led to deposits of considerable amounts of starch in the explants (Martin, 1954). Conversely, if the carbohydrate reserves are rapidly depleted, abscission is usually initiated. For example, when cotton was growing under favorable conditions, only 20 percent of the young fruit abscised, but when such plants were placed in total darkness for forty-eight hours, 90 percent of the young fruit abscised. The abscission was correlated with greatly reduced translocation of carbohydrate into the young fruit (Rabey and Bate, 1978).

As was already mentioned, the abscission zone characteristically deposits starch, which provides sufficient carbohydrate to enable abscission to take place at any time the signal may come. In other words, the substrate requirements for abscission are programmed and automatically provided, and we must turn to some other function of carbohydrate for the explanation of abscission retardation by high levels of carbohydrate. The explanation appears to lie in the general accumulation of carbohydrates in the abscission zone, especially in cell-wall components. It is well known that high carbohydrate plants have heavier cell walls, and low carbohydrate plants have lighter, less heavily deposited walls. Some of the high levels of carbohydrates will be in the form of sugars and other soluble compounds in the cells of the abscission zone. Such high levels would tend to prevent hydrolysis of wall materials in accordance with the laws of mass action. It is a reasonable interpretation, therefore, that abscission will be more difficult and proceed more slowly in plants with large carbohydrate reserves. High carbohydrates in a plant will contribute to the vigor of fruits and leaves, and in general will enable such organs to more readily synthesize the hormones required for growth, development, and inhibition of abscission.

2. Nitrogen Metabolism

It is common knowledge in horticulture that plants which are supplied with an abundance of nitrogen retain their leaves much longer than deficient plants and set more fruit than deficient plants. The tissues of plants that are high in nitrogen will be well supplied with amino acids and the other nitrogenous compounds essential to active metabolism. High-nitrogen plants were early identified as high-auxin plants (Avery et al., 1937; Avery and Pottorf, 1945). In view of the high nitrogen content of the CK molecule, high-nitrogen plants seem certain to be high-CK plants, although I am not aware of the publication of direct evidence on this point. Certainly also, the vigor of high-nitrogen plants is a reflection of other factors, such as an ample supply of building materials for the synthesis of DNA, RNA, and protein.

The inhibiting effect of high nitrogen upon abscission cannot be readily related to effects in the abscission zone. Rather, the action appears to be indirect through the promotion of the vigor of leaves and fruits and the delaying of their senescence. Vigorous organs delay their own abscission by exporting to their abscission zones greater amounts of auxin and lesser amounts of abscisic acid than do normal or senescent organs. It is also possible that the abscission zone is sensitive to the amount of carbohydrate passing through it. High levels of carbohydrate could be a signal indicating that all is well in the subtended leaf, or that the subtended young fruit is a vigorous sink. Very likely, there are additional ways in which a vigorous organ helps to prevent abscission; but as yet, these have not been identified.

3. Mineral Nutrition

Each of the mineral elements required by plants is necessary for the functioning of one or more essential biochemical reactions. Plants weaken and die if the available amount of an essential element falls below the required minimum. It is not surprising, therefore, that as a mineral deficiency develops abscission of leaves, fruits, or other structures occurs. Abscission of leaves and sometimes fruits is one of the recognized symptoms of deficiencies of the following elements: N, P, K, S, Ca, Mg, Fe, Zn, and B. With most deficiencies, plants maintain the nutrient status of the abscission zone at a level sufficient to enable it to function even when the rest of the plant is suffering. In these circumstances the deficiency has its first effect on the subtending organ. As that weakens, information of either a positive or a negative sort is transmitted to the abscission zone, and abscission is initiated. An interesting example is the case of zinc deficiency, which as it develops brings about considerable leaf abscission. Skoog (1940) showed that zinc is required for normal auxin levels in plants, and one of the first symptoms of zinc deficiency is lowered auxin levels. The reduced amounts of auxin moving from a young leaf, for example, reduces the normal inhibition of abscission and permits the process to commence. Other mineral deficiencies— e.g., Cu and Mn—can influence auxin levels, but only in a secondary way after the overall vigor of the leaf has declined. Whether there are other agents besides auxin whereby the abscission zone is informed of the vigor of the subtending organ is not known. It is possible that there are a number of related changes. In some of the mineral deficiencies, also, leaf abscission does not occur. In those cases, the deficiency may weaken the abscission zone as much as it does the rest of the plant.

Calcium's involvement in abscission deserves special mention. It has long been known that Ca-pectate is a major constituent of the cell wall (Jones and Lunt, 1967), particularly of the middle lamella. During abscission, Ca essentially disappears from the separation layer and adjacent cells (Sampson, 1918; Stösser, 1969). Thus, it is not surprising that Ca deficiency promotes abscission and that application of Ca^{2+} can retard or inhibit it (Poovaiah and Leopold, 1973; Stösser, 1975).

4. Preabscission Changes in Organs

When an organ approaches the end of its life as a result of either normal senescence or injury, or for whatever reason, it undergoes a number of chemical and physiological changes, and the pattern and amounts of chemicals translocated from the organ change drastically. Some of the changes in the chemicals reaching the abscission zone are obviously sensed and serve to trigger the onset of abscission. Since our understanding of this mechanism is far from complete, a careful examination of the available information on preabscission changes is entirely appropriate. Certain of the mineral elements are translocated from leaves in increased amounts as they senesce. These elements include N, P, K, Fe, Mg, Zn, and Cu (Williams, 1955; Woodwell, 1974). There is a gradual decline and disappearance of chlorophyll. This makes the carotenoid pigments more conspicuous. In some species there is a temporary and sometimes striking increase in anthocyanins. Eventually, all of the pigments break down. Hydrolytic enzymes become active, resulting in a decline in DNA, RNA, polysaccharides, and proteins generally. There is a parallel increase in amino acids and other soluble nitrogenous compounds, as well as sugars. Most of the enzymes associated with respiration decline in activity during senescence, but enzymes such as polyphenol oxidase increase (Leinweber and Hall, 1959b; Spencer and Titus, 1972). Similar changes take place during senescence of rose petals (Weinstein, 1957). The net effect of the decline in respiratory enzymes is a rapid lowering of the amount of ATP and other energy-rich phosphates (Komiya and Misawa, 1976) that are available for synthetic processes.

The preabscission senescent (ripening) changes in fleshy fruits have been studied extensively. These include a decline in organic acids, starch, and pectin, and a parallel increase in levels of sugars and soluble pectins, among numerous other changes (Biale, 1960; Varner, 1961). Associated with the senescent changes is an increased rate of respiration, the "climacteric," which is shown by almost all organs during senescence (Biale, 1960; Siegelman, 1951; Leinweber and Hall, 1959a).

A number of hormonal changes occur in senescent and aborting organs. There is a large decline in auxin (Shoji et al., 1951; Jacobs, 1962), and indications of decline in GA and CK as well (Smith, 1969; Sandstedt, 1971; Rodgers, 1977). Increases in ABA are common in aborting young fruits and leaves subjected to stress (Davis and Addicott, 1972; van Steveninck, 1957; Wright, 1978). Somewhat similarly, ETH increases in young fruit and mature fruit at the time of abscission, but there is only limited evidence of ETH involvement in the abscission of mature leaves. From what is now known, hormones are the primary agents whereby information on the decline or senescence of an organ is transmitted to its abscission zone. But it is entirely possible that some of the other preabscission changes, particularly those that involve export of chemicals from the organ, influence the abscission zone. The recent evidence on the ability of Ca^{2+} to retard abscission (Poovaiah and Leopold, 1973) suggests that monovalent cations such as K^+ and NH_4^+ that tend to increase permeability could be expected to promote abscission. Further, the changing levels of amino acids and sugars reaching the abscission zone could conceivably convey a message and have an additional promotive effect. The physiology of the hormones in relation to abscission and the possible action of some of the other factors are discussed in subsequent sections.

D. HORMONAL FACTORS

1. Auxin (IAA)

a. Auxin and Abscission

After the discovery of the growth hormone, auxin, little could be done by way of further investigation of its effects until a supply of the hormone could be found and developed. In 1932, Laibach found that pollen contained growth-promoting substances, and in collaboration with Maschmann (1933), determined that the active substance in pollen was probably identical chemically and physiologically with the auxin Went had discovered in coleoptiles. Also, they determined that the pollen of orchids was an extremely rich source of auxin. This enabled a number of investigations, including those of Mai (1934), who showed that the pollen-auxin would strongly retard abscission when applied to debladed petioles. In 1936, La Rue confirmed Mai's work, using crystalline IAA. The foregoing experiments and others in which auxin preparations were used in attempts to stimulate the development of parthenocarpic fruits led to the discovery that auxin could retard the abscission of mature fruits (Gardner et al., 1939). These early discoveries, which provided methods for investigation and suggested important agricultural uses, gave impetus to the abscission research of the subsequent decades.

b. Correlative Occurrence

In a pioneering investigation, Shoji et al. (1951) followed the levels of extractable auxin through the life of a bean leaf. The levels, which were very high in the young leaf, fell to a steady level after the leaf was fully expanded, and then dropped rapidly as the leaf blade yellowed and abscission approached (Fig. 4:6). In a somewhat similar investigation of the leaves on a *Coleus* shoot, Wetmore and Jacobs (1953) noted that the time to abscission of the leaves on a shoot decreased with the age of the leaf. The length of time for abscission was correlated with the amount of diffusible auxin obtainable from the leaf (Fig. 4:7). The older the leaf the less auxin it produced and the shorter the time until it absciscd. In investigations with auxin and fruit development, Luckwill (1953) found that the auxin content of the seeds of apples was low during periods of fruit abscission and relatively high during periods of fruit growth where there was little abscission. Similar changes have been identified in the black currant (*Ribes nigrum*) by Wright (1956), and the subject was reviewed generally by Luckwill (1957). These observations have been confirmed repeatedly in a variety of species in subsequent years by a number of workers (e.g., Storey, 1957; Rodgers, 1977).

c. Abscission Responses

Since IAA is the natural auxin, discussion in this section will be focused on responses to it. Responses to the synthetic auxin regulators, such as NAA and 2,4-D, are discussed in Chapter 10. While the synthetic auxins have yielded valuable information in experiments and are valuable agricultural chemicals, it is obvious that in many situations plants do not respond to the regulators as they do to the natural auxin, IAA. Indeed, both NAA and 2,4-D are toxic, although NAA is less frequently so. An early investigation of responses of the date (*Phoenix dactylifera*) showed that NAA inhibited car-

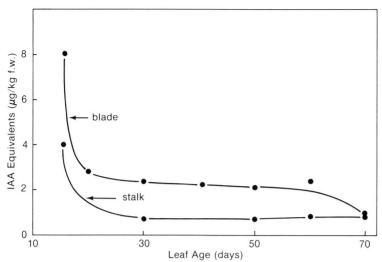

Fig. 4:6.
Changes in extractable auxin with age of bean leaflets and leaf stalks. At 70 days the leaflets were yellow and about to abscise from the stalks. (Redrawn from Shoji et al., 1951)

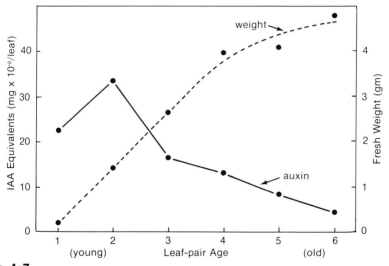

Fig. 4:7.
Changes in diffusible auxin with leaf age of *Coleus blumei*. (Redrawn from Wetmore and Jacobs, 1953)

pel development and led to the degeneration of the carpels (Nixon and Gardner, 1939). Further, in its action as a fruit-thinning agent, NAA and related chemicals induce a number of toxic responses, particularly embryo abortion (Teubner and Murneek, 1955). Functioning of the style and stigma, growth of the pollen tube, and translocation of nutrients are also affected (see Addicott, 1976).

In subsequent discussions the designation IAA will be used whenever the pure substance was employed in the research. If a preparation was used that contained natural auxin of unknown purity, the designation auxin will be used.

(1) Retardation Laibach's student Mai (1934) applied auxin (in the form of orchid pollen) to the debladed petioles of four species of *Coleus* and more than a dozen other genera. In most cases the auxin doubled or quadrupled the time to abscission of the debladed petioles. With three of the *Coleus* species, abscission took eleven times as long when auxin was applied. Shortly after this work, La Rue (1936), using IAA on debladed petioles of *Coleus blumei* and *Ricinus communis*, confirmed the strong retardations. And since that time, there have been innumerable confirmations of the ability of IAA to retard abscission when applied in a position distal to the abscission zone (Figs. 4:8, 4:9), usually to debladed petioles or pulvini of explants (see Addicott, 1970). In addition, general applications, such as brief immersion of an explant (Livingston, 1950), or sprays to branches or entire plants frequently retard abscission. However, sprays are sometimes ineffective (see, e.g., Nixon and Gardner, 1939). The lack of response could be due to any of a number of factors, such as poor penetration or rapid inactivation of the IAA (see Chapter 10).

Further, proximal applications, while they usually accelerate abscission, have retarded it in certain circumstances (Fig. 4:9). The retardation is correlated with a high dosage of IAA and susceptible research material, such as small explants (Portheim, 1941; Chatterjee and Leopold, 1963).

(2) Acceleration The discovery that IAA applied proximal to the abscission zone of an explant would accelerate abscission (Addicott and Lynch, 1951) identified a new role for auxin in the control of abscission. Subsequent investigators found that applications of auxin in positions proximal to the abscission zone almost invariably accelerate abscission. (Two exceptions are mentioned in the paragraph just above.) Accelerating proximal applications are commonly made to the petiole side of leaflet explants and to the stem or hypocotyl tissues of explants of cotton cotyledonary nodes (Fig. 4:9). An active shoot-tip tends to promote the abscission of the older leaves on the shoot (Louie, 1963; Rosas et al., 1976). With young plants of

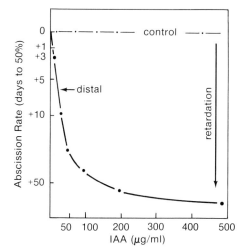

Fig. 4:8.
Retardation of abscission by distal applications of IAA to cotton explants. (Redrawn from Louie and Addicott, 1970)

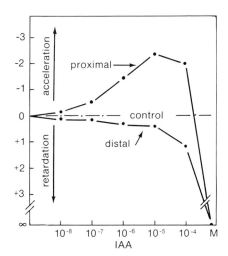

Fig. 4:9.
Acceleration and retardation of abscission by proximal and distal applications of IAA to bean leaflet explants. (Redrawn from Chatterjee and Leopold, 1963)

Coleus, W. P. Jacobs (1955) showed that auxin in the stem coming from nearby intact leaves accelerated the abscission of debladed petioles. Young leaves above the debladed petioles were the most effective in the acceleration, but older leaves below debladed petioles also had some accelerating effect. Jacobs showed that these accelerations were the result of proximal auxin. These observations were confirmed by Louie (1963) on cotton plants in the field. He found that removal of the apical bud and young leaves delayed the abscission of debladed petioles on the lower parts of the plant. The abscission accelerating influence of the intact apical bud could be replaced by application of IAA. (Other investigations that found evidence of abscission acceleration by proximal auxin include Laudi, 1956; Terpstra, 1956; Vendrig, 1960; Chatterjee and Leopold, 1963; Louie and Addicott, 1970; and see the reviews: W. P. Jacobs, 1962; and Addicott, 1970.)

Under some circumstances, distal applications of IAA can accelerate abscission. A response curve for such acceleration is shown in Figure 4:12. Lyon (1964) found that distal applications of low amounts of IAA would accelerate abscission in cotton explants if the explants had been taken from seedlings fourteen days old or younger. If the seedlings from which the explants were taken were sixteen or more days old, all amounts of IAA retarded abscission.

Abscission acceleration can also result from application of IAA or auxin regulators distal to the abscission zone if the application is delayed. Rubinstein and Leopold (1963) discovered that application of NAA to bean explants which was delayed for nine or more hours after the explants were cut would accelerate abscission. This response was confirmed with IAA (Fig. 4:10) by Chatterjee and Leopold (1963). Their results showed that IAA applied during the first several hours retarded abscission, as had been shown many times before; but by delaying the application, abscission was actually accelerated. This was a finding of some significance, as is discussed in later sections.

(3) Stages I and II A *critical period* in abscission responses to auxin regulators was recognized by Barlow in 1950. Working with debladed petioles on apple trees in the field, he observed that if application of NAA was delayed for four days or more, the application was without effect in retarding abscission. Application delayed one or two days gave some retardation, but application made immediately after deblading resulted in the most prolonged retardation. Barlow concluded that there was a critical period after removal of the natural hormone supply by deblading within which abscission was initiated. To be effective in retardation, the applied auxin-regulator should reach the abscission zone before the end of the critical period.

Two phases in abscission responses to ETH were detected by S. Yamagu-

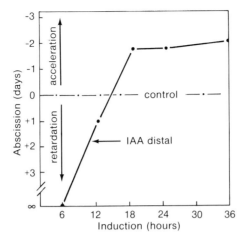

Fig. 4:10.
Acceleration of abscission by delayed applications of IAA to bean leaflet explants. (Redrawn from Chatterjee and Leopold, 1963)

chi in 1954. From a series of experiments with explants from the primary leaves of bean, he found evidence for *phase I*, cessation of the inhibition imposed by the subtending organ; *phase II*, the period during which the chemical and physical processes of separation occur. He estimated that, in his bean explants, phase I lasted about 18 h, because short exposures to ETH during that period were relatively ineffective in accelerating abscission. After 18 h, exposure to ETH strongly promoted abscission.

Rubinstein and Leopold (1963), from their experiments with delayed applications of NAA to bean explants, independently of Barlow and Yamaguchi, described two stages of abscission. *Stage I* was a period during which application of auxins tended to retard or inhibit abscission. It appears identical with Barlow's "critical period" and Yamaguchi's "phase I." *Stage II* followed stage I, and was identified as a period during which auxins accelerated the processes of abscission. It appears identical with Yamaguchi's "phase II." The terms *stage I* and *stage II* are now the accepted usage. Subsequent investigations have given further information on the roles of auxin and ETH in the stages.

Stage I appears to be of indefinite length, commencing from at least the time the leaf is fully mature and persisting until the terminal physiological activities of the separation layer are initiated. Jackson and Osborne (1972) describe stage I as an auxin-dependent stage of relative insensitivity to ETH, supporting Yamaguchi's observations. All indications are that stage I is a period during which the normal auxin inhibition of abscission gradually diminishes and is lost. The length of time required for an explant to accomplish the changes of stage I decreases with the age of the leaf from which it is taken. With very old leaves, the length of time for stage I can be quite short (Chatterjee and Leopold, 1965).

Stage II is characterized by the ability of the explant to respond to both IAA and ETH with accelerated abscission (Yamaguchi, 1954; Chatterjee and Leopold, 1965). During stage II, the active biochemical and ultrastructural changes of separation take place, and the break strength of the abscission zone declines rapidly. However, the line between the end of stage I and the beginning of stage II is not sharply demarked; the transition appears gradual. As far as can be estimated from the relatively early decline in break strength of explants, biochemical changes and the secretion of hydrolytic enzymes appear to be initiated before the end of stage I (Morré, 1968; dela Fuente and Leopold, 1969).

(4) Response Curves When the abscission responses to a series of concentrations of IAA are plotted, the curves have shown three forms: monophasic, biphasic, and multiphasic. Monophasic response curves are by far the most common. They indicate little or no response at low dosages, and retardation or acceleration at high dosages. Such curves appear in Figures 4:8–4:10, and are characteristic of most abscission responses to concentrations of auxin regulators, and the other hormones as well. Occasionally, biphasic response curves are obtained. These have been found most frequently following application of NAA to relatively large explants (Gaur and Leopold, 1955; Biggs and Leopold, 1958), but they also are obtained from proximal applications of IAA (Fig. 4:9) (Chatterjee and Leopold, 1963). A biphasic response to distal applications of IAA to young cotton explants is also obtained (Fig. 4:12) (Lyon, 1964).

The physiology of the biphasic response is not clear, but it is significant that, for the most part, the biphasic curves are characteristic of large explants. This suggests that the response involves a reaction between the auxin and the tissue distant from the abscission zone, rather than a direct effect on the abscission zone itself. Various suggestions have been made as to the reaction at the distant site of application. IAA and the regulators could (i) promote the mobilization of nutrients to the site of application and (ii) stimulate the production of ETH; and (iii) in the case of the auxin regulators, a mildly toxic reaction could lead to an acceleration of abscission (see Chapter 10). Acceleration of abscission by distal application of low dosages of IAA to young cotton explants is difficult to analyze. The explants would certainly be in stage I at the start of the experiment, and distal applications would retard under all other circumstances. Our understanding of auxin physiology must be considerably expanded before a reasonable explanation for that effect can be formulated.

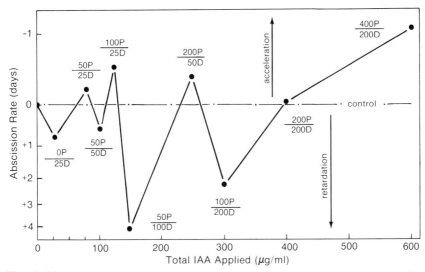

Fig. 4:11.
Multiphasic abscission-response curve to various combinations of distal and proximal applications of IAA. (Redrawn from Louie and Addicott, 1970)

When Louie and Addicott (1970) applied a wide variety of concentrations both distally and proximally to cotton explants, they obtained a multiphasic response curve (Fig. 4:11). The total amounts applied varied widely. The important aspect of the response was that whenever the amount applied distally exceeded the amount applied proximally, abscission was retarded; and whenever the amount applied proximally exceeded the amount applied distally, abscission was accelerated. It is obvious that any number of such curves could be produced, depending on the combinations of simultaneous distal and proximal applications. Perhaps the most significant aspect of these results is their demonstration of the importance of the gradient of applied auxin as a controlling factor in abscission. In this response the total amount of auxin applied to the explant had no influence on the rate of abscission.

d. Auxin Physiology in Relation to Abscission

The hormone relations of any tissue are influenced by three major aspects of hormone physiology: production, transport, and inactivation. With respect to auxin, these aspects of its physiology have been studied in great detail and reviewed many times. For our purposes, a few of the more salient features should be mentioned. Auxin production is high in young tissues, and drops off as the tissues age (e.g. Went and Thimann, 1937; Wetmore and Jacobs, 1953). Further, auxin levels are influenced by a number of en-

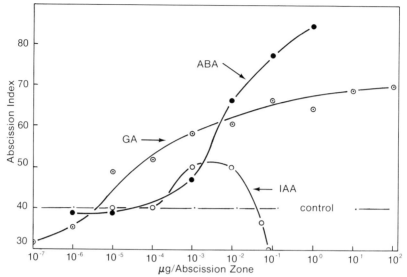

Fig. 4:12.
Abscission-response curves to abscisic acid, gibberellic acid, and indoleacetic acid applied to cotton explants. (Data of J. L. Lyon)

vironmental factors, particularly the supply of N and Zn. The abscission behavior of organs closely parallels their production of auxin.

Mai (1934) demonstrated that the ability of auxin to retard abscission was correlated with the ability of petioles to transport auxin. Subsequent workers have found that as leaves age, the ability of the petioles to transport auxin declines (see, e.g., Storey, 1957; Beyer and Morgan, 1971). Further, chemicals that lower the auxin-transport capacity of tissues also promote abscission. These include TIBA, ETH and DPX1840 (Beyer, 1972).

Most plant tissues have the ability to inactivate IAA, and this ability tends to increase with age.

e. Auxin Gradient

As is evident from even a cursory perusal of this book, auxin is the internal factor that has the most important influence(s) on the control of abscission. This situation is in part an artifact stemming from the fact that auxin was the first of the major hormones to be identified; and in consequence, the auxin relations of abscission have been under investigation for nearly fifty years. But it is significant that as the other factors involved in the control of abscission are investigated, these factors are soon found to affect the auxin physiology of the plant as well. From consideration of all of these influences, it is impossible to escape the conclusion that auxin has a central role in the control of abscission. The biochemical aspects of this role await future refinements and developments. The physiological aspects of the role, however, now seem very clear and have been summarized in the preceding section. While no single concept can encompass all that is known or has been claimed for the role of auxin in abscission, the concept of an auxin gradient across the abscission zone summarizes the significant aspects of auxin physiology in a very reasonable way. The auxin-gradient concept was conceived in terms of the abscission zone of the leaf. This is exposed to two auxin influences: a flow of auxin from the leaf, and a flow of auxin down the stem. Auxin coming from the leaf and approaching the abscission zone from a distal direction tends to delay and slow abscission. Auxin moving down the stem tends to accelerate abscission. This general concept is supported by a variety of investigations of the production and movement of auxin in the plant (see especially Jacobs's 1955 "auxin-auxin balance" evidence), and by experiments in which the auxin gradient has been manipulated by applications distal or proximal or both. The conclusion is inescapable that the auxin gradient across the abscission zone is a primary controlling factor in abscission. Many of the other factors that influence abscission are now known to influence auxin physiology, at least as a part of their effects.

When the concept of an auxin gradient as a controlling factor in abscission was enunciated (Addicott et al., 1955), it appeared to be the central and perhaps the overriding mechanism in the control of abscission. Since then, the involvement of other hormones in the control of abscission, notably ABA and ETH, has been demonstrated; and at times, each of these seems able to initiate and precipitate abscission. Yet it is significant to note that neither ABA nor ETH is completely able to overcome the influence of auxin (see Figs. 4:13, 4:14, and later sections). Although there have been a few experiments with IAA or auxin-regulators, the results of which have been interpreted as being inconsistent with an auxin gradient, for the most part, those experiments involved treatment that prohibited the functioning of the gradient, and hence, really do not bear on the validity of the concept. The auxin gradient remains, therefore, as a central factor in the control of abscission, and perhaps the single most important one. It is recognized, of course, that many other physiological factors also influence abscission, and most of these factors can modify to some extent the function of the auxin gradient.

f. Mechanism of Auxin Action

The involvement of auxin in a specific biochemical reaction basic to abscission has not yet been demonstrated. However, the recent identification of biochemical interactions of auxin with the plasma membrane is an encouraging advance (Morré and Cherry, 1977). Once the biochemical role of auxin in growth is understood, determination of the precise reactions in auxin's control of abscission should become feasible.

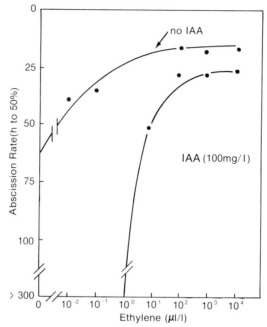

Fig. 4:13.
Interaction of ETH and IAA in abscission of cotton explants. Note that ETH cannot completely overcome the retardation of IAA.

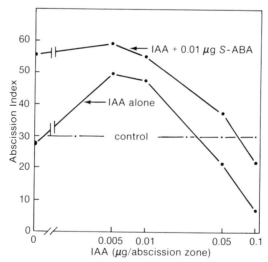

Fig. 4:14.
Interaction of ABA and IAA in abscission of cotton explants. (Redrawn from Smith et al., 1968)

In contrast to the lack of specific biochemical information, the function of auxin in the control of abscission is quite clear in terms of general physiology. Auxin approaching the abscission zone from a distal direction inhibits abscission. The inhibition is released by a decline in the amount of distal auxin, setting in motion the biochemical activities that culminate in the secretion of hydrolytic enzymes and the separation of the distal organ. However, the triggering decline is a relative one. An increase in auxin approaching the abscission zone from a proximal direction will also promote abscission, emphasizing the importance of the auxin gradient in abscission control. It is of value and importance in the life of plants that abscission is sensitive to auxin gradients rather than to absolute amounts of auxin. The auxin gradient gives an indication to the abscission zone of the relative importance to the plant of the subtending organ. If a plant is weak and poorly supplied with nitrogen, the overall auxin levels will still be low. In such circumstances it is likely to be advantageous to retain an organ, particularly a leaf, as long as possible. In other circumstances, when a plant is vigorous and the auxin levels are high, it still is desirable to maintain homeostasis and abscise the least vigorous of leaves and fruits.

The auxin accelerations of abscission likewise can only be explained in general physiological terms. The accelerations that are related to proximal auxin are understandable in terms of strong sink activity of vigorous organs that are high in auxin. The high auxin levels moving away from a strong sink serve as a signal to the plant to translocate nutrients to the sink. Thus, there is a withdrawal of nutrients from weaker organs in response to the signal of auxin from more vigorous organs. It is reasonable that the abscission zone can sense both auxin gradients and the levels of nutrients passing through the zone. Possibly, ETH is involved in the experiments where IAA and auxin-regulators applied in proximal locations accelerate abscission. Both wounding and application of auxin stimulate the release of a burst of ETH. Under some experimental conditions, this ETH could add to the promoting influences. However, in many situations, and especially in those where the promoting influence is coming from an intact bud or fruit, it is unlikely that ETH is involved; and indeed, Jacobs (1955) eliminated the possibility of such involvement in the leaf abscission of *Coleus* shoots.

The abscission promotion by delayed applications of IAA is a surprising response in light of the retardations following most other applications. No real explanation is at hand, but this and other responses to applied IAA suggest that IAA tends to support and augment whatever the ongoing processes may be at the time of application. Thus, IAA tends to promote and continue the inhibition of abscission when that is still in effect, and it tends to promote the processes of separation once they have begun.

The acceleration induced by low dosage of IAA applied to the petiole stumps of young cotton explants also has no ready explanation. Such material is relatively high in auxin and has a relatively slow rate of abscission. Because of the high levels of endogenous auxin, wound ETH from application of IAA would not be expected to be effective. Clearly, much remains to be learned about the auxin relations of abscission.

2. Abscisic Acid (ABA)

a. Discovery and Hormonal Nature

The research that led to the isolation and identification of ABA had its beginning in a study of the auxin relations of developing cotton fruit. A substance that interfered with the growth response in the *Avena* bioassay was present in the young fruit, and was present in highest amounts as young fruits approached abscission. The initial program was led by H. R. Carns; and among numerous associates, J. L. Lyon participated continuously and K. Ohkuma was responsible for the chemical isolation and identification of ABA. The major investigation (Addicott et al., 1964) extended over a period of ten years. The abscission of cotton explants was used as a bioassay to guide the isolation procedures. At about the same time, van Steveninck (1959b) obtained evidence of a substance produced in developing fruit of lupines which induced the abscission of flowers and young fruit. This substance was eventually shown to be ABA (Porter and van Steveninck, 1966). In 1963, it became apparent that a substance present in senescent leaves of *Acer* could inhibit the growth of terminal buds (Wareing et al., 1964). The early experiments showed that the active substance induced considerable leaf abscission, as well as growth inhibition (Eagles and Wareing, 1964), and eventually it was demonstrated that the active substance from leaves of *Acer* was ABA. It is now apparent that several other investigators working with growth-inhibiting or dormancy substances were also, in fact, working with extracts and materials that were rich in ABA (Milborrow, 1967). The result has been the impressive spectrum of physiological and hormonal functions that are now known for ABA.

Early in their research, Carns and his associates wondered if ABA might have a hormonal function in the abscission of young cotton fruit. They demonstrated that such was indeed the case: ABA increases in the young fruit at the time of their heavy abscission, and when reapplied either to young fruit or to defruited pedicels, the ABA promoted abscission, a demonstration that parallels the classical demonstration that auxin is a hormone promoting plant growth. The early work of van Steveninck (1959a,b) and of Eagles and Wareing (1964) strongly suggested that in both lupine fruit abscission and leaf abscission of *Fraxinus*, ABA was functioning as a hormone. (Since the early experiments, it has been discovered that ABA is possibly the most important hormone prolonging seed dormancy, and that it has a central role in the action of the stomatal guard cells.)

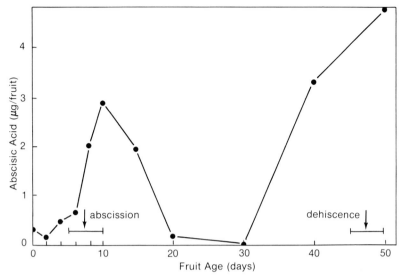

Fig. 4:15.
ABA and the development of the cotton fruit. Note the correlation of high ABA with young-fruit abscission and with fruit dehiscence. (Redrawn from Davis and Addicott, 1972)

General recognition and acceptance of ABA's role in abscission was slow in coming. This was due in part to an early desire to exploit ABA's properties as a dormancy-inducing and -prolonging substance; hence, its abscission-inducing properties were sometimes minimized. Also in ABA's early years, abscission researchers found it more convenient to work with ETH, which could be easily assayed by gas chromatography, whereas assays for ABA were slow and laborious. In that period the possible involvement of ABA in abscission often was overlooked or ignored. Gradually, however, an impressive number of researches have been published that support, confirm, or extend the early demonstrations of ABA as an abscission-accelerating hormone.

b. Correlative Occurrence

Since the pioneering investigations mentioned above, ABA has been demonstrated in the leaves, flowers, and fruits of a wide variety of higher plants. In most cases, ABA has been found to increase as maturity and abscission approach (see, e.g., Fig. 4:15). Many of these studies have, in fact, established or strongly indicated a hormonal role of ABA in the promotion of abscission, although usually that has not been a direct object of the investigation.

A significant example of correlative occurrence is found in the recent

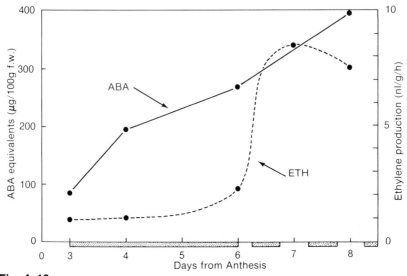

Fig. 4:16.
Changes in ABA and ETH during darkness-induced abscission of young cotton fruit. Treatment was started three days after anthesis. Dark treatment and subsequent light periods are indicated by the bars at the base of the figure. Average time to abscission was five days (actually 118 h) from the start of the dark treatment. (Redrawn from Vaughan and Bate, 1977)

work of Vaughan and Bate (1977). They induced abscission of young cotton fruit by placing plants in complete darkness for three days. Plants were then returned to the greenhouse, and abscission occurred approximately two days later. ABA concentrations in the young fruit doubled during the first day in darkness, tripled by the end of the third day, and were four times the initial concentration at the time of abscission on the fifth day. In the experiment, ETH production showed little change during the first three days, but increased sharply during the two days before abscission (Fig. 4:16). The results showed ABA to be correlated with the initiation of abscission (and quite possibly the later stages as well), while ETH production was correlated with the terminal separatory activity.

c. Abscission Responses

The isolation of ABA from extracts of young cotton fruit was guided by an abscission bioassay where those fractions containing ABA promoted the abscission of petiole stumps of explants of the cotyledonary node of cotton (Fig. 4:12). When applied to explants either proximally or distally, ABA has almost invariably accelerated abscission. In one of the early field trials where a low dosage of ABA was applied to vigorous cotton plants, there was no response (van Overbeek, unpubl.). This led to the widely distributed misunderstanding that ABA was ineffective as an abscission accelerator except for explant abscission. Actually, when a moderate dosage is applied to mature cotton plants, leaf abscission is indeed induced (Johnson and Addicott, unpubl.), and single applications have completely defoliated *Citrus* (Cooper and Henry, 1968) and *Olea* (Hartmann et al., 1968). (Numerous other abscission responses are listed in the Appendix.)

d. Mechanism of ABA Action

The physical and chemical properties of ABA are remarkably similar to those properties of GA and IAA. In extracts, both GA and IAA are contaminants that must be removed in the course of purification of ABA. In the original isolation of ABA, a gibberellin was the final impurity to be removed before ABA was obtained in completely pure form.

Carns's earliest experiments showed that ABA prevented coleoptiles from responding to IAA (see Carns et al., 1958). Later Chrispeels and Varner (1967) showed that ABA could prevent the GA-induced synthesis of α-amylase. Upon careful investigation, however, neither of the two antagonisms were biochemically competitive, nor was the antagonism with CK in the growth of coleoptiles (Wright, 1968; Smith et al., 1968; and see Addicott, 1972). Investigations of effects of ABA on various aspects of plant biochemistry have given results that sometimes seemed inconsistent, but varied with the nature of the biochemical system or presumed biochemical system being investigated. In general, however, ABA tends to inhibit the nucleic acid changes that are promoted by IAA, GA, and CK (e.g., Varner and Ho, 1977), such changes as occur during germination and growth. In contrast, ABA promotes the nucleic acid changes associated with senescence (Sacher, 1973; and see Addicott, 1972). As yet, a specific role for ABA in connection with nucleic acid metabolism has not been established, but it is apparent that in one way or another, ABA promotes protein synthesis, at least the synthesis of certain proteins (Milborrow, 1974). In various systems, treatment with ABA has promoted a synthesis or increased the activity of acid phosphatase, phenylalanine ammonia-lyase, cellulase (Fig. 4:17), invertase, IAA-oxidase, ribonuclease, and PEP-carboxylase. In other systems it has inhibited synthesis or reduced activity of α-amylase, protease, ribonuclease, dipeptidase, fatty acid synthetase, transaminase, invertase, and nitrate reductase.

As is the case with IAA, for the present, the action of ABA is most meaningfully and usefully described in general physiological terms. ABA counteracts IAA, GA, and CK in almost all of their actions, but especially in those actions that promote growth and inhibit or retard abscission. This counter-

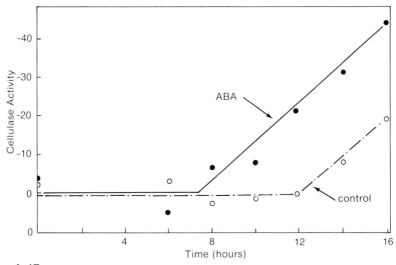

Fig. 4:17.
ABA acceleration of cellulase activity in the presence of saturating (10 μℓ/ℓ) ETH, the control. (Redrawn from Craker and Abeles, 1969)

action is apparent from abscission investigations, as well as investigations of interactions in growth and in some biochemical systems—e.g., the counteractions of GA-induced α-amylase in aleurone preparations. As will be obvious by the end of this chapter, the interactions of the five major hormones in the whole plant can be viewed as a complex balance, with the action of IAA, GA, and CK tending to inhibit and retard abscission, and the action of ABA and ETH tending to initiate and accelerate abscission. (The Appendix includes a check list of the literature on ABA in relation to abscission.)

3. Gibberellin (GA)

a. Correlative Occurrence

Many plant organs and tissues have been examined for GA activity, using bioassays. While no single bioassay gives but a fraction of the possible information on the more than fifty naturally occurring gibberellins that have been identified, still even a single assay method gives some indication of activity. In leaves, GA is high in young developing leaves and falls during senescence (Fletcher et al., 1969; Chin and Beevers, 1970; see Sheldrake, 1973). In developing young cotton fruit, Rodgers (1977) found a rapid increase in GA, whereas GA declined in the abscising young fruit. The amounts in the developing fruit were about ten times greater than in the abscising fruit. In *Citrus*, Goldschmidt et al. (1972) found that GA decreased and almost disappeared from the flavedo of senescent fruits.

b. Lack of Response to General Applications

When preparations of GA became available for extensive application to horticultural and agricultural plants, it was used in hundreds of experiments under a wide variety of conditions. It is noteworthy that among the many responses observed there was very little abscission (see Stowe and Yamaki, 1957; Wittwer and Bukovac, 1958). These observations indicate that GA is seldom a limiting factor in the control of abscission. This aspect of GA physiology is in contrast to that of IAA, as applications of IAA (and especially the synthetic auxins) regularly affect abscission, although the effects are sometimes transitory.

c. Responses to Localized Applications

When applied to individual fruits or branches, GA often, but not always, acts to retard abscission. In early experiments, Brian et al. (1959) applied GA at weekly intervals to branches of a number of deciduous trees in late summer. Little effect was observed on five species, but on seven other species, GA delayed development of autumn-leaf color and leaf abscission. On one species, *Taxodium distichum*, GA accelerated branchlet abscission (cladoptosis). Applied to individual young fruit of tomatoes, cotton, apples, and *Citrus*, GA stimulates fruit development and reduces the amount of young fruit abscission (Rappaport, 1957; Walhood, 1957; Hield et al., 1958; Crane, 1969). In an experiment with cotton, Walhood (1957) applied GA to each young fruit as it appeared on the plant; each fruit was retained and not abscised, whereas two-thirds of the control fruit did abscise. The GA did not, however, affect the yield of the plant. The total amount of seed and lint was the same on treated and untreated plants. Similar results are found where only a portion of the fruit on a plant are treated; while the GA-treated fruit are prevented from abscising, untreated fruit on the same plant do abscise, with the result that the yield is unaffected. These observations support the established view that the fruitfulness of a plant is limited primarily by the ability of the plant to provide nutrients to the developing fruit. In one way or another such observations indicate that the abscission of supernumerary young fruit is induced by deficiency (sometimes localized) of available nutrients (see Chandler, 1951). Apparently, an abundant supply of GA in a particular fruit makes that fruit a stronger sink and better able to attract nutrients when sources are limited.

d. Applications to Petioles

The early experiments of Laibach (1957) and Brian et al. (1955) detected no effect of applications of GA to debladed petioles of *Coleus*. It is likely that in those experiments there *was* some effect that escaped unnoticed, as the careful work of Jacobs and Kirk (1966) with debladed petioles of *Coleus* detected an acceleration from GA applications of about 18 h that was "statistically highly significant." Abscission of petioles of oldest leaves on the *Coleus* shoots were unaffected by GA. Further, the abscission-retarding effect of IAA was greatly reduced by simultaneous application of GA. Again, petioles of older leaves were unaffected by the GA (Jacobs and Kirk, 1966). Debladed petioles of *Morus alba* and *Robinia pseudoacacia* responded to applications of GA in a similar fashion (Trippi and Boninsegna, 1966).

Application of GA to the stem apex of *Coleus* accelerated abscission of the debladed petioles, as does application of IAA. Muir and Valdovinos (1970) found that such proximal applications increased auxin levels in the stem, similar to the GA-induced increases of auxin that had been found earlier in peas and tomatoes (Kuraishi and Muir, 1962; Sastry and Muir, 1963). Muir and Valdovinos concluded that the abscission-promoting effect of GA applied to the stem apex was due to an increase in the level of endogenous auxin.

e. Responses of Explants

Applied to explants of cotton or *Citrus*, GA strongly promotes abscission (Fig. 4:12). In contrast to the lack of response in older petioles of *Coleus*, GA accelerated abscission in cotton explants from seedlings as old as twenty-six days. With cotton explants, also, GA strongly accelerated petiole abscission when applied to the stem stump, proximal to the abscission zone (Carns et al., 1961). Further, applications of moderate to high amounts of GA to one of the pair of petiole stumps of a cotton explant induced abscission of the untreated petiole stump as rapidly as the treated stump (Lyon and Smith, 1966).

As with *Coleus* petioles, simultaneous application of GA with high amounts of IAA could overcome the abscission retardation of the IAA both in cotton explants (Greenblatt, 1965) and in *Citrus* leaf explants (Lewis and Bakhshi, 1968a). In cotton explants, Greenblatt (1965) found that combined application of GA and low amounts of IAA induced much greater acceleration than the maximum possible from either substance alone. Essentially, the two accelerations were additive (not synergistic) (Fig. 4:18). This response indicates that GA accelerates abscission in a manner different from the acceleration by IAA. If the action of GA in this situation were merely to increase the levels of IAA, the combination should have reduced the acceleration of IAA rather than increased it.

Low dosages of GA applied either distal or proximal to the abscission zone of cotton explants retard abscission (Fig. 4:12) (Carns et al., 1961; Addicott, 1970). There is no immediate explanation for such a biphasic re-

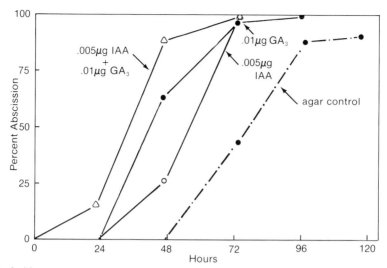

Fig. 4:18.
Abscission acceleration from application of GA and IAA to cotton explants. The combination is additive, not synergistic. (Redrawn from Greenblatt, 1965)

sponse to GA, although at low dosage GA may possibly increase auxin levels, thereby adding to the retardation of abscission. Applied to young cotton plants, GA augments the leaf abscission induced by near-saturating levels of ETH (Morgan and Durham, 1975). Further, as might be expected, GA counteracted the abscission retardation of IAA in experiments with cotton explants in the presence of ETH (Morgan, 1976). High levels of GA may be overriding auxin inhibitions by stimulating the synthesis of hydrolytic enzymes.

Lewis and Bakhshi (1968b) investigated protein synthesis in leaflet abscission zones of *Citrus* and found that GA stimulated protein synthesis in the abscission zone. This synthesis is undoubtedly related to the synthesis of the hydrolytic enzymes of abscission. In contrast, application of IAA to the explants maintained the pre-existing rate of protein synthesis in the abscission zone.

Anatomical and cytological aspects of GA-induced abscission of petiole and stem stumps in cotton explants was investigated by Bornman et al. (1967, 1968). The cytological changes were essentially the same in petiole and stem abscission, involving a limited amount of cell division, followed by hydrolysis of cell-wall constituents. Considerable cellular hypertrophy developed in the petiole abscission zones following application of higher dosages of GA to the stem stump.

f. Mechanism of GA Action

Although the amount of research on GA and abscission has been relatively small, at least three important actions are now apparent. The first is that applied GA intensifies the ability of an organ to which it is applied to function as a nutrient sink. The net result of this effect is to inhibit or at least delay the abscission of such organs.

A second action of GA that can influence abscission is its ability to increase the synthesis of IAA in plant tissues. In favorable circumstances, as in the stem of *Coleus*, this GA-induced increase in auxin has a promotive effect on abscission (Muir and Valdovinos, 1970). This effect may be responsible for the abscission retardation of low dosages of GA applied to cotton explants. Further, it is quite likely that GA-induced increases in auxin intensify the activity of nutrient sinks.

The third action involves accelerated synthesis of hydrolytic enzymes as exemplified by GA's strong abscission acceleration in cotton explants. This response resembles the classical action of GA-induced synthesis of α-amylase and other hydrolytic enzymes in aleurone cells.

As was mentioned at the beginning of this section, GA seldom appears to be a limiting factor in abscission in the field. Nevertheless, the experiments with explants and debladed petioles discussed here indicate that GA indeed has important functions in the control of abscission.

4. Cytokinin (CK)

a. Correlative Occurrence

Bornman (1969) found that by autumn the concentrations of cytokinin in the distal portions of the leaf of *Streptocarpus molweniensis* drop to one-ninth of the summer value. Similar changes have been found in the leaves of *Salix babylonica* (van Staden, 1977) and in abscising young fruit of cotton (Rodgers, 1977). Lowering levels of CK in the xylem sap of *Salix viminalis* were also correlated with autumnal leaf abscission (Alvim et al., 1976). In *Ginkgo*, van Staden (1976) found CK to be largely in the form of glucosides by autumn, rather than in the form of zeatin and zeatin riboside, which were high in the young leaves. In addition to the foregoing, it will be recalled that the classical (rich) sources of CK were the developing fruits of coconut and maize, so that it is likely that future investigations will disclose numerous correlations between fruit abscission and lowering levels of CK, as Rodgers (1977) already has demonstrated for cotton.

b. Abscission Responses

When a CK (usually kinetin or benzyladenine) is applied to young fruits, it stimulates growth, and it can induce parthenocarpic development and consequently decrease young fruit abscission (see Crane, 1964, 1965; Letham, 1967). CK is also one of the most effective retardants of leaf senescence. It is very effective as a promotor of sink activity of either organs or tissues. For example, when CK is applied to a portion of a leaf, senescence of the treated area is delayed while it progresses at the normal or even accelerated rate in adjacent areas (Leopold, 1964; Mothes, 1964).

As might be expected, CK is an effective retardant of abscission (Gorter, 1964). However, when attention is given to site of application, CK applied to sites distant from the abscission zone tends to accelerate abscission, whereas application directly to the abscission zone inhibited abscission (Osborne and Moss, 1963). Similar results were obtained by Carr and Burrows (1967) with applications to the leaflets (promotion) or the pulvini (retardation) of lupine. As with auxins, Chatterjee and Leopold (1964) found that low concentrations of CK induced a small acceleration of abscission, that all higher concentrations retarded it, and that delayed applications of CK tend to accelerate abscission.

c. Mechanism of CK Action

From the meager information available on CK and abscission, two modes of action are already apparent. The first is the ability of CK to enhance the sink strength of organs or tissues to which it is applied. Thus, CK tends to promote the development and retard the senescence and abscission of fruits and leaves. However, when CK is applied to an explant at even a short distance from the abscission zone, it has the tendency to promote sink strength at the precise site of application, and consequently to accelerate abscission changes in the abscission zone. In this respect its action resembles that of the synthetic auxins such as NAA.

The second effect of CK, that of retarding senescence, is observed following applications directly to the abscission zone, or if at a small distance from the abscission zone, applications of sufficiently high dosage that substantial amounts of CK reach the abscission zone promptly. In these cases the action appears to be similar to that of IAA, maintaining the status quo of the abscission zone and delaying the onset of abscission.

5. Ethylene (ETH)

a. Occurrence

The deleterious effects of coal gas on growth and development of plants, including the induction of abscission, have been known for more than a hundred years. At the turn of the century, ethylene was identified as the component that has the most adverse effects on plant growth and development (Neljubow, 1901). Here the matter rested until it became clear that plant tissues themselves, particularly tissues that were injured or infected with certain organisms, could produce appreciable amounts of ETH (see Burg, 1962). Then with the development of sensitive assay methods, it became possible to investigate the ETH physiology and biochemistry of plants (Pratt and Goeschl, 1969). Now the relative ease with which ETH can be assayed by gas chromatography has given great impetus to the study of ETH in plants. The current methods for ETH are so simple, sensitive, and efficient that research with other hormones affecting abscission has been neglected because methods for their assay remain laborious and expensive. (It would be a mistake to measure the significance of ETH in abscission by the number of publications touching on the relationship!)

It is now apparent that many, but not all, fruits produce considerable amounts of ETH in correlation with ripening and abscission (Walsh, 1977; and see Pratt, 1974). Also, aborting, injured, infected, or detached organs, such as leaves, petals, and fruits, produce appreciable amounts of ETH (Adato and Gazit, 1977; Burg, 1962; Aharoni, 1978; Lipe and Morgan, 1972). All plant cells probably have the ability to produce ETH (Burg, 1962), but the amounts produced appear related primarily to the nutrient levels of the organ. The leaves of young cotton plants produce considerable ETH, rather more in the petioles than in the blades (McAfee and Morgan, 1971), and cotton cotyledons more than doubled their rate of ETH production as they senesced and approached abscission (Beyer and Morgan, 1971). However, it is important to note that Hall et al. (1957) found that mature, green leaves of cotton produced small amounts of ETH, while the amount of ETH produced by senescent chlorotic leaves was below the sensitivity of their method. As vegetative tissues of peas mature, Burg (1968) found a more than tenfold decrease in the ability of tissues to produce ETH. In contrast, Jackson and Osborne (1970) found that the petioles of *Prunus* and *Parthenocissus* showed increased ETH production as abscission approached. However, with bean leaflet explants that were kept in flasks and vented at 12-h intervals, they found that abscission came at the end of 80 h, marked by very low ETH production by the explants. Shortly after abscission, ETH production rose in the pulvinus, and still later in the petiole (Fig. 4:19). The explant results hardly support the assertion that ETH is "the natural regulator of abscission" (Jackson and Osborne, 1970). In this connection, Abeles' (1973) summary-comment should be noted: "The observation that some plants are readily defoliated by ethylene, or others less so, and still others not at all, suggests that in some plants ultimate control depends on factors other than ethylene."

Many chemicals that induce abscission either in the field or in explants in the laboratory also induce the production of ETH (Morgan and Hall, 1964; Jackson, 1952), but such ETH production cannot always be readily correlated with abscission. For example, Lewis et al. (1968) in their study of the effects of various chemicals on ETH production during abscission of *Citrus* fruit found that, whereas ETH production usually increased when abscission was accelerated, some treatments with iodoacetic acid increased ETH

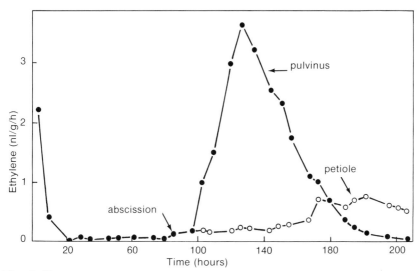

Fig. 4:19.
ETH production by pulvinus and petiole of bean leaflet explants. (Redrawn from Jackson and Osborne, 1970)

production but did not accelerate abscission. They found treatments with GA and coumarin that accelerated abscission but did not increase ETH production. From the foregoing, it appears clear that while ETH is usually associated with abscission, it is not a universal requirement or prerequisite for abscission.

Early studies of ETH production sometimes failed to take into account wound ETH that is released upon injury, such as that which occurs with the excision of an explant or with the application of any of a number of chemical agents. Such ETH production can persist from a few hours to more than a day, depending on the species and condition of treatment (see Abeles, 1967). Under most conditions, wound ETH dissipates and may have little influence on abscission, but if it is confined with the explants, it can have an accelerating effect (Jackson and Osborne, 1970).

It is important that the ability of a tissue or an organ to produce ETH is correlated with its supply of substrates and other nutrients. This relationship appears to result from the fact that methionine is the major precursor of ETH. Hence, tissues that are rich in amino acids and well supplied with energy-yielding substrates can be active producers of ETH. Conversely, as tissues become depleted, particularly of amino acids, their ability to produce ETH will drop. From this argument it is unlikely that a moribund organ in the advanced stages of senescence would be able to produce sufficient amounts of ETH to affect its abscission. However, as is noted elsewhere, the abscission zone maintains a high level of nutrients until it has

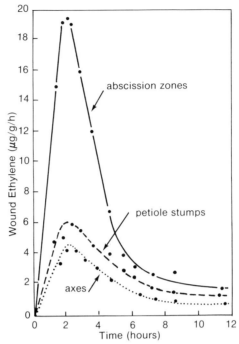

Fig. 4:20.
Wound ETH production by portions of cotton cotyledonary-node explants. (Redrawn from Marynick, 1977)

functioned; and it can be expected, therefore, that ETH produced in the cells of the abscission zone could participate in the processes of separation. Evidence in support of this point had been obtained from *Parthenocissus* (Jackson and Osborne, 1970), where the most basal portion of the petiole, which presumably included part of the abscission zone, produced twice as much ETH after excision as a segment of the petiole that was 2 cm distant from the abscission zone. In an analysis of wound ETH production of cotyledonary node explants of cotton, Marynick (1977) found that the abscission zones also produced approximately twice as much wound ETH as either the petiole stumps or the proximal, axial tissues (Fig. 4:20). After the subsidence of wound ETH production, the abscission zones continued to produce ETH at a higher rate than either the proximal or distal tissues.

b. Abscission Responses

As was already mentioned, abscission is one of the conspicuous responses to ETH. In particular, leaf, petal, and fruit abscission is commonly accelerated by ETH. Dehiscence also can be promoted by ETH, and the separation of fruit hulls of *Juglans* and *Carya* (Sorber and Kimball, 1950; Lipe and Morgan, 1972; Morgan, 1976). Chemicals that release ETH, or ETH itself, have been used to promote leaf abscission, to thin young fruit, and to promote the abscission of mature fruit to facilitate harvest (see Chapter 10).

The concentration of ETH required to induce abscission varies from 0.01 to 1.0 µl/l (see Abeles, 1973). The length of time to which plant materials must be exposed to ETH in order to induce abscission varies greatly. For susceptible flowers (e.g., *Geranium* spp.), petal abscission follows after a few minutes' exposure (to coal gas, Fitting, 1911). With their bean explants, Jackson and Osborne (1970) found that the explants were essentially insensitive to ETH for the first 24 h after excision, but abscission followed a second 24 h of exposure. The period of insensitivity coincides with stage I of auxin responses, and appears to be a period in which the auxin levels in distal tissues are declining. At the end of the period, abscission can proceed normally, and in most situations the abscission zone becomes susceptible to acceleration by ETH. The period of insensitivity can be prolonged indefinitely by the application of IAA.

ETH does not induce leaf abscission of some species as was already noted (Abeles, 1973). Further, treatment with ETH (as CEPA) retards fruit abscission of the oil palm (Chan et al., 1972) and substantially reduced activity of cell-wall-degrading enzymes on preparations of prune mesocarp tissues (Weinbaum et al., 1979).

Anatomical responses to ETH have received some attention. In explants of bean leaflets, ETH-induced abscission proceeds rapidly, and separation is accomplished without the cell divisions that precede separation in untreated explants (Brown and Addicott, 1950). ETH does not seem to merely accelerate the entire process of abscission. After exposure to ETH, the bean explants showed few indications of either cell divisions or hypertrophy that occur in the untreated explants. Further, in appropriate materials, high dosage of ETH can induce adventitious separation layers (Webster and Leopold, 1972). Application of 1,000 µl/l to stem explants of bean usually induced a separation layer at the base of the internode, and sometimes induced one or more well up in the internode. In such cases, after one separation layer was initiated, others developed below it sequentially. This adventitious internodal abscission appears similar to that induced by TIBA in intact bean plants (Whiting and Murray, 1948). The similarity of these two responses is probably related to the fact that both ETH and TIBA drastically alter auxin physiology.

In recent years it has become fashionable to induce abscission by treatment with ETH prior to cytological study. Such investigations assume that ETH treatment is merely a convenience to facilitate the timing of sample collection. But in fact, several anatomical and ultrastructural differences between ETH-induced abscission and natural, untreated abscission have already been described (see, e.g., Valdovinos and Jensen, 1974). Future workers have the responsibility of demonstrating whether or not ETH-induced abscission is identical with natural abscission.

c. Interrelation with Hormones and Regulators

The responses to ETH are modified in various ways by interactions with the nonvolatile hormones and regulators. An early investigation followed the interaction of simultaneous application of IAA and ETH to cotton explants (Addicott, 1965). While the two hormones tended to counteract each other in abscission (Fig. 4:13), it was noteworthy that even the highest concentrations of ETH were unable to completely overcome the retardation of IAA. In the cotton explants the interaction is clearly biochemically noncompetitive.

Production of ETH is stimulated by wounding, as was already mentioned. It is also stimulated by stress of various sorts, including the stress from applied chemicals. Such chemicals include the hormones and regulators that influence abscission; and it is interesting that ETH production is stimulated by chemicals that retard abscission, such as IAA and NAA, as well as by chemicals that accelerate abscission (see, e.g., Palmer et al., 1969). Abeles (1967) examined the ability of a selected series of abscission accelerants and retardants to stimulate production of ETH in explants of *Cassia*, *Coleus*, and cotton. The results showed no consistent relation between the amount of ETH produced by the chemicals and the rate of abscission. The rates of abscission that he observed were clearly influenced by the applied retardants and accelerants as much or more than they were by the ETH produced. A similar conclusion was reached by Palmer et al. (1969) from experiments in which GA and ABA were applied to *Citrus* leaf explants. The amounts of the two hormones applied gave identical rates of abscission, but the ABA applications produced considerably more ETH than did the GA applications. The results suggested that under these conditions, the levels of ETH were not a determining factor in the abscission rate. In further experiments, Palmer et al. also found that pretreatment of *Citrus* explants with GA or ABA for 16 h before application of ETH accelerated abscission. They interpreted these results as indicating that prior to the addition of ETH, the two hormones had initiated and accelerated the process of abscission.

Cycloheximide (CHI) is a well-known inhibitor of protein synthesis and has been a useful aid in biochemical experiments. When applied to *Citrus* fruit, it induces the flavedo of the fruit to produce substantial amounts of wound ETH. The amounts produced appear to account for the abscission acceleration of the CHI (Cooper and Henry, 1974). The response is all the more interesting because *Citrus* fruit normally produce relatively little ETH as they ripen. In their experiments, Cooper and Henry found that GA tends to inhibit the *Citrus* fruit's responses to CHI, while simultaneous application of ABA promoted both ETH production and abscission.

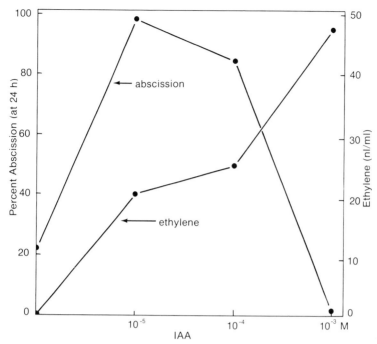

Fig. 4:21.
ETH production in relation to IAA acceleration and retardation of abscission in *Coleus* explants. (Redrawn from Abeles, 1967)

Perhaps the most important interrelation between ETH and the other hormones is its interaction with IAA. As was already mentioned, applications of IAA and ETH tend to counteract each other. Beyond that, application of IAA stimulates production of ETH by tissues. Abeles (1967) examined ETH production by *Coleus* explants when they were subjected to proximal application of a range of IAA applications. ETH production rose steadily with increasing concentrations of IAA. However, abscission was accelerated by the lower concentrations, whereas the highest concentration inhibited abscission (Fig. 4:21). Similar results were obtained by Marynick (1977) with distal applications of IAA to cotton explants. Such data make it difficult to ascribe a controlling role to ETH in abscission. In this case at least, the inhibiting action of IAA overrides any promotive effect of ETH. Further, it seems entirely possible that in situations like that shown in Figure 4:21, IAA is having a promotive effect over and beyond that of increased ETH production. This possibility has been given only limited attention, but must be thoroughly explored before the mechanism of IAA accelerations can be understood.

Because some organs, such as marcescent leaves and some fruits, abscise under conditions where little or no ETH appears to be produced, it is important to learn if abscission is feasible under conditions where the external ETH is reduced to as low a level as is technically possible. Such levels can now be maintained well below the threshold for ETH stimulation of abscission. The two principal methods involve (i) reduced atmospheric pressures (so-called hypobaric pressures), and (ii) use of mercuric perchlorate to absorb ETH. In extensive experiments with *Citrus*, Cooper and Horanic (1973) found that hypobaric conditions would prevent the ETH-induced fruit abscission that otherwise would follow from application of CHI. Interestingly enough, when hypobaric *Citrus* fruits were supplied with ABA, abscission was accelerated. In related experiments with explants of cotton cotyledonary nodes where the explants were confined in small chambers in the presence of mercuric perchlorate, Marynick (1977) found that the abscission responses to IAA, ABA, and GA were unaffected by the absorption of external ETH. ABA and GA accelerated abscission, and IAA accelerated at low levels and retarded at high levels, as had been previously described for explants in petri dishes (Fig. 4:12). Also, bean explants confined with mercuric perchlorate eventually absciced (Jackson and Osborne, 1970). The above results show that abscission can take place in the essential absence of external ETH. Unfortunately, such experiments tell us nothing about the internal ETH in the abscission zone, which conceivably could contribute to abscission in a limited way. The *Citrus* experiments suggest that ETH was, indeed, not produced in sufficient quantities in the fruit abscission zones, while the question remains open for cotton and bean explants.

Auxin physiology has been found to be greatly affected by ETH and in ways that in turn modify auxin's influences on abscission. For example, ETH lowers levels of IAA in the plant by stimulating the activity of IAA-oxidase (Morgan et al., 1968; Hall and Morgan, 1964). Also, monophenols that are cofactors for IAA-oxidase also accelerate abscission (Schwertner and Morgan, 1966). Further, ETH inhibits the transport of IAA in petiole and stem tissues of a number of species, such as those of *Gossypium, Hibiscus, Pisum, Phaseolus,* and *Vigna* (Morgan and Gausman, 1966; Morgan et al., 1968). Auxin transport in other species, such as *Lycopersicon* and *Helianthus*, was little affected by ETH. It is interesting that those two genera have poorly developed leaf-abscission mechanisms. Auxin transport normally declines as leaves become senescent and approach abscission (Beyer and Morgan, 1971). Moreover, treatments with auxins such as NAA tend to prevent the ETH-induced decline in auxin transport (as well as to prevent abscission) (Beyer, 1973). Other transport inhibitors, such as TIBA and DPX-1840, augment the overall effect of ETH in promoting abscission (Morgan and Durham, 1972). Experiments that involved the application of ETH either to the blade of leaves of cotton or to the remainder of the plant showed that the major effect of ETH in promoting abscission of intact leaves is on the blade of the leaf. Exposure of the abscission zone to ETH has a promotive effect, but only after the auxin from the blade, and passing through the petiole, has been reduced below inhibitory levels (Beyer, 1975). In these experiments, application of ETH to the leaf blade alone was not sufficient to accelerate abscission, indicating that there was little, if any, translocation of ETH through the petiole to the abscission zone.

ETH has a number of interesting interrelations with ABA. Early experiments showed that ABA, like many other substances, would stimulate the production of ETH by plant tissues such as explants of cotton (Abeles, 1967) and bean (Craker and Abeles, 1969). In segments of *Citrus* fruit, ABA doubled the rate of ETH production (Hyodo, 1978). In other materials the response can be much smaller. Applied to bean explants, ABA induced only low levels of ETH production (Osborne et al., 1972). With her cotton explants, Marynick (1977) found only a slight enhancement of ETH production over a wide range of increasing amounts of ABA applied to the explants, yet the higher rates of ABA strongly accelerated abscission. In contrast, ETH has induced a rapid increase of ABA in *Citrus* flavedo (Goldschmidt, 1974). This was the earliest change detected following ETH treatment; other senescent changes came subsequently. Since increased levels of ABA commonly are detected following stress, it is likely that other instances of ETH promotion of ABA synthesis will soon be reported. An interesting paradox has recently been described by Kondo et al. (1975). While ABA commonly stimulates some production of ETH by susceptible plant tissues, investigation of its interaction with IAA in the induction of ETH production showed that ABA could strongly inhibit the IAA-induced production of ETH.

d. Mechanism of ETH Action

As with the other hormones, a precise biochemical site of action has not yet been established for ETH. Although many biochemical changes have been detected and measured after ETH treatment, most such responses appear to be secondary consequences. For the present, the general physiological responses to ETH appear to have the most meaning. ETH's influences on abscission fall into two categories: first, those influences that have their primary effect on the organ being abscised; and second, those influences that directly affect the abscission zone. In the first category, we find ETH commonly (but not always) increasing with senescent changes or ripening of the organ, changes that usually precede abscission of the organ. A major consequence of increased ETH levels in leaves and fruits is increased inactivation of IAA and decreased transport of IAA by petioles and pedicels. The net effect of this action is to reduce the amounts of IAA reaching the abscission zone and the intensity of IAA's inhibition of abscission.

The second group of effects of ETH are directly within the abscission zone. Experiments with explants have shown that the abscission zone is usually sensitive to ETH and responds with rapid abscission. More important to considerations of abscission under natural conditions is the production of ETH by the nutrient-rich cells of the abscission zone. Although the amounts of ETH produced may seldom be enough to bring about rapid rates of abscission, even relatively low amounts would be somewhat promotive, according to our present view of the processes.

In its action within the abscission zone, ETH has a special role. Its production does not go up until appropriate signals have reached the abscission zone. The most likely of these signals are: (i) a lowering of IAA levels, (ii) an increase in ABA levels, or (iii) a lowering of the ratio of IAA-ABA reaching the abscission zone. Other signals, such as the changing levels of GA and CK, are very likely involved in at least some kinds of abscission. The net result of the translocated signals is an increased activity in the abscission zone, including increased synthesis of ETH, which in turn functions to promote the subsequent changes of abscission. Thus, in such cases, ETH can be described as a *secondary hormone* analogous to the "second messengers" of animal endocrinology. This concept is entirely consistent with the emerging view that ETH is principally a hormone of local coordination (Varner and Ho, 1976).

Although it is not yet possible to assign to ETH a specific biochemical activity or site of action, there are two general effects of ETH on cell physiology that have been observed repeatedly and that suggest a general biochemical role for ETH, possibly in contrast to a specific one. One of the earliest recognized responses to ETH was a stimulation of oxidative metabolism. This appears to be a widespread, if not universal, response (see Burg, 1962), and has the effect of accelerating any respiration-dependent changes in progress. This response alone can account for ETH's acceleration of abscission. If, in fact, abscission is initiated by a decline in the ratio of IAA to ABA reaching the abscission zone, ETH would have the role of accelerating the enzyme syntheses triggered by the nonvolatile hormones. The other response to ETH by cells is a general increase in permeability, particularly the permeability of the plasmalemma. Such increases have been noticed for some time and have received increasing attention in recent years (Sacher, 1973; Suttle and Kende, 1978). The nonvolatile acidic hormones (IAA, ABA, GA) and CK are chemically closely similar to the active components of known enzyme systems. Thus, it appears likely that in due course the specific biochemical actions and sites will be identified for those four hormones. On the other hand, ETH is so different chemically and physically from the other hormones that it is difficult to imagine a similar role for it. Consequently, the idea of a different kind of function, that of a general increase in permeability, especially in permeability to O_2, which could then lead to an increase in oxidative respiration, is an attractive explanation of the role of ETH. Such a role appears entirely compatible with what is known of the action of ETH in growth processes generally, and particularly in abscission. The observations of increased enzyme synthesis following application of ETH, according to this explanation, would not be due to ETH's forming a biochemically specific molecule but would be due rather to a more general effect, that of increasing the amount of energy-rich compounds available for enzyme synthesis. Such an explanation is strongly supported by the observations that IAA, ABA, and GA in appropriate circumstances accelerate abscission to rates significantly above those inducible by ETH alone.

6. Miscellaneous Abscission Substances, Possibly Hormonal

The previous five sections have been restricted to the five major hormones of higher plants and their abscission physiology. There are, of course, numerous other plant hormones, putative hormones, and naturally occurring substances for which hormonal functions may someday be demonstrated. Some of these substances influence abscission, and it is appropriate to give brief mention to them at this point. Possible candidates for new abscission hormones could well be found among the substances translocated from leaves during senescence. One of the more conspicuous groups of translocated substances includes the amino acids. Several of these were found to accelerate abscission of bean explants or tobacco flowers; they included alanine, glutamic acid, serine, glycine, aspartic acid, phenylalanine, methionine, glutamine, leucine, and valine (Rubinstein and Leopold, 1962; Yager and Muir, 1958a). However, with her bean explants, Osborne (1959) detected no abscission acceleration from glycine, aspartic acid, glutamic acid, valine, histidine, leucine, tyrosine, cysteine, or methionine. In this case the bean explants may have been too vigorous—i.e., too high in auxin to respond to the amino acids. There are at least three ways in which amino acids might promote abscission. The first was suggested by Yager and Muir (1958a), who noted that methionine can serve as a donor of methyl groups to pectic substances and weaken them structurally by replacing the calcium bridges. A second function of methionine is that of the principal precursor of ETH. When supplied in sufficient amounts, methionine is a strong accelerant of abscission (Yager and Muir, 1958b). A third possible function is that the amino acids reaching the abscission zone could contribute to the synthesis of enzyme protein.

Sugars that are also exported from senescent organs have, in general, been found to have little effect of a hormonal nature. Absence of abscission acceleration has been reported following application to bean explants of glucose, fructose, and sucrose (Osborne, 1959), and arabinose, mannitol, and rhamnose (Morré, 1968). Morré made the significant observation that while other sugars are inactive, galactose and galacturonic acid strongly promote abscission. He suggested that they act by a mechanism of substrate induction of pectinase.

The phenolic substances are widely distributed in plants and influence the activity of IAA-oxidase: polyphenols tend to inhibit the oxidase, and monophenols tend to enhance its activity. Thus, phenolic substances can modify auxin levels, and could thereby influence abscission. Tomaszewska (1964, 1968) confirmed this possibility in experiments with debladed petioles of *Deutzia*. She found that polyphenols prolonged the abscission retardation of applied IAA, while monophenols reduced the retardation (and see Schwertner and Morgan, 1966).

Ascorbic acid is distributed widely in higher plants. It has long been known to be essential to the metabolism of animals, and is able to influence a number of plant processes (see Audus, 1972). In young cotton fruit, retained fruits have much higher levels of ascorbic acid than the abscising fruits (Varma, 1977). In *Crataegus*, Tomaszewska and Tomaszewski (1970) noted a large decline in ascorbic acid levels of leaves immediately preceding abscission. They pointed out that since ascorbic acid counteracts the oxidation of IAA and of phenols, ascorbic acid levels may indirectly control leaf abscission through these influences on IAA. On the other hand, ascorbic acid (at rather high dosages) has promoted the abscission of *Citrus* fruit, and sometimes of leaves as well (Cooper et al., 1968).

The literature contains a number of references to abscission acceleration or retardation by diffusates or extracts from plant materials. In general, little can be said about such extracts and diffusates until the active substances are chemically isolated in pure form. The chemical complexity of such preparations is exemplified by a progress report of research on abscission retardants and accelerants from young cotton fruit (Lyon et al., 1972). The investigation examined abscission activity of various fractions prepared in the course of the isolation of ABA. Several substances were identified, some only tentatively. These included (i) ethyl-indoleacetate, which like IAA, accelerated abscission at low concentrations and retarded it at high concentrations; (ii) a bound form of ABA, which when hydrolyzed, strongly accelerated abscission; (iii) phaseic acid, a relative of ABA; (iv) at least two gibberellins; (v) β-bisabolol, a somewhat toxic substance that retarded abscission of explants; (vi) a new substance tentatively designated "acetate X," and (vii) two strong abscission retardants designated "retardant A" and "retardant B" which could not be isolated because of chemical instability. The foregoing summary gives some indication of the complexities that can be encountered in attempts to isolate and chemically identify new substances.

Abscisin I is a moderately active abscission accelerant isolated from extracts of the mature-fruit walls of cotton. It was isolated in pure form, and an excellent infra-red spectrum was obtained (Addicott et al., 1964). However, in spite of extensive efforts, we have been unable to reisolate it (Addicott and Lyon, 1972).

Another abscission substance(s) has also defied isolation and identification. Osborne (1955) found that the diffusate from senescent leaves of bean and other species had the property of accelerating abscission. The diffusate could be collected in agar, and the agar reapplied to bean explants, where it accelerated abscission. Diffusates from young leaves were ineffective in accelerating abscission. She concluded that the diffusate from senescent leaves contained an "abscission factor." Subsequent research showed that diffusates from the leaves of a number of species also accelerated abscission, and the effect was ascribed to an "abscission-accelerating substance" (Osborne, 1958b). A study by Jacobs, Shield, and Osborne (1962) of the ability of diffusates from *Coleus* petioles to accelerate the abscission of debladed petioles showed the presence of abscission-accelerating activity from the petioles of naturally senescent leaves, whereas diffusates from mature leaves did not affect petiole abscission, and diffusates from young leaves retarded it. In that paper the authors renamed the abscission-accelerating factor(s) "senescence factor." In a subsequent paper, Chang and Jacobs (1973) compared the physiological properties of senescence factor (diffusates from senescent petioles of *Coleus*) with ABA. They found that the senescence factor diffusates and ABA were similar in each of the several properties examined. Both accelerated abscission and both decreased basipetal movement of IAA through petiole sections, but had no effect on acropetal movement. Both increased the IAA-aspartate extracted from the sections. Böttger (1970b) identified ABA in *Coleus* and found that a diffusate fraction rich in ABA increased thirtyfold as the adult leaves became senescent. More recently, Dörffling, Böttger, and co-workers (1978) extended these investigations to an analysis of the abscission-accelerating substances (senescence factor) in petioles of *Coleus*, *Phaseolus*, *Acer*, and pedicels of *Malus*, using a *Coleus* explant bioassay. From each material they identified two abscission accelerants: ABA and xanthoxin, and an abscission retardant: IAA. A third abscission accelerant was not identified. From this work, it is apparent that senescence factor could well be a mixture of substances having the net effect of accelerating abscission. The investigation of Osborne et al. (1972) involving chromatography of diffusates and extracts of leaves of beans and *Euonymus* found evidence of several abscission accelerants in addition to ABA. Various fractions they studied differed in ability to stimulate production of ETH in their bean explant assay. In their discussion, the authors redefined senescence factor as that portion of the extract mixture which is distinct from ABA and IAA and has the ability to stimulate ETH production. However, such a redefinition seems unjustified and is likely to be confusing. Dörffling, Böttger, et al. (1978) consider that senescence factor is a complex consisting of at least three substances: ABA, xanthoxin, and the unidentified accelerant. In addition, the several abscission substances described by Lyon et al. (1972) must also be considered components of the complex. Thus, it appears likely that the senescence factor phenomenon is the result of a mixture of chemicals, and that the original active diffusates will ultimately be shown to contain a number of substances influencing abscission in perhaps several ways.

7. Summary of Hormone Action

In the last three decades, the physiology of hormones has been given more attention than any other aspect of abscission. Yet, our knowledge must still be considered incomplete, and indeed, in most respects fragmentary. In order to synthesize an overview of the mechanism of the hormonal controls, it is necessary to draw on the results of experiments done with diverse species under a wide variety of conditions. Even the most thorough of recent investigations, such as Rodgers' (1977) study of the hormones in cotton fruit, could attack only a small portion of the questions to which answers are needed. Nevertheless, the broad picture of the physiology of abscission is emerging, and some of the relationships are now sufficiently clear that the following brief summary can be made with a reasonable confidence that future discoveries will not alter it greatly.

a. In the Abscission Zone

The hormones that influence events in the abscission zone may arise within it, as is very likely the case with ETH, and in part with IAA, and possibly even ABA. However, the most important influences are clearly those that reach the abscission zone by transport from the subtended organ, such as the leaf or the fruit. Secondarily, some influences may come from a proximal direction.

Figure 4:22 summarizes the present evidence on the action of the hormones in the abscission zone. The figure assumes that the critical biochemical reactions center on the equilibrium between pectins of the middle lamella (and other cell-wall substances) and soluble sugars that function either as building blocks or breakdown products of the pectins and other wall materials. Abscission is brought about by the activity of hydrolytic enzymes, synthesized by the chain of reactions from DNA to hydrolases, specific for pectins. For this synthesis, energy is required and is provided by the usual oxidative respiration, accounting for the absolute requirement of O_2 for abscission. There is increasing evidence that synthetic processes, in particular cell-wall synthesis, are in progress, albeit at a low level, throughout the life of the abscission zone, and even through the period of abscission and separation. These processes are symbolized by the lower chain from DNA to synthetases.

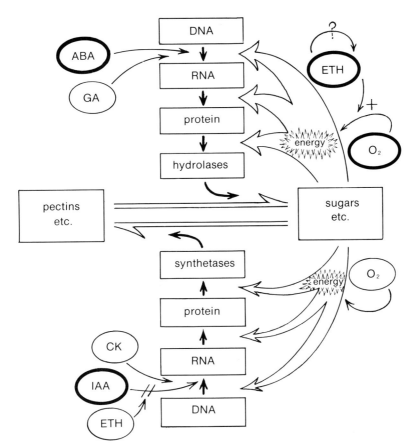

Fig. 4:22.
Summary scheme of hormonal action in the abscission zone (see text for explanation).

While the precise sites of action of the hormones in these chains of events are still uncertain, the hormones do in one way or another affect the pattern or intensity of enzyme synthesis. The powerful action of IAA in delaying abscission is interpreted as influencing the pattern of RNAs continuing and promoting production of the synthetases. In contrast, ABA appears to promote a pattern of RNAs leading to the production of hydrolases. A lowering of the level of IAA or an increase in the level of ABA leads to a preponderance of hydrolase synthesis that culminates in breakdown of the pectins and subsequent abscission. ETH's ability to increase permeability and to promote oxidative respiration suggests that its major role is to increase the availability of energy-rich compounds for enzyme synthesis. It is probably more than a coincidence that the increased levels of ETH in the abscission zone usually (but not always) are correlated with the shift to increased synthesis of hydrolases. Thus, ETH is frequently in the role of a promotor of abscission. It is probable, however, that ETH is not essential to abscission; in a number of instances abscission has responded to changes in IAA and ABA, and proceeded in the virtual absence of ETH.

While GA and CK have been shown to have some significant influences on abscission when applied to explants and abscission zones, we know almost nothing of these hormones as they may be translocated to abscission zones. Hence, in the figure they are given only a light emphasis. From the cytological responses of explants, GA appears to have a strongly promotive effect on the synthesis of hydrolases, parallel to ABA; but the evidence suggests that it probably stimulates a somewhat different pattern of hydrolase synthesis. On the other hand, CKs in the abscission zone probably act in ways that augment IAA in maintaining ongoing processes.

b. In the Subtended Organ

Plant hormones are intimately involved in the development, vigor, and ultimate senescence of plant organs. The hormone levels are a reflection of the vigor of the organ and the agency whereby the rest of the plant, and in particular the organ's abscission zone, is informed of the status of the organ. A vigorous organ contains high levels of IAA, CK, and GA. When the levels decline, senescence usually follows. The ability of each of these hormones to delay or reverse the processes of senescence strongly supports the view that they are controlling factors. In contrast, ABA and ETH often rise when vigor is lost and when the processes of abortion and senescence are in progress. In abscission of young fruit, ABA rose quickly with initiation of the process, whereas ETH production changed little at first, but peaked as separation approached (Fig. 4:16). However, in other preabscission situations, levels of ABA do not change greatly, but the levels of IAA, CK, or GA fall, with the consequence that the ratio of ABA to IAA, for example, goes up. Increases in the production of ETH are characteristic of ripening and senescent changes in many fruits, and such increases are also found in the abortion of young fruits. However, mature leaves, such as those that are abscised by deciduous trees in the autumn, apparently produce little if any ETH in their blades. While there is abundant evidence that IAA and ABA from the leaf and fruit reach the abscission zones, the evidence indicates that ETH produced in the organ has little direct effect on the abscission zone at the base of the petiole or pedicel. Such ETH does affect the abscission zone indirectly through its inactivation of IAA and reduction of the transport of IAA. There appears to be some translocation of GA and CK from the organ to its abscission zone, but there is still only limited evidence about it. In overview, we can no longer link abscission to the levels of a single hormone in a leaf or

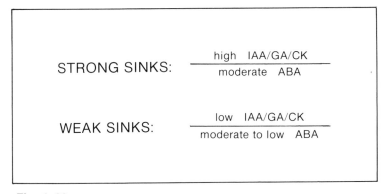

Fig. 4:23.
Hormone relationships in sink strength.

a fruit, which could be done so conveniently thirty years ago; but we must think in terms of a balance of hormonal messages interacting to indicate to the abscission zone and the remainder of the plant, the relative vigor or senescence of an organ (Fig. 4:23).

It is of interest to note in passing that physiologists concerned with the role of hormones in the control of fruit ripening are now inclined to the view similar to that expressed just above. IAA, and to some extent GA and CK, are being regarded as juvenile hormones associated with growth, development, and the maintenance of vigor, while ABA is being increasingly recognized as a nonvolatile fruit-ripening hormone interacting with ETH in the promotion of ripening in ways that are still not completely understood (Pratt, 1974).

c. In Source-Sink Relations

In considering the competition among leaves, fruits, and similar organs on a plant, it is helpful to view each organ as a potential "sink" attracting water and nutrients from the available supplies. Among the fruits on a plant, for example, some fruits will be stronger sinks than others; the weaker sinks will lose out in competition with stronger sinks, and will typically be abscised. The strength of a sink is a reflection of its ability to produce IAA, GA, and CK (e.g., Gersani et al., 1980), the hormones of vigor, as was discussed in the preceding section. Similarly, when the sinks weaken and abort, levels of ABA may increase, and sometimes levels of ETH as well. In any event, the evidence suggests that strong sinks have a relatively high ratio of IAA, GA, and CK, to ABA, and weak sinks have a relatively high ratio of ABA to IAA, GA, and CK (Fig. 4:23). The important point here is that the same signals that indicate to the abscission zone the vigor of the subtending organ are also transmitted widely within the plant, and interact to regulate the total number of leaves or fruits that the plant will carry.

The discussions in this chapter have emphasized the variety of physiological factors involved in abscission, including the underlying metabolic and respirational factors, and the importance of carbohydrate and mineral (especially nitrogen) metabolism. The hormones appear to be the principal messengers in the coordination of abscission behavior. We can no longer think of a single hormone as being the controlling factor in abscission, but must think in terms of a balance of hormonal influences reflecting the physiological status on each side of the abscission zone. As our knowledge now stands, O_2 is the central physiological factor that is absolutely essential to the process of abscission. Beyond that, IAA appears to have an overriding ability to inhibit abscission in the face of the promotive factors. There is increasing evidence of the involvement of ABA, once levels of IAA have fallen sufficiently to permit it to act. ETH can promote abscission in many situations, but abscission can proceed in the virtual absence of ETH.

If the reader is still tempted to accept a simplistic view of the physiological control of abscission, he is encouraged to peruse the discussions of the ecological factors in Chapter 6.

5. Biochemistry and Ultrastructure of Abscission

Chapter Contents

A. BIOCHEMISTRY   153
  1. Introduction   153
  2. Cellulases   156
  3. Pectinases   158
  4. Lignase and Other Enzymes   160
  5. Are the Cell Walls of the Abscission Zone in a Dynamic State?   161
B. CYTOLOGY: INVOLVEMENT OF CYTOPLASMIC ORGANELLES   162
C. CYTOCHEMICAL LOCALIZATION OF ENZYMES   170
D. CYTOLOGY: CELL-WALL CHANGES   170
  1. Middle Lamella   170
  2. Primary Cell Wall   177
  3. Lignin and Secondary Cell-Wall Separation   177
  4. Cutin and Suberin   184

A. BIOCHEMISTRY

1. Introduction

The present view is that the central biochemical activity in abscission is the production, secretion, and action of enzymes which attack and degrade the middle lamellae and cell walls. Evidence of a second biochemical activity has recently been recognized. It appears to involve the deposition of wall materials, and is going on simultaneously with the degrading activity. The two would appear to be the biochemical manifestation of von Mohl's (1860a) observation that abscission consists of two functions: separation and protection.

Molisch (1886), from his many observations of abscission, suggested the probability that separation was the result of enzymatic activity. However, knowledge of enzymes accrued slowly, and even by 1918 Hodgson, from his anatomical study of leaf abscission in *Citrus*, was only able to suggest that pectinases were responsible for wall dissolution during abscission. At that time and for many years after, botanists and plant physiologists still had such a fragmentary knowledge of enzymes that it was difficult to conceive how they might be involved with cell-wall metabolism. As knowledge of enzyme synthesis expanded and the physiological and biochemical activities of the abscission zone became known, it was possible during the last two decades to construct the present concept. One of the important bodies of information on the involvement of oxidative respiration with abscission was discussed in Chapter 4. The evidence presented there established that O_2 is an absolute requirement for physiological separation, and that the participation of the recognized enzymes of oxidative respiration is also necessary (Carns, 1951). Those observations have been extended by the histochemical demonstration of several other respiratory enzymes in abscission zones as separation approaches. These enzymes include peroxidase, succinic dehydrogenase, malic dehydrogenase, acid phosphatase, and esterase (Sutcliffe et al., 1969; Poovaiah et al., 1973; Henry, 1975).

As biochemists slowly discovered the steps whereby the DNA of the nucleus directs and guides the synthesis of enzymes, plant tissues were examined and found to contain closely similar mechanisms to animals and microorganisms. Changes in RNA and protein synthesis precede abscission. Almost all investigations in relation to abscission have been investigations of senescence or ripening activities in leaves and fruits (Sacher, 1973). In a few situations, such as sessile leaves and fruit abscission between the ovary and the pedicel, the changes in the abscission zone leading to separation probably go forward simultaneously with changes in the leaf or fruit. The usual morphological relationship is for a petiole or a pedicel to lie between the leaf blade or the fruit and the main abscission zone. Therefore, biochemical changes in the leaf blade or fruit do not necessarily indicate the changes that will take place subsequently in the abscission zone some distance away. Senescence and ripening changes do, however, result in a change in the signals reaching the subtending abscission zone. Consequently, the activities of DNA and RNA in senescent organs are of significance in the control of abscission, since changes in these activities lead to changes in hormone levels being exported from the organ. As was indicated in Chapter 4, hormonal changes appear to be the principal signal for abscission from the organ to the abscission zone. Thus, as far as abscission is concerned, the significant preabscission changes in organs convey the message to the abscission zone that the time has come to commence the process of abscission.

The ripening and softening of a fruit appears to involve biochemical and cytological changes that are closely similar to the changes of separation in the abscission zone. Fruits have now been studied in some detail because material for investigation is available in large quantities when desired. Changes in fruit include the following: degradation of chlorophyll, and synthesis of carotenoids and anthocyanins (in some species), conversion of starch to sugar and interconversion of sugars, increased respiration, decrease in IAA, and often increase in ABA and ETH, changes in patterns of DNA and RNA activity, production of secondary substances such as flavor constituents, and development of enzymes that attack cell walls (Pratt, 1974; Knee, 1978a,b; Pesis et al., 1978; Zauberman and Schiffmann-Nadel, 1972; Awad and Young, 1979). These activities commonly occur in correlation with abscission.

In the abscission zone changes in pigmentation are often conspicuous. More important, there is an increase in the synthesis of RNA (Holm and Abeles, 1967), and a parallel increase in protein synthesis (Abeles and Holm, 1967; Lewis and Bakhshi, 1968b). These changes precede and accompany increased activity (presumably production) of the numerous enzymes that become conspicuous at the time of abscission. The mechanism of such a changeover in pattern of enzyme synthesis is far from clear, but it can be triggered in explants by various hormone applications, and in nature by changes in the levels or patterns of hormones reaching the abscission zone. The investigation of the mechanism of hormonal regulation of RNA and protein metabolism has presented extremely complex and difficult problems for the biochemist. However, the challenges are being met; recent developments have been described by Jacobsen (1977). Although specific relationships have been difficult to pin down, there is voluminous information that leads to the interpretation that modification of RNA metabolism is a specific and necessary event for the initiation and perpetuation of biochemical responses to hormones. Now, in at least two cases, hormonal controls have been demonstrated. One is by GA for mRNA translatable for α-amylase, and the other is by auxin for translation of mRNA for cellulase (Jacobsen, 1977). While responses to hormones that extend over some period of time regularly involve a broad spectrum of RNA metabolism, only changes in minor species of RNA seem indispensable to hormone action. For example, when ABA inhibits the synthesis of α-amylase in barley aleurone, the synthesis of most of the other proteins remains unchanged (Ho and Varner, 1976).

In the abscission zone, enzyme changes of several sorts have been detected. The increase in respiratory hormones has already been mentioned, as well as the inhibition of abscission by inhibitors of the respiratory hormones. Noteworthy, also, are the increases in enzymes affecting the hormones, in particular enzymes such as "peroxidase" (Henry, 1975; Gahagan et al., 1968; Hinman and Lang, 1965) that can inactivate IAA (Fig. 5:1). However, by far the greatest attention has been given to enzymes that might possibly be involved in the degradation of cell walls, particularly the cellulases and pectinases. Enzymes capable of degrading compounds other than cellulose and pectins have as yet received little attention.

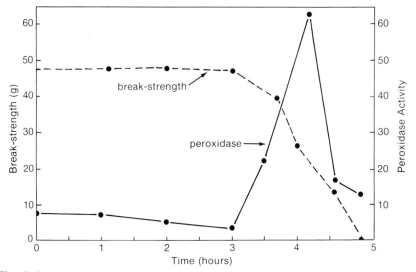

Fig. 5:1.
Peroxidase and abscission in explants of tobacco flower-pedicel abscission zones. (Redrawn from Henry et al., 1974)

2. Cellulases

In bean explants, Horton and Osborne (1967) found increased cellulase activity to be correlated with abscission. Similar correlations were observed in cotton and *Coleus* (Abeles, 1969), *Citrus* (Ratner et al., 1969; Greenberg et al., 1975), and other species (Osborne, 1973). Lewis and Varner (1970) established that the increased cellulase activity in bean-leaflet abscission zones was, indeed, a *de novo* synthesis, and detected two forms of cellulase in the abscission zones. Subsequent investigations disclosed the existence of several isozymes of cellulase in abscission zones (Lewis et al., 1972; Reid et al., 1974; Goren and Huberman, 1976). Only one of the isozymes, however, appears closely correlated with abscission (Sexton et al., 1980; Fig. 5:2). Others are present and are active when abscission is not in progress.

Because cellulose is recognized as the major constituent of the fibrils of the primary cell wall, it was once assumed that cellulase was required for cell separation (Abeles, 1969), and that the increased cellulase activity promoted the cytolysis and cell disintegration (Horton and Osborne, 1967) that sometimes accompanies separation. Other evidence would question those assumptions. Many of the early anatomical studies (reviewed by Pfeiffer in 1928) described the cell separation in the abscission zone with little indication of breakdown of cellulose in the primary wall. In contrast, there are numerous descriptions of separation achieved by the degradation of the pectic substances in middle lamellae. The latter observations have now been confirmed repeatedly with the electron microscope. From EM studies, it is clear that separation is achieved almost entirely by the dissolution of the middle lamella, and that there is little cellulose breakdown by the time of separation (see sec. D, this chapter). Further, the extensive investigations of the molecular organization of the cell wall indicate that the cellulases that have been investigated in connection with abscission have little effect on the strength of the cellulose fibrils in the cell wall (Albersheim et al., 1977). Their main effect appears to be on the hemicellulose fraction, and they seem to be involved with a steady-state turnover of wall materials that takes place without weakening the wall. Such a turnover permits slippage among the fibrils and enables growth to take place without disrupting the organization of tissues. Possibly, some of the cellulase activity is correlated with the development of protective layers that takes place immediately after separation. Under favorable circumstances, the cells of the primary protective layer that are exposed upon separation enlarge considerably. Increased cellulase activity would facilitate this enlargement. This interpretation is supported by the observation of Ratner et al. (1969) that by far the greatest cellulase activity associated with *Citrus* leaflet abscission occurred in the 0.2 mm of tissues immediately proximal to the separation layer (Fig. 5:3). Other support comes from the results of Hänisch ten Cate et al. (1975), who found in *Begonia* pedicel abscission that cellulase activity did

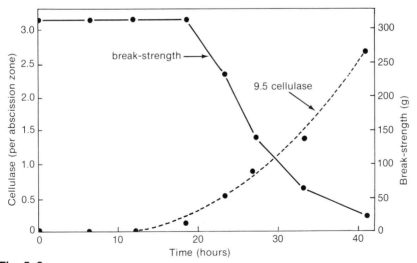

Fig. 5:2.
Correlation of cellulase isozyme 9.5 with abscission in the bean-leaf abscission zone. (Redrawn from Sexton et al., 1980)

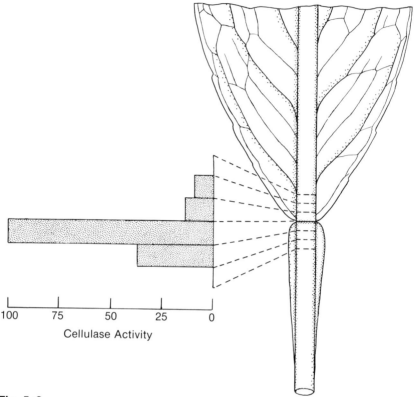

Fig. 5:3.
Distribution of cellulase in and near the *Citrus* leaflet abscission zone. (Redrawn from Ratner et al., 1969)

not develop in the abscission zone until after separation was essentially complete (Fig. 5:4). From studies of plant growth, there is increasing evidence of cellulase activity in cell-wall expansion of growth (Maclachlan, 1977). Thus, it appears likely that some form(s) of cellulase functions in abscission, particularly in the development of the protective layers. However, the involvement of cellulase in cell separation remains in question. It will not be easy to resolve this problem. Rigorous methods will be required to identify the isozymes of cellulase and the chemical and cytological forms of cellulosic materials which they could attack in the cells of the separation layer.

3. Pectinases

The possibility that an enzyme might be involved in the dissolution of the middle lamella was first suggested in 1886 by Molisch, who pointed out that

Wiesner's recently described "Gummiferment" (a mucilage-forming enzyme) could well be involved. The suggestion lay dormant for decades, awaiting the development of techniques and the acceptability of the idea that cells could secrete enzymes through their cell walls to the middle lamella. Even when the first experiments with pectic enzymes and abscission began, the feasibility of enzyme secretion was not apparent.

Commencing in 1958, evidence of changes in activity of pectic enzymes in correlation with abscission began to appear. Osborne (1958a) found a decrease in pectin methylesterase (PME) correlated with abscission of bean explants. Similar correlations have been found in *Coleus* (Moline et al., 1972) and in flower abscission of *Nicotiana* (Yager, 1960a). Yager (1960b) compared PME activities in the flower abscission zones of two varieties of *Nicotiana*. The variety that retained its flowers longer also had the higher levels of PME. These investigations suggest that abscission is in some way facilitated by reduced activity of PME; that is, by increased esterification of pectic substances. The information on PME and abscission is far from clear, as a number of workers have reported their inability to find changes in PME correlated with abscission (Ratner et al., 1969; Abeles et al., 1971; Moline et al., 1972; Hänisch ten Cate et al., 1975). The reasons for these failures are difficult to understand, but such determinations are confronted with many problems. *In vitro* assays do not necessarily reflect the activity within the plant, and extraction procedures have varied greatly. Sampling methods,

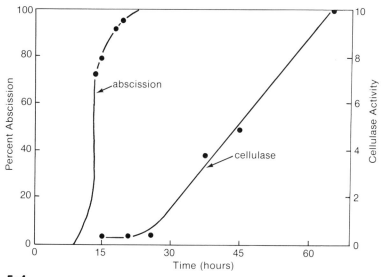

Fig. 5:4.
Abscission and cellulase in the *Begonia* flower-pedicel abscission zone. (Redrawn from Hänisch ten Cate et al., 1975a, with permission of the publisher.)

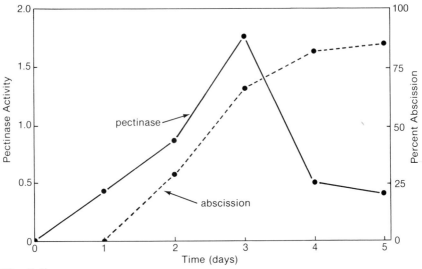

Fig. 5:5.
Abscission and pectinase (polygalacturonase) in the bean leaflet explant. (Redrawn from Morré, 1968)

likewise, have varied widely, and some may have missed the critical stages (cf., Moline et al., 1972; Osborne 1958a).

Another pectic enzyme, polygalacturonase (PGU), occurs in the abscission zones and increases shortly prior to separation. Using a bioassay, specific for PGU (Mussell and Morré, 1969), Morré (1968) found a rapid increase in PGU immediately prior to abscission of bean-leaflet abscission zones (Fig. 5:5). A similar PGU has been identified in leaflet abscission zones of *Citrus* (Riov, 1974) and in the fruit abscission zones of *Citrus* (Greenberg et al., 1975). Application of 2,4-D (synthetic auxin) delayed, and application of ETH accelerated, the increase in PGU activity in *Citrus*. Further, CHI, a protein synthesis inhibitor, inhibited the development of PGU activity and abscission (Riov, 1974). As with PME, some workers have been unable to confirm PGU involvement in abscission (e.g., Berger and Reid, 1979; Hänisch ten Cate et al., 1975), but as before, this may be due to differences in materials and sampling methods.

4. Lignase and Other Enzymes

The foregoing discussion has centered on cellulases and pectinases, in part because cellulose and pectic substances have been easily detected constituents of parenchyma cell walls, and in part because assay methods for the enzymes have been available. Other important constituents of the cell wall are slowly becoming known, constituents such as the matrix materials that bind the cellulose fibrils. While the abscission zone cytology of some species indicates that dissolution of pectic substances between adjacent cells is all that is necessary to bring about separation, in others, such as *Citrus* (Hodgson, 1918), the entire primary wall is affected during abscission. It is likely that the matrix materials, which can include xyloglucan, arabinangalactan, and rhamnogalacturonan, among other polysaccharides (Albersheim et al., 1977), are attacked enzymatically during abscission in plants like *Citrus*. In addition, there is considerable evidence that the constituents of secondary cell walls, in particular the lignin of the secondary thickenings in xylem, are attacked and at least softened during abscission (see sec. D.3., this chapter). Thus, it is likely that lignases are also produced in the course of abscission. There is no evidence as yet, from cytological or other investigations, that the waxes and related substances of the cuticle are attacked by enzymes during abscission. The cuticle merely appears to break mechanically.

5. Are the Cell Walls of the Abscission Zone in a Dynamic State?

Two series of observations support the view that the biochemical machinery for cell-wall synthesis is present and potentially active in the abscission zone at the same time that the machinery for cell-wall degradation is functioning. These are observations of the "retightening phenomenon."

In the course of testing chemicals for their effectiveness as cotton defoliants, more than one researcher has observed that some of the weaker chemical treatments seemed only able to initiate an incomplete process of abscission. After two or three days, the treated leaves were noticeably "loose," so that they could be dislodged easily by the touch of a finger. However, on the following day, the separation layer had retightened, and thereafter the leaf could no longer be removed by a finger touch. Apparently, the abscission zones returned to their previous intact condition (V. L. Hall, p.c.; V. T. Walhood, p.c.).

In the tests of chemicals as fruit-loosening (abscission) agents to assist in the harvest of oranges, partial loosening followed by retightening has been observed repeatedly (W. C. Wilson, p.c.). By the use of appropriate instruments, the decrease and subsequent increase in fruit-removal force is readily measured (e.g., Holm and Wilson, 1977).

The retightening phenomenon is a matter of only passing interest to the field researcher; the more effective treatments are at his disposal to meet agricultural needs. What is remarkable about the phenomenon is that it demonstrates the presence, in relatively mature plant tissues, of the machinery for cell-wall deposition, and this in the same cells and essentially at the same time that cell-wall degradation is occurring. The interplay between the factors making for cell-wall synthesis and those making for cell-wall degradation offers an attractive field for cyto-morphogenetic research.

B. CYTOLOGY: INVOLVEMENT OF CYTOPLASMIC ORGANELLES

The synthesis and secretion of macromolecules, such as RNA and enzymes, is a function of the endomembrane system of the cell. The results of painstaking and meticulous investigations of ultrastructure following the development of the electron microscope have led to the recognition that many of the organelles of the cell are closely related and, indeed, seem to be derived from the nucleus in a flow of membranes with physical and physiological connections and interrelations stemming from the nuclear membrane. As membrane material is synthesized, it contributes to the endoplasmic reticulum, which in turn forms vesicles; in some situations the vesicles give rise to provacuoles and the vacuole, and at other times they give rise to the dictyosomes, which in turn release vesicles that can contribute to the vacuole(s) or to the plasmalemma (Reinert and Ursprung, 1971; Matile, 1975). The current concept is that much of the ribosomal RNA is assembled in the nucleolus (Clowes and Juniper, 1968). The endoplasmic reticulum appears to be the principal site of protein synthesis. As macromolecules accumulate in the vesicles arising from the ER, in some situations the vesicles will fuse and join to form the vacuole, which is increasingly recognized as a storage organelle rather than a "dust-bin." In other situations the vesicles fuse and become active in the dictyosomes, which accumulate substrates as well as enzymes, with the result that the dictyosomal vesicles, when they move to the plasmalemma and fuse with it, bring both enzymes and substrate for wall deposition. The involvement of dictyosomal vesicles in such secretion is now well documented in higher plants (Mollenhauer and Morré, 1966)—e.g., deposition of wall materials in pollen tubes (Van der Woude et al., 1971). In other situations, vesicles from the ER appear to move directly to the plasmalemma and fuse with it. The foregoing summarizes the current and widely accepted interpretation of the role of the endomembrane system. The biochemical evidence in support of this role in plants is consistent but still limited in extent (Chrispeels, 1976).

Studies of the ultrastructural changes in the abscission zone have now been undertaken in several laboratories and in a variety of plant materials. Because of the demanding requirements of ultrastructural microtechnique and the time-consuming procedures, no one investigation has given more than a few glimpses of the changes. Nevertheless, collectively, a picture is emerging that is completely consistent with the overall concept of the structure and function of the endomembrane system summarized above. The discussion will comment on some of the more noteworthy observations and is based mainly on the following publications: Addicott and Wiatr, 1977; Gilliland et al., 1976; Iwahori and van Steveninck, 1976; Jensen and Valdovinos, 1967, 1968a,b; Osborne and Sargent, 1976a,b; Sexton and Hall, 1974; Sexton et al., 1977; Valdovinos et al., 1974; Webster, 1968; Webster and Chiu, 1975; and Wiatr, 1978.

The nuclear envelope can be interrupted by outfoldings of the membrane that connect with the endoplasmic reticulum. The nucleolus becomes conspicuous as activity in the abscission zone commences, and RNA increases rapidly in the abscission zone, as staining with pyronin Y indicates. Investigators are unanimous in observing a considerable increase in ER as abscission gets under way. Both smooth and rough ER develop and become conspicuous (Figs. 5:6, 5:7). Likewise, dictyosomes can increase greatly in number, particularly in the cells of leaf abscission zones (Fig. 5:8 and Table 5:1). Vesicles become more numerous; they may be simple or show up as "multivesicular bodies." The immediate origin of the vesicles appears to vary. Some are clearly dictyosomal in origin (Figs. 5:7A, 5:9). In the species where the dictyosomes become numerous, they may be the major source. In other situations, such as petal abscission zones of *Linum lewisii* (Wiatr, 1978) where dictyosomes are inconspicuous, ER is abundant and appears to be the source of most of the vesicles (Fig. 5:10). Some of the vesicles may be cut off from the vacuole; the multivesicular bodies may possibly arise in that manner. Vesicles can become numerous near the plasmalemma (Fig. 5:11).

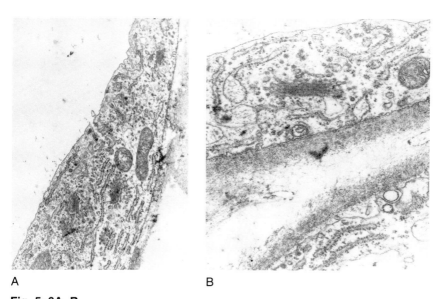

A   B

Fig. 5:6A, B.
Electron micrographs of cells of the bean-leaf separation layer. Note dictyosomes and endoplasmic reticulum, showing dilations. (Courtesy of Sexton and Hall, 1974)

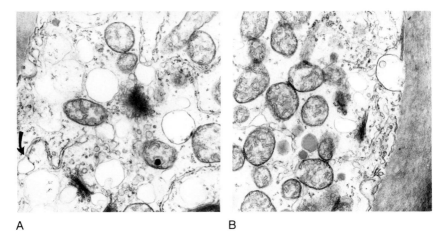

Fig. 5:7A, B.
Electron micrographs of pedicel separation-layer cell of *Bruguiera gymnorhiza*. Note vesiculated endoplasmic reticulum and vesicles fused with the plasmalemma. (Courtesy of P. Berjak and J. P. Withers, unpubl.)

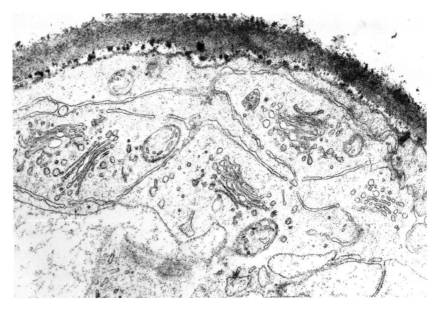

Fig. 5:8.
Electron micrograph of a cell on the distal separation surface immediately after abscission of a *Coleus* leaf. Note extensive endoplasmic reticulum and numerous dictyosomes. (Courtesy of Sexton et al., 1977)

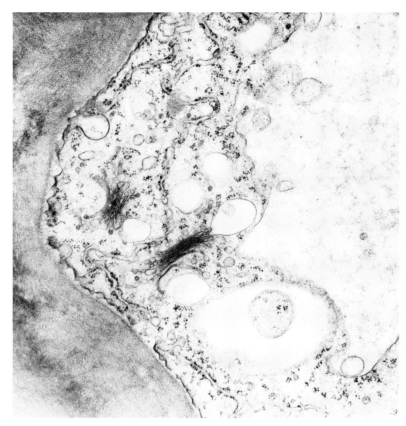

Fig. 5:9.
Portion of a cell from the pedicel separation layer of *Bruguiera gymnorhiza*. Note vesicle fusion with plasmalemma. (Courtesy of P. Berjak and J. P. Withers, unpubl.)

The crenulations and invaginations that are seen in the plasmalemma (Figs. 5:9, 5:11, 5:12B) appear to be the consequence of vesicles fusing with plasmalemma. Intact vesicles are occasionally found in the periplasmic region (Fig. 5:12A), and more rarely in the primary wall or in the middle lamella.

Sexton and co-workers (Sexton and Hall, 1974; Sexton et al., 1977) made quantitative studies of the changes in cytoplasmic organelles as the cells of the bean, *Coleus*, and *Impatiens* leaf separation layers became active (Tables 5:1, 5:2). In bean they found a doubling of the amount of ER and a fivefold increase in the number of dictyosomes. There was no change in the amount of mitochondria and only a small increase in the number of chlo-

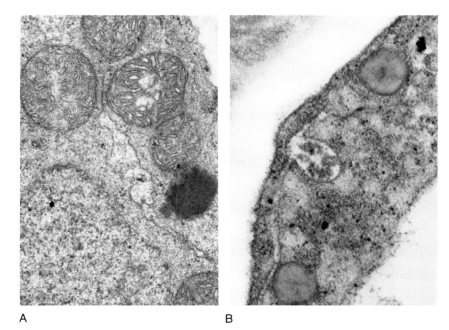

Fig. 5:10A, B.
Dilations and blebs of the endoplasmic reticulum in separation-layer cells of the petals of *Linum lewisii*. (Courtesy of Wiatr, 1978)

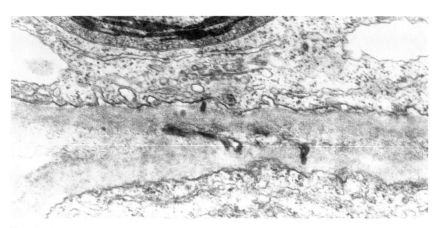

Fig. 5:11.
Endoplasmic reticulum and vesicles in a cell of the distal separation surface of a *Coleus* leaf. (Courtesy of Sexton et al., 1977)

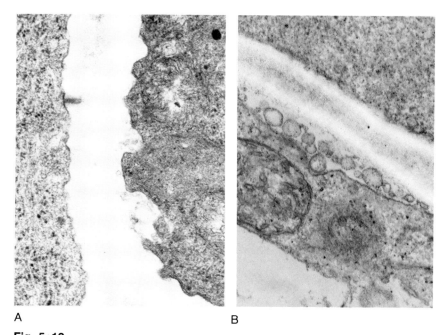

Fig. 5:12.
Vesicles in petal separation layer of *Linum lewisii*. **A.** Fused with the plasmalemma. **B.** Between plasmalemma and cell wall. (Courtesy of Wiatr, 1978)

roplasts. A considerable increase in the number of microtubules was noted, but no counts were made.

The foregoing observations make it clear that the endomembrane system is very much involved in the activities that bring about separation. In other tissues the system is responsible for the synthesis of enzymes, particularly hydrolases (Matile, 1978). Although the number of observations is limited, it is likely that the immediate source of vesicles carrying enzymes to the periplasmic region varies with species and the kind of abscission zone. In leaf abscission zones (Sexton and Hall, 1974; Sexton et al., 1977) and in pedicel abscission zones (e.g., Gilliland et al., 1976) the dictyosomes appear to be an immediate source of the vesicles, and possibly of the vacuole as well. However, in the petal abscission zone of *Linum lewisii*, Wiatr (1978) found little increased activity of the dictyosomes, but considerable ER and evidence that ER vesicles were the ones moving to the periplasmic region.

The numbers of mitochondria and chloroplasts change very little in the course of abscission. Also, their appearance remains normal with no indication of deterioration. Aside from the increased amounts of organelles of the endomembrane system, the cells of the abscission zone and separation layer change remarkably little throughout the course of abscission. The plas-

Table 5:1. *Percentage of Increase in Area Occupied by Organelles in Separation-Layer Cells of Bean Leaves after Abscission Induction*

Endoplasmic Reticulum	102.0
Dictyosomes	488.0
Mitochondria	1.2
Chloroplasts	26.4
Microtubules	++

Source: Sexton and Hall, 1974.

Table 5:2. *Organelle Frequency in Cells of the Leaf-Separation Layer during Abscission*

Plant	Stage	Mitochondria	Dictyosomes	RER	Cytoplasmic Vesicles
Coleus	uninduced	717	165	685	101
	abscised	889	534	879	246
Impatiens	uninduced	1,294	281	314	468
	abscised	1,078	382	798	838

Source: Sexton et al., 1977.

malemma remains intact and the cells remain capable of plasmolysis (Sexton et al., 1977).

Starch deposits increase conspicuously in the abscission zone as abscission approaches, but largely disappear by the time of separation. In all likelihood, the starch is utilized as a source of energy, as Lloyd suggested in 1916. Also, some of it may well contribute to the increased deposition of cell-wall materials of the protective layers. In the pedicel abscission zones of *Nicotiana* and *Lycopersicon*, Jensen and Valdovinos (1967) found crystalloidal microbodies which were proteinaceous and which diminished during the course of abscission (Fig. 5:13). They suggested (1968a) that the microbodies may consist of latent enzymes involved with cell separation.

Microtubules have been observed in some studies of abscission. These may have some relation to the increased cellulase activity that is commonly associated with abscission. Two of the significant observations appear to be related to wall deposition: Sexton and Hall (1974) found increased numbers of microtubules in the cells of the proximal fracture surface of bean-leaf abscission zones, cells developing into the primary protective layer. Wiatr (1978) found oriented microtubules in areas where cells of the petal abscission zone were undergoing localized thickening, as if the cells were in an early stage of becoming transfer cells. Such observations support the view

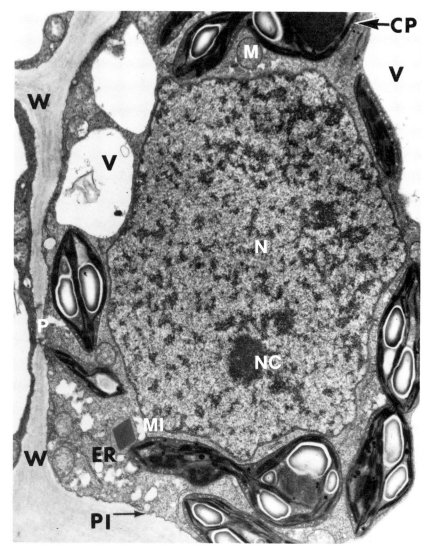

Fig. 5:13.
Cortical cell of the pedicel abscission zone of *Nicotiana*. Note crystalline microbody (MI) and numerous plastids with starch grains. (Courtesy of Jensen and Valdovinos, 1967)

that while the processes of separation are under way, other processes of primary wall growth may be going on simultaneously.

Hormones accelerate or retard the normal ultrastructural changes more or less in direct correlation with their effects on the rate of abscission. However, they often induce other cytological and anatomical changes as well (Bornman, 1967a; Bornman et al., 1967; Valdovinos et al., 1974; Osborne and Sargent, 1976b).

C. CYTOCHEMICAL LOCALIZATION OF ENZYMES

While increased activity of a number of enzymes has been detected in abscising tissues, most of the observations have been made using histochemical stains and the light microscope. Methods for the ultrastructural localization of enzymes have been developed for only a few enzymes. To date, only "peroxidase" and "acid phosphatase" have been localized in EM studies of abscission. The results are of considerable interest, but not completely satisfactory. Each of the enzymes is known to occur in several forms. Thus, a simple, positive reaction for "peroxidase" or "acid phosphatase" gives little indication of the precise biochemical function taking place in the cell. Nevertheless, there is undoubtedly some significance in the fact that positive reactions for "peroxidase" and "acid phosphatase" are widely distributed in the endomembrane system, particularly in the membranes of ER, the dictyosomes, the vacuole and their vesicles, as well as the plasmalemma (Figs. 5:14, 5:15). Positive reactions are also localized in the middle lamella and on the surfaces of the primary wall after separation (Figs. 5:14, 5:16) (Henry and Jensen, 1973; Hall and Sexton, 1974; Gilliland et al., 1976; Webster et al., 1976).

The functions of peroxidase and acid phosphatase in abscission are still unknown. Both enzymes increase in activity in correlation with many metabolic functions. "Peroxidase" is, of course, a group of enzymes with various specificities. Some of the peroxidase activities that may be related to abscission are those found (i) in cell walls during growth, (ii) in the oxidative inactivation of IAA, and (iii) in the biosynthesis of ETH. Further, Hagemann (1971) has suggested that the increased phenoloxidase activity that he has observed functions to promote the peroxidase mediated reactions. Acid phosphatases are, likewise, commonly found in regions of metabolic activity. They are hydrolases, and their presence is considered indicative of increased hydrolase synthesis generally.

D. CYTOLOGY: CELL-WALL CHANGES

1. Middle Lamella

The early plant anatomists described a wide range of patterns of separation, but in most instances, they found that separation resulted from the

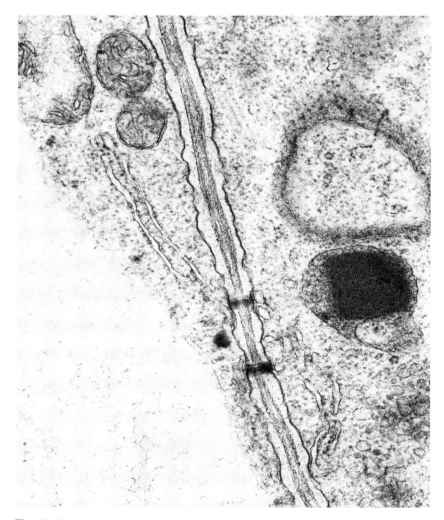

Fig. 5:14.
Localization of peroxidase in endoplasmic reticulum, plasmalemma, and middle lamella in flower-pedicel abscission-zone cell of *Nicotiana*. (Courtesy of Henry, 1979)

dissolution of middle lamellae. Lee (1911) summarized the consensus in his comments on leaf abscission in *Rhus typhina*: "Separation occurs in the usual way. The primary walls of the Separation-layer become slightly swollen, the middle lamellae become mucilaginous and disappear, leaving the neighbouring cells quite free." This description applies to separation of the majority of leaves, flowers, and fruits (Figs. P:2, P:3). In a few species, notably *Citrus*, the entire primary wall swells during the process of abscis-

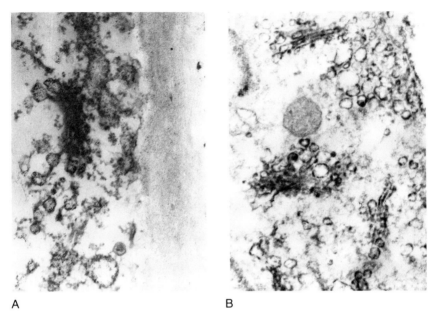

Fig. 5:15.
Enzyme localization in dictyosomes of cells on the proximal surface of the bean-leaf separation layer. **A.** Acid phosphatase. **B.** Peroxidase. (Courtesy of Hall and Sexton, 1974)

sion (Fig. 5:17) (Hodgson, 1918). Such species probably have an unusually large amount of pectic and other branching polysaccharides in the primary wall.

In most of the cases that have been studied, the middle lamella is the first part of the cell wall to be affected. It swells considerably, starting in the corners between cells and about the intercellular spaces if such are present (Fig. 5:18). Soon the entire middle lamella weakens and separates. Middle lamellae of two or more cell layers may be involved, with the result that some cells separate completely and become isolated from the abscission zone tissues (Fig. 5:16A) (Bornman, 1967b, 1969; Valdovinos and Jensen, 1968; Sexton, 1976; Osborne and Sargent, 1976a). If freshly separated surfaces are examined with the scanning electron microscope, the cells on each surface are found to be intact and rounded; some are completely free (Figs. 5:19, 5:20) as a consequence of middle-lamella dissolution on all sides. In many species the exposed cells remain rounded (abgerundeten in the German literature). In other species they expand and become somewhat clavate or merely elongate (schlauchförmig). In some species the rounding or elongation of cells commences early in abscission and is apparent before actual separation (Pfeiffer, 1928).

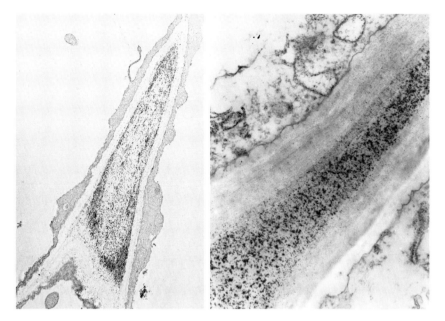

Fig. 5:16.
A. Localization of peroxidase in an intercellular space of the flower-pedicel abscission zone of *Nicotiana*. (Courtesy of Henry and Jensen, 1973) **B.** Acid phosphatase in the middle lamella, plasmalemma, and vesicles of separating cells in the bean-leaf abscission zone. (Courtesy of Hall and Sexton, 1974)

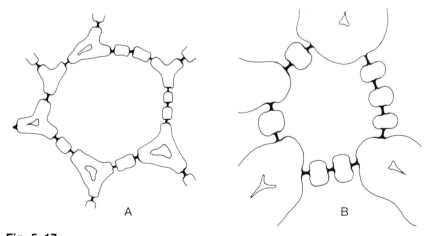

Fig. 5:17.
Changes in the walls of cortical cells of the abscission zone during leaf abscission of *Citrus sinensis*. **A.** Before abscission. **B.** An advanced stage of wall gelatinization. (Redrawn from Hodgson, 1918)

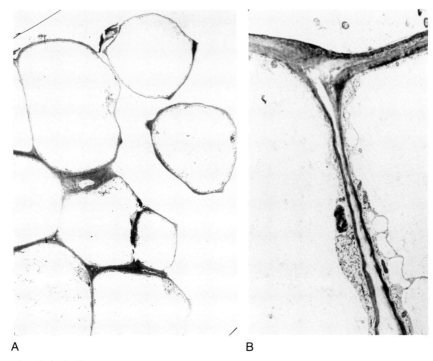

Fig. 5:18A, B.
Electron micrographs of separation in the leaf abscission zone of *Impatiens sultani*. (Courtesy of Sexton, 1976)

Between the time of the early anatomical investigations that utilized fresh sections and the commencement of the ultrastructural investigations in 1967, anatomical and cytological studies utilized paraffin sections almost exclusively. As the micrographs from these investigations are examined and compared with electron micrographs, on one hand, and the drawings of the early anatomical studies of fresh materials, on the other hand, it is apparent that the paraffin-embedded materials often show evidence of artifacts. These include poor fixation with plasmolyzed or distorted protoplasts; shrinkage that led to irregular cell walls; compression or tearing in sectioning; and stains that either failed to show or obliterated cytoplasmic details. Further, physical pressure on the separation layer, either from manipulation or from tissue tensions, often distorted the very cells of greatest interest. The consequence is that the cytological details of studies using paraffin methods are often blurred and must be reexamined in the light of what we have learned from electron microscopy. Such studies are now in progress.

Several lines of evidence indicate that the middle lamella is composed al-

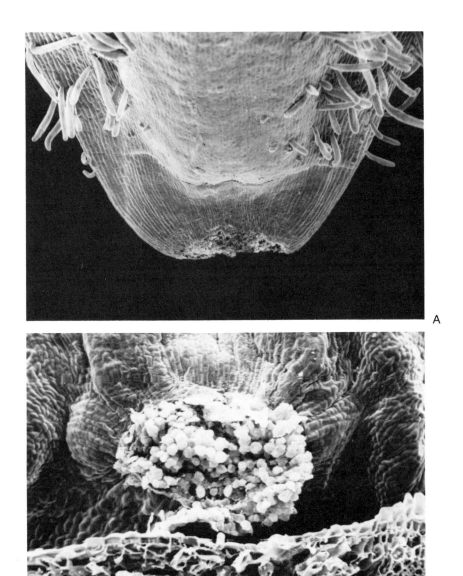

Fig. 5:19.
Scanning electron micrographs of the surfaces exposed by petal abscission of *Linum lewisii*. **A.** Distal surface (petal base). **B.** Proximal surface (receptacle). (Courtesy of Wiatr, 1978)

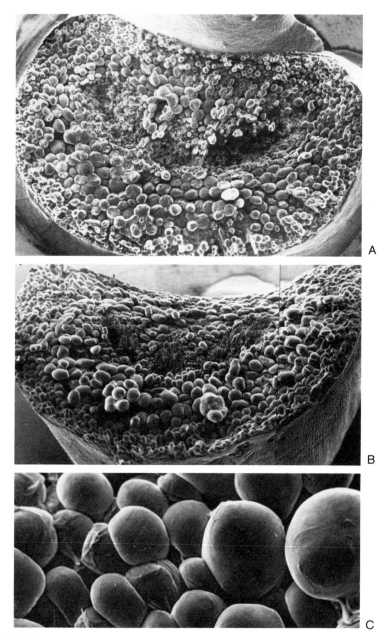

Fig. 5:20.
Scanning electron micrographs of the surfaces exposed by leaf abscission of *Impatiens sultani*. **A.** Proximal surface (leaf scar). **B.** Distal surface (petiole base). **C.** Proximal surface. (Courtesy of Sexton, 1976)

most exclusively of what are called pectic substances. Using the most specific reagent now recognized for use in electron microscopy, alkaline hydroxylamine-ferric chloride (Albersheim and Killias, 1963), Bornman et al. (1969) demonstrated that the middle lamellae of the separation layers of *Coleus* and *Gossypium* are composed almost entirely, if not completely, of pectic substances. These are now recognized as being composed basically of polygalacturonides. Other sugars may be linked covalently to the branching chains of galacturonic acid residues. The acid groups are methylated in varying degrees. For some time, it was believed that calcium ions served as bridges between acid groups and thereby contributed to the strength of pectic substances. Some recent work has questioned this function of calcium, and clarification awaits the results of further research.

2. Primary Cell Wall

The ultrastructural studies of abscission zones have disclosed very few changes in the primary wall by the time the middle lamella has disintegrated and separation is achieved. Usually, a small amount of swelling is all that is detected (Sexton, 1976; Valdovinos and Jensen, 1968). However, ultrastructural studies to date have been concerned with a limited number of species. The picture will undoubtedly broaden as a wider range of anatomical types of abscission are studied. The swelling and gelatinization of the primary wall in *Citrus* leaflet abscission observed by Hodgson (1918) (Fig. 5:17) is clearly in a different category from what has been studied ultrastructurally to date, although a few EM micrographs, including some of Wiatr's 1978 preparations (Fig. 5:21), appear somewhat similar to *Citrus*. Several investigators have found irregular thickenings in the primary walls of some cells in the abscission zone, thickenings that indicate that those cells were functioning as transfer cells (Sexton, 1976; Wiatr, 1978; Valdovinos and Jensen, 1968). Although present evidence indicates that separation is usually achieved with little or no breakdown of the primary cell wall, such breakdown has been observed in ETH-treated bean leaflet explants (Webster, 1973).

Quantitative chemical analysis of the changes in abscission zones is difficult because of the very small amount of tissue involved. The most thorough chemical work to date, however, confirms and supports the EM observations (Morré, 1968).

The ultrastructural changes described in the preceding section are summarized diagrammatically in Figure 5:22).

3. Lignin and Secondary Cell-Wall Separation

The involvement of lignin and lignified tissues in abscission has been recognized from the time of the early anatomical investigations. Lignin is often deposited in the protective layers and, in some instances, in the cells imme-

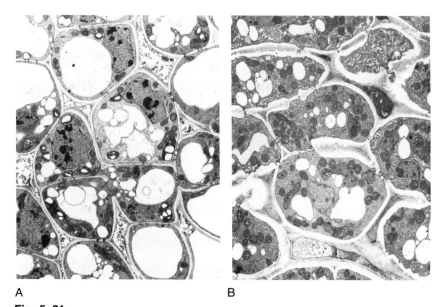

Fig. 5:21.
Electron micrographs of the cortical parenchyma of the petal abscission zone of *Linum lewisii*. **A.** An early stage showing middle-lamella breakdown at the corners of cells. **B.** A later stage showing primary-wall swelling. (Courtesy of Wiatr, 1978)

diately distal to the separation layer. The identification of lignified cells has often been inaccurate, however, because of the lack of specificity of the usual histochemical reagent (phloroglucinol-HCl). Further, the separation of lignified xylem elements has rarely been studied in detail, and the precise methods of separation of xylem elements are only now being given the careful study they deserve. These aspects will be discussed in the following paragraphs.

In leaf abscission of woody species, "ligno-suberization" of the protective layers is very common. Deposition of the secondary-wall materials may commence some time before separation or only afterwards. In almost all cases the proximal ligno-suberized tissues become continuous with the periderm (Lee, 1911). In a few species what appears to be authentic deposition of lignin occurs in the cells immediately distal to the separation layer (e.g., *Tilia europaea*, *Baccharis halimifolia*, *Gleditsia triacanthos*, *Rhus typhina*, and *Cornus sanguinea*). As far as it can be determined from the evidence presented (Lee, 1911), those positive reactions for lignin were localized to recently thickened secondary walls.

Not long ago, it became apparent that in many, if not all, cases of leaf abscission in both woody and herbaceous species, the parenchyma of the distal portions of the abscission zone stain positively (red) with phloroglucinol-HCl shortly before abscission. In a survey of several dozen ornamental species growing in the vicinity of the University of California at Los Angeles, every species examined developed a positive reaction in the abscission zone parenchyma as abscission approached (K. C. Baker, unpubl.). These observations confirmed and extended those of Tison (1899) in which he reported that all 105 species which he studied gave a positive reaction. In most cases the positive reaction was general in the cells and not limited to secondary-cell-wall thickenings. Further, it is difficult to imagine that lignification of an abscising leaf would have any value to the plant. This led Bornman et al. (1966) to reexamine the specificity of the phloroglucinol-HCl reagent and other reagents that had been developed for detection of lignin. In his review of the chemical literature, he found that phloroglucinol-HCl is not considered specific for lignin, and that pentoses, polysaccharides containing pentoses, and uronic acids also give a positive reaction. On the other hand, the chlorine-sulfite test and Maule's reagent do, indeed, appear to be specific for lignin. He observed that in abscising cotton explants posi-

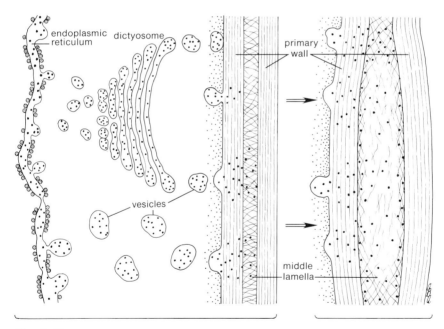

Fig. 5:22.
Diagram of ultrastructural activity correlated with abscission. Both ER and dictyosomes contribute vesicles that fuse with the plasmalemma and release their contents to the cell wall. Swelling and disintegration is usually far more extensive in the middle lamella than in the primary wall.

tive reactions to phloroglucinol-HCl were given by the secondary walls of xylem elements and by parenchyma immediately distal to the abscission zone. The reaction of the parenchyma was negative before abscission, but became positive and intensified as abscission progressed. Only the secondary walls of the xylem gave positive reactions to the chlorine-sulfite test and to Maule's reagent. From these considerations and observations, Bornman concluded that phloroglucinol-HCl could not be considered specific for lignin.

It is likely that many reports of the presence of lignin, even in the recent literature, must be considered false positives. Such reports are especially suspect where the positive reaction develops over a short period of time and in cells where there is no evidence of secondary-wall deposition (e.g., Poovaiah, 1974).

The separation of lignified elements is usually said to be the result of mechanical breakage after the parenchyma has separated (see, e.g., Lee, 1911; Webster, 1974; Eames and MacDaniels, 1947). The implication of the assertions is that the xylem elements are unaffected by the activities in the parenchyma that lead to separation of middle lamellae. Actually, there is little in the literature to indicate that separation of the xylem has been carefully studied. The statements in the literature do indicate that a number of workers have observed some xylem elements stretched out and still connecting the two sides of a separated abscission zone. These stretched elements eventually break to permit complete separation. The observations do not include any evidence as to what proportion of the xylem stretched and broke mechanically, and what proportion may have separated by enzymatic or other physiologic activity. That some softening of lignified secondary walls can occur during abscission has been recognized for some time. In the course of development of the protective layer in the abscission zone of *Parthenium argentatum*, xylem vessels are crushed and practically obliterated. Crushing does not produce discrete fragments; the thickenings seem to melt and distort as if the secondary walls had become a soft plastic (Addicott, 1945). Similar distortion and plastic deformation of xylem thickenings appear in the figures of several sources (e.g., Webster, 1974), but are usually ignored by the author, who goes on to speak of mechanical xylem breakage. A few have recognized something more than breakage. Parkin (1898) described the lumens of xylem vessels and raphid-cells as "being obliterated by the pressure of the dividing cells" of the separation layer in leaves of *Narcissus* (Fig. 5:23). In floral-cup abscission of peach the xylem elements in the separation layer undergo drastic deformation (Fig. 5:24, Simons, unpubl.). In commenting on style abscission of *Citrus medica*, Goldschmidt and Leshem (1971) noted that in the separation of the vascular strands, the helical thickenings of the tracheary elements "became flexible and stretchy" in the processes of separation. In petal abscission of *Linum lewisii*, Wiatr (1978) ob-

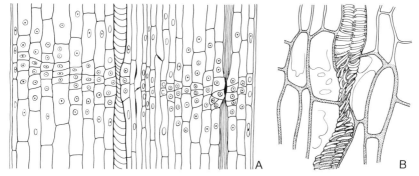

Fig. 5:23.
Changes in lignified cell walls during leaf abscission of *Narcissus*. **A.** Constriction of a xylem vessel and a raphide-cell by developing cells of the abscission zone as diagrammed by Parkin, 1898. **B.** Distorted tracheids in the leaf abscission zone, filled with "a gum-like substance." (Redrawn from Namikawa, 1926)

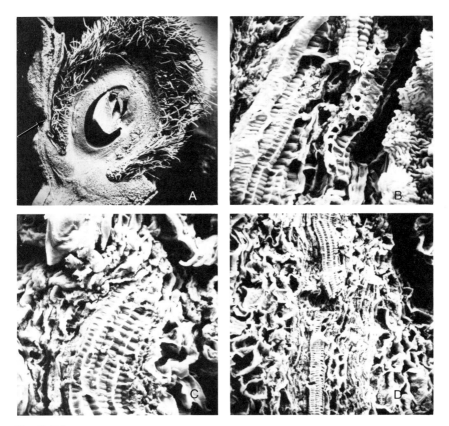

Fig. 5:24.
Scanning electron micrographs of distortion of lignified xylem elements during floral-cup abscission of peach. **A.** Location of floral-cup abscission zone. **B–D.** Distorted xylem elements at the separation layer. (Courtesy of R. K. Simons)

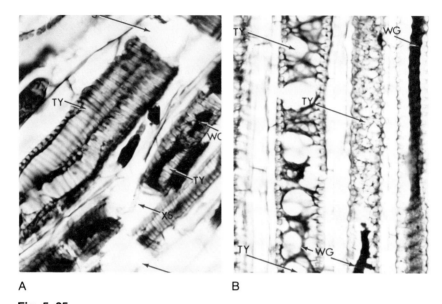

Fig. 5:25.
Changes in xylem vessels during abscission in the cotton cotyledonary-node explant. **A.** Photomicrograph suggests the separation between end walls of vessel elements. **B.** Disruption of secondary wall by penetration of tyloses. (Courtesy of Bornman et al., 1967)

served middle-lamella separation of the phloem and xylem parenchyma elements, and intrusion of tyloses into xylem vessels. Such intrusions can be extensive (Fig. 5:25) (Bornman, 1967a), and could well contribute to weakening of the xylem.

Separation of lignified xylem and other non-living elements of the abscission zone could also be accomplished by the pectinases secreted by adjacent parenchyma of the separation layer. Such enzymes attacking the middle lamellae between end walls of xylem elements would certainly facilitate separation. Some published micrographs suggest that possibility (see, e.g., Fig. 5:25A). Recent work of Wiatr (1978) and Sexton (1979) gives further support to the view that wall separation of lignified elements is a regular part of the process of abscission (Fig. 5:26).

In an extensive study of branch abscission in *Eucalyptus*, Ewart (1935) found that in many species branch abscission takes place in two stages. Typically, the branch first breaks away in the weaker region at its base, leaving a stump of a few to several cm. In the course of the next three years, the new tissues of the main stem developing around the dead stump grip it tightly and exert an outward force. At the same time, the outer layers of the branch's woody lignified tissues within the trunk gradually soften, disintegrate, and become replaced by a layer of gum containing "no wood fibres or

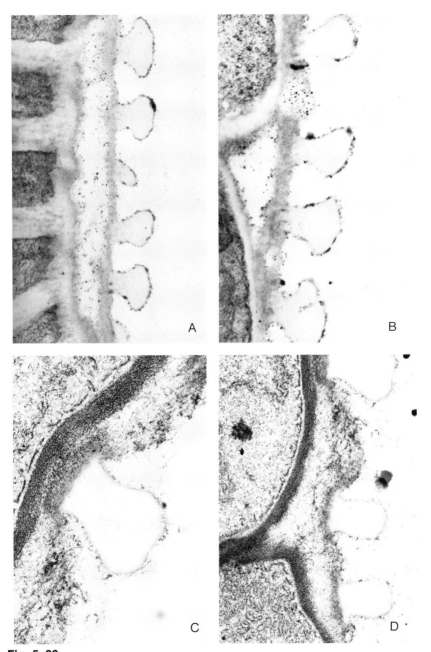

Fig. 5:26.
Electron micrographs of tracheary (xylem) elements in the petal abscission zone of *Linum lewisii*. **A, B.** Separation of middle lamella adjacent to xylem vessels. The secondary thickenings appear as clavate protrusions in these sections. **C, D.** Disintegration of the primary walls of tracheary elements. Note that the primary walls of adjacent parenchyma are still intact. (Courtesy of Wiatr, 1978)

organized structural elements." By the end of three years, the direct xylem connections between the branch and the main stem have been repaced completely by the layer of gum, and the branch stump loosens and falls from the tree, primarily from its own weight (Fig. 3:9). Ewart (1935) observed twenty-seven species that showed the foregoing type of branch separation, and a number of other species in which branch stumps did not separate. He considered that the most likely way in which separation was accomplished was by the production of "lignase or other enzyme," which could attack the wood at the base of the branch.

The foregoing discussion gives the evidence for believing that considerably more than simple mechanical breakage is involved in xylem separation. (i) Pectinases from adjacent parenchyma can hydrolyze the middle lamellae of the end walls of xylem vessels. (ii) The penetration of tyloses can weaken the primary wall of vessels and possibly distort the secondary wall sufficiently to weaken it also. (iii) The action of lignases, which to date has been suggested in only a few instances, is likely to be found in many more cases, possibly universally in connection with the development of protective layers. Investigation of these and possibly other factors in the separation of lignified elements offers a promising field for future research.

4. Cutin and Suberin

Little attention has been given to the waxy and fatty materials and their involvement with abscission. There is yet no evidence of enzymatic degradation of the cuticle. Although the epidermal cells separate in the same manner as parenchyma of the abscission zone, the cuticle appears merely to break mechanically.

There are few publications touching on separation of suberized tissues. Of course, much of the bark that is abscised consists of suberized cells, and suberin is a regular constituent of the protective layers.

6. Ecology

Chapter Contents

A. PHYSICAL AND CHEMICAL FACTORS AND THEIR PHYSIOLOGICAL EFFECTS   186
   1. Light   186
   2. Other Radiation   188
   3. Temperature   188
      a. Cold   188
      b. Heat   189
      c. Fire   189
   4. Water   190
      a. Moisture Stress   190
      b. Rain and Mist   193
      c. Flooding   193
   5. Wind   194
   6. Soil Factors   195
      a. Mineral Deficiencies   195
      b. Toxicities   197
      c. Salinity and Alkalinity   197
   7. Atmospheric Pollutants   198
   8. Ecological Effects of Abscised Parts   201

B. PHENOLOGY: SEASONAL ABSCISSION   202
   1. Autumnal Defoliation   202
   2. Vernal Leaf Abscission   203
      a. Abscission of Marcescent Leaves   203
      b. Abscission of a Portion of the Leaves of Evergreen Trees   204
      c. Abscission of All of the Leaves of "Evergreen" Trees   205
      d. Physiology of Vernal Leaf Abscission   205
   3. Drought (and Summer) Leaf Abscission   205
   4. "Hygrophobic" Leaf Abscission   207

C. BIOTIC FACTORS IN ABSCISSION   207
   1. Pathogenic Microorganisms   207
   2. Insects and Mites (Acarina)   210
      a. "Defoliator" Insects   210

      b. General Abscission Effects of Insect Feeding   211
      c. *Lygus*   211
      d. *Anthonomus*   211
      e. Mites   213
      f. Insect Galls   214

A. PHYSICAL AND CHEMICAL FACTORS AND THEIR PHYSIOLOGICAL EFFECTS

1. Light

The most important effect of light is on photosynthesis and consequently on the supply of carbohydrate. The involvement of carbohydrate nutrients in abscission has already been mentioned. Moderate levels of carbohydrate are necessary for the accomplishment of the changes of abscission. Responses to light intensity are most conspicuous in leaves. Low light intensity tends to promote abscission. When the canopy of an entire tree and the competition among its leaves is considered, it is the shaded, disadvantaged leaves that are first abscised (Meyers, 1940; Schaffalitzky de Muckadell, 1961). Leaves exposed to full light intensity are retained for the full period of their "life expectancy." Nevertheless, when individual plants that have received adequate illumination are changed to conditions in which they receive much reduced illumination, they typically abscise at least a portion of their leaves. Also, it is common for flower buds that may have been initiated under adequate light intensities to be abscised if light intensities become reduced in the course of bud development. An example is the flower bud abscission of *Lilium* in the greenhouse during winter (Durieux, 1975). For growers of greenhouse tomatoes, flower abscission can be a source of heavy losses.

With cotton plants, three days of darkness induced abscission of 100 percent of the young fruit five days after start of the experiment. During the three days of darkness, the carbohydrate level in the young fruit dropped to half the original value while ABA rose to three times its original level. ETH production was low at first, but rose to several times its original level during the two days immediately before abscission (Vaughan and Bate, 1977).

Light intensities high enough to promote accumulation of carbohydrates tend to prolong the retention of leaves and fruits (see, e.g., Roy and Chatterjee, 1967). The effect of high levels of carbohydrate can become evident months after the actual photosynthesis. Carbohydrate reserves deposited in the twigs of deciduous fruit trees are directly related to the amount of young fruit retained the following season (Chandler, 1951).

A shortening photoperiod is one of the important triggers for autumnal defoliation of many trees. This was first suggested by Garner and Allard

(1923) on the basis of their pioneering experiments on photoperiodism. Botanists noticed some time ago that autumnal leaf fall was usually delayed on trees in the vicinity of street lights (e.g., Matzke, 1936; Schroeder, 1945). In experiments with *Acer saccharum*, Olmsted (1951) found that young trees kept on a 16 h photoperiod retained their leaves up to five months longer than controls under the natural autumnal photoperiod (at Chicago). Murashige (1966) was able to completely prevent the autumnal leaf fall of *Plumeria* (in Hawaii) by interrupting the longer nights of autumn and winter with artificial light. The response to photoperiod can be very finely tuned. For example, "Several trees growing at Peradeniya (7° N), Ceylon (e.g., *Hevea brasiliensis, Bombax malabaricum, Manihot glaziovii*, and *Erythrina velutina*), drop their leaves between December and March, but at Buitenzorg [Bogor] (7° S), Java, those same species shed their leaves from June to August" (Alvim, 1964).

The fact that leaf abscission (of some species) is sensitive to photoperiod suggests that it may be under the control of phytochrome, an interpretation supported by recent experiments with leaf abscission in *Phaseolus aureus*. Cuttings kept in the dark showed enhanced leaf abscission, but that abscission could be inhibited by red light, and the inhibition was reversible by far red light (Curtis, 1978).

The mechanism of the induction of leaf abscission by shortening days has not been investigated. However, the levels of hormones in leaves are often correlated with photoperiod. In long days IAA, GA, and CK are often high, and typically drop as days shorten and leaves become senescent (Wareing, 1957; Nitsch, 1963; Bornman, 1969; Zeevaart, 1976). Further, levels of ABA either increase or remain constant as days shorten (Eagles and Wareing, 1964; Rudnicki et al., 1968; Bornman, 1969). The net result of the hormonal changes is a shift in the balance between the growth-promoting hormones and ABA such that the ratio of the growth promotors to ABA lowers, a change promotive to abscission.

In contrast to leaf abscission induced by shortening photoperiods is flower bud and young fruit abscission induced in some species by long photoperiods. In an investigation of several varieties of soybeans in photoperiods of 12, 16, and 20 h, van Schaik and Probst (1958) found increased flower and young fruit abscission as the photoperiod was lengthened. Ojehomon et al. (1968) studied flower bud abscission of short-day varieties of *Phaseolus vulgaris* that originated in Colombia and Peru near the Equator. In contrast to most varieties of *Phaseolus vulgaris*, which are completely day-neutral and which flower and fruit in either short or long days, the equatorial varieties under controlled long-day conditions in the greenhouse grew vigorously and initiated flowers, but abscised most of their flower buds. Thus, on the basis of their flowering (in contrast to flower initiation) the equatorial varieties were short-day types.

2. Other Radiation

Radiation with wave lengths outside the range of visible light has important biological effects, and abscission is one of the responses to such radiation. Abscission of flower buds and leaves has been induced by UV-B radiation (Hall and Liverman, 1956; Carns and Christiansen, 1975). Low dosage of gamma radiation when sufficiently prolonged appears to accelerate the natural processes of senescence. For example, trees of *Pinus taeda* that received 4,000 rads over a period of a month showed excessive needle abscission. Exposure of deciduous hardwood species to 20 R/day for six months of winter and spring brought on typical autumnal leaf-color changes and abscission by June. In other experiments, 4,000 rads over a period of several weeks in the summer accelerated the autumnal color changes and abscission by about a week, whereas 15,000 rads initiated these changes seven weeks earlier than normal (Krebs, 1965). In laboratory experiments petiole stem explants of beans and of *Impatiens* showed accelerated abscission with moderate dosages of gamma radiation. High dosages were inhibitory (Dwelle, 1975). The physiological and biochemical effects of such radiation on plants are only beginning to be investigated. It is noteworthy that low to moderate doses of radiation sufficient to induce considerable response produced little or no outward signs of injury; the abscission zones are able to function in a normal manner after the radiation. In some ways the responses are very similar to those following a mild frost. The present view that radiation primarily accelerates the changes of senescence (Krebs, 1965) is plausible. Dwelle (1975) has shown that just as abscission induced by normal senescence can be retarded by auxin and Ca^{2+} and accelerated by ETH, so can the abscission induced by gamma radiation. Similarly, UV-induced branch abscission in *Spirodela oligorhiza* can be greatly reduced by pre-incubation with IAA or 0.1 M or 0.2 M sucrose (0.3 M and 0.4 M sucrose promoted abscission) (Witztum and Keren, 1978a, b).

3. Temperature

The temperature characteristics of the abscission zone have been discussed in Chapter 4. In this section the effects of temperature on abscission in general will be discussed, primarily conditions under which the entire plant or portions of the plant, such as leaves, react to unusual temperature with an abscission response.

a. Cold

Leaf abscission is one of the common responses to cold that develops over a short period of time. For example, a light frost can induce considerable leaf abscission of plants such as *Citrus* and *Gossypium*. With such plants, a severe freeze will kill leaves and actually prevent leaf abscission. The light frost appears to bring about changes in leaf metabolism that induce abscission, but does not injure the abscission zone. Thus, abscission is initiated and is able to proceed to separation. Severe cold injures all tissues, including the abscission zone, and if the cold is sufficiently severe, the cells of the abscission zone are killed and physiological separation is not possible. Depending on the species and the conditions under which the plant is growing, widely varying degrees of cold are sufficient to induce abscission. Under some circumstances, cold appears to be the factor initiating autumnal defoliation of deciduous trees.

b. Heat

In a somewhat similar way, excessive heat can contribute to abscission (Wiesner, 1904c). It should be borne in mind in reading the following comments that high temperatures almost invariably induce water stress, and some of the effects of heat may be the result of that stress rather than of heat itself. Well-watered trees show little or no leaf abscission from heat (Wiesner, 1904c). Following a period of hot weather, however, many trees do abscise a portion of their leaves. Such abscission is characteristic of species of *Eucalyptus*, *Citrus*, and other subtropical broad-leaved evergreens. Typically, the flush of leaf fall comes a day or two after the close of the hot weather, and it is the older leaves that are abscised. Wiesner (1904c) has pointed out that it is something of a paradox that in response to the heat of the sun, it is not the leaves that receive the most direct rays that are abscised, but it is the lower and innermost leaves. This observation also supports the view that it is water deficit rather than heat itself that induces the abscission. Another conspicuous response to hot weather is abscission of flower buds, flowers, and young fruits, which has been noted many times in crop plants such as beans and the tree fruits (Wittwer, 1954). Fitting (1911) accelerated petal abscission of several genera by brief exposure of the flowers to temperatures of 33°–40°C. Leaf abscission of mature cotton plants has been induced by brief exposure to air heated to 150°–300°C (Batchelder et al., 1971).

c. Fire

Fire is such a drastic agent that injuries from it are seldom sufficiently mild to permit the abscission zone to function. Obviously, fire is a very potent defoliating agent and is sometimes so used in forestry (Hough, 1968). Further, fire can have a significant pruning effect on branches. In forest regions that are subject to periodic understory fires, such as the yellow pine (*Pinus ponderosa*) forest of western America, trees whose bark is resistant can survive an understory fire with the loss of only their lower branches (see

Daubenmire, 1967). *Sequoiadendron gigantea*, the Big Tree, is the prime example of a tree whose architecture at maturity has been determined largely by fire pruning.

As far as leaf abscission is concerned, the intensity of heat to which an individual leaf is exposed will determine whether the leaf is injured and abscised, killed and not abscised, or completely destroyed.

Fire is also an important factor in the shedding of seeds of serotinous species (Daubenmire, 1967). For example, cones of *Pinus attenuata* can remain closed for many years and seldom open to release seeds unless they have been heated by a forest fire. Thus, a burned forest of *P. attenuata* reseeds itself promptly and heavily (Jepson, 1925). The scales of serotinous cones are held together firmly by a resinous bond. Temperatures above 50°C are required to melt the resin. Then the scales can separate and the seeds disperse (see Teich, 1970). In *Banksia ornata* the follicles are quite woody, and the sutures weaken under the intense heat of a fire. More important is the effect of such heat in bringing about the desiccation of the woody follicle walls. After the suture is weakened, dehiscence is possible, and the desiccating walls curl back, opening the follicle (Gill, 1976).

4. Water

a. Moisture Stress

Abscission is one of the principal defenses of a plant against the injury of water deficiency. Under moisture stress most plants will abscise leaves, flowers, or fruit and bring the plant back into balance with the environment. For most species, the extent of abscission is in direct proportion to the severity of the moisture stress (e.g., Vieira da Silva, 1973).

Many trees that are exposed to seasonal drought respond by defoliation, abscission of all of their leaves. For these deciduous species, moisture stress is the trigger for the annual or periodic defoliation (Figs. 6:1, 6:2). (Various patterns of the drought deciduous habit are discussed in sec. B.3.) Leaf abscission responses to moisture stress are not always as simple as those mentioned above; there is considerable variation among species and with leaf age (see Neger and Fuchs, 1915; Sifton, 1965). In *Citrus*, for example, leaves that are injured by moisture stress are not abscised promptly; they remain attached until water relations improve materially (as from a rain or irrigation), and then they are abscised.

The physiological effects of drought on plants have been studied for many years (see Crafts, 1968; Vaadia et al., 1961). Moisture stress induces numerous chemical and biochemical changes, many of which appear to assist the plant in adapting to moisture stress (e.g., Darbyshire, 1971b). Some of these changes tend to promote abscission, particularly the changes affecting hormone levels and activities. With respect to auxin, drought brings about increased activity of IAA-oxidase (Darbyshire, 1971a), a decrease in

Fig. 6:1.
Fouquieria splendens in leaf. Note the desert habitat.

diffusible auxin (Hartung and Witt, 1968), and a decrease in auxin-transport capacity (Davenport et al., 1977). A decrease in CK activity has also been observed (Itai and Vaadia, 1965, 1971). Increased ETH has also been found (McMichael et al., 1973; Guinn, 1976; Davenport et al., 1976). One of the most impressive responses to moisture stress is the rapid rise in levels of ABA that develops within an hour or two of the onset of the stress (Fig. 6:3) (Wright, 1978; Zeevaart, 1971; Mizrahi et al., 1971). Each of the foregoing hormonal changes tends to promote abscission; in combination their effect is a very strong stimulus to abscission. The changes would be expected to be minimal in the younger leaves that are stronger sinks, and to be greater and more effective in the older leaves with limited nutrient reserves.

It may be of interest to comment briefly on plants whose leaf abscission is resistant to moisture stress, especially those that can be described as "desiccation tolerant." A number of higher plants are able to withstand virtually complete desiccation involving prolonged exposure to relative humidities from 15 to 0 percent (Gaff, 1971). Broad-leaved species, such as *Myrothamnus flabellifolius* of the arid regions of southern Africa, are known locally as "resurrection plants." Under moisture stress, leaves are not abscised, but fold and become appressed to the stem and remain in that position until moisture again becomes available. At that time the leaves unfold and appear to function without damage. The leaves of *Xerophyta plicata*, a rupicolous

Fig. 6:2.
Shoots of *Fouquieria splendens*. **A.** With leafy short shoots that develop after a rain. **B.** After drought-induced leaf abscission. The spines are persistent parts of the primary, long-shoot leaves (see Fig. 3:3).

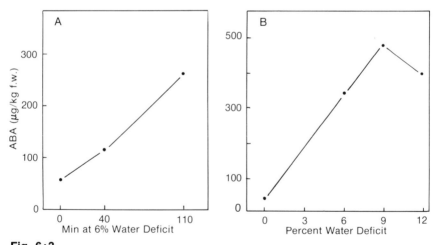

Fig. 6:3.
A. Rapid increase of abscisic acid in detached wheat leaves held at 6% water deficit. **B.** Abscisic acid content of detached wheat leaves held at 0, 6%, 9%, and 12% water deficit for 4 h at 23° C. (Redrawn from Wright, 1978)

plant of Brazil, were studied in some detail by Meguro et al. (1977). They could withstand loss of 90 percent of their water without injury. When they were rehydrated, chlorophyll reappeared in two to three days, and normal photosynthetic activity was resumed.

b. Rain and Mist

Moisture in the form of rain or mist has relatively little influence on abscission. Most species appear well adapted to the atmospheric water to which they are exposed. However, prolonged exposure to more than the normal amounts may have an effect. Lee and Tukey (1972) found that when *Euonymus alatus* was grown under intermittent mist during the autumn months, the development of red foliage color was inhibited, and leaf abscission was delayed for more than four weeks. The treatment modified the levels of a number of chemicals within the leaves and leached considerable sugars, K, and soluble N. Leaves of plants under mist contained much less ABA than non-misted leaves (Hemphill and Tukey, 1973). Also, the hydration of the leaves may have promoted a buildup of IAA, a change that could easily delay abscission. Shoji and Addicott (1954) found that immersion of bean petioles in water led to the accumulation of auxin.

c. Flooding

When the roots of a plant are flooded for more than a very brief period, abscission of leaves, flowers, or fruits is one of the symptoms that develops (Kramer, 1951; Treshow, 1970; Reid and Crozier, 1971; Ladiges and Kelso, 1977). The primary effect of flooding is O_2 deficiency in the soil. The symptoms can be induced by replacing the soil O_2 with N_2 gas as readily as by flooding (Bradford and Dilley, 1978). A study of flood-sensitive and flood-tolerant species of *Senecio* showed that the flood tolerant *S. congestus* was able to transport sufficient O_2 from the shoots to the roots to maintain normal rates of O_2 utilization (Lambers et al., 1978).

One of the principal effects of the O_2 deficiency from flooding is impaired uptake of mineral nutrients. The symptoms include those of mineral deficiency, particularly deficiency of nitrogen, which tends to develop first. Consequently, chlorophyll formation and photosynthesis are affected, as well as respiration. Growth and development are retarded, and if soil flooding is at all prolonged, there is considerable abscission of leaves, as well as of flower buds and young fruits (see, e.g., Ladiges and Kelso, 1977; Lloyd, 1920b), similar to that induced by the mineral deficiencies.

Several hormonal responses to flooding are known. In *Helianthus*, flooding induced an increase in auxin levels that reach three times the control level after two weeks (Phillips, 1964b). Further, the symptoms of epinasty and adventitious root development described by Kramer (1951) suggest that auxin also increases, at least in some species, such as the tomato. In other species, such as *Liriodendron tulipifera* and cotton, such symptoms are slight (Kramer, 1951; Lloyd, 1920b). Presumably, non-reacting plants have few reserves of nitrogenous substances from which to synthesize auxin after the nitrogen supply from the soil is cut off. This view is supported by the ease with which cotton sheds its buds and young fruit after flooding (Lloyd, 1920b). High amounts of soil moisture, short of flooding, can decrease auxin levels. Hartung and Witt (1968) found that soil moisture levels of about 60 percent of field capacity greatly reduced the amount of diffusible auxin in the stems of *Helianthus*. In the leaves of *Anastatica* diffusible auxin was reduced by soil moisture above 60 percent of field capacity and reached 0 at 90 percent of field capacity. Such changes would be strongly promotive of leaf abscission. Levels of other hormones including CK (Burrows and Carr, 1969) and GA (Reid and Crozier, 1971; and see Phillips, 1964a) also fall after soil flooding. Further, ETH increases after flooding, particularly in the portions of the plant that are made anaerobic (Kawase, 1976, 1978; Clemens and Pearson, 1977; Bradford and Dilley, 1978).

The reported hormonal responses to flooding are somewhat inconsistent, and it is difficult to interpret their possible significance to abscission. Perhaps the inconsistencies are understandable in view of the variety of materials that have been studied. What appears to happen is that soon after flooding ETH is produced and IAA accumulates if the plant is reasonably well supplied with the nitrogenous precursors at the time of flooding. Such hormone buildup would account for the development of epinasty and adventitious roots. Leaf abscission would follow at a later stage in flooding when nutrients and levels of IAA became depleted.

5. Wind

Wind can affect abscission by amplifying the effect of such factors as cold and low humidities. Further, through its mechanical action, wind can bring about earlier separation of leaves, branches, fruits, and other organs. The latter effect of wind is particularly noticeable in the abscission of branches of species where the physiological processes of separation are incomplete. A strong wind will bring about the abscission of such branches months or even years sooner than otherwise. Another interesting involvement of wind is in the separation of tumbleweeds where it effects the separation of the mature shoot and the subsequent dissemination of the seeds.

Nearly one hundred years ago, Darwin (1888) recorded his observations of the rapid corolla abscission of the flowers of *Verbascum* spp. "when the stem is jarred or struck by a stick." Fitting (1911) extended those experiments and noted that corolla abscission of *Verbascum* usually occurred 1–3 min after shaking, and that petals of flowers of *Veronica* spp. and *Cistus* spp. were also abscised rapidly after shaking.

6. Soil Factors

a. Mineral Deficiencies

Abscission of leaves, buds, flowers, and fruits is a common response to mineral deficiency. Essential elements whose deficiency can lead to abscission include N, P, K, S, Ca, Mg, Zn, B, Mo, and Fe (Sprague, 1964; Cook et al., 1960; Seeley, 1950; Smith and Reuther, 1949; Treshow, 1970; Oberly and Boynton, 1966; Cook, 1975). It is not surprising that these deficiencies lead to abscission, as each of the elements is essential to the physiology and survival of the plant. What is surprising is that leaf abscission has not been observed with deficiencies of the other essential elements. Presumably, those deficiencies affect all parts of the plant and impair the function of the abscission zone from the onset. In the case of the elements whose deficiency does induce abscission, it seems likely that the deficiency is first felt in the subtended organ, promoting the degeneration of the organ while the abscission zone is able to conserve sufficient of the essential element to function and bring about separation.

The involvement of nitrogenous substances in abscission has already been reviewed (Chapter 4). In the plant that is adequately supplied with nitrogen (and other nutrients) a full complement of amino acids is synthesized and serves as a pool of intermediate compounds from which many of the more specialized compounds of the plant are synthesized. Among the products that influence abscission are the hormones IAA, CK, and ETH. The biosynthesis of IAA proceeds directly from the amino acid tryptophan (Gruen, 1959; Muir and Lantican, 1968), and the levels of IAA in the plant have long been known to be correlated with the availability of nitrogen in the soil (Avery et al., 1937). Conversely, nitrogen deficiency leads to reduced levels of IAA in the plant and favors abscission and defoliation (Table 6:1) (see, e.g., Nightingale and Farnham, 1936). Presumably, the levels of CK in the plant are very sensitive to the supply of nitrogen, as five amino acids enter into the synthesis of each molecule of CK. The immediate precursor of ETH is the amino acid methionine. It is noteworthy that the tissues which are most active in producing ETH contain high levels of amino acids. And conversely, mature tissues low in amino acids are poor sources of ETH (Abeles, 1973). Ordinarily, there appear to be adequate reserves of amino acids in the typical abscission zone to enable it to release considerable ETH, but an old, senescent leaf can produce little or none (Hall et al., 1957).

Zinc is a micronutrient whose deficiency reduces growth and promotes leaf abscission. Its mode of action stems from the fact that it is essential to the synthesis of IAA (Skoog, 1940). When zinc is deficient, tryptophan and tryptamine accumulate, and the plant is unable to convert them to IAA (Takaki and Kushizaki, 1970). Skoog (1940) also observed that plants defi-

Table 6:1. *Applied Nitrogen, Petiole Auxin, and Defoliation Response in Cotton*

As soil nitrogen was increased, petiole auxin increased and the leaf abscission response was greatly reduced.

Soil Nitrogen (pounds per acre, applied as NO_3^-)	Petiole Auxin (μg IAA equiv. per kg fresh weight, 60 h after defoliant application)	Percent Defoliation (Percent leaf abscission from 20 pounds per acre of calcium cyanamid)
0	0.1	95
75	0.4	80
105	0.8	70
150	1.4	50

Source: H. R. Carns (unpubl.).

cient in Cu and Mn became low in auxin, but the general symptoms of the deficiencies had to be severe before auxin levels decreased. In the case of these three deficiencies, leaf abscission induced by Zn deficiency appears to be the result of an early and rapid lowering of auxin levels. In the case of Cu and Mn, the plants appear to be so weakened by the time auxin levels are reduced that they are unable to abscise leaves.

Boron deficiency also leads to abscission in susceptible species. It is essential to normal carbohydrate metabolism, and the abscission induced by B deficiency appears related to that role. IAA metabolism is also affected by boron, but a clear relationship between B deficiency and auxin physiology has not yet emerged (see Coke and Whittington, 1968, however). Boron is essential to the growth of the pollen tube and is normally present in stigmatic fluid in relatively high concentrations (see Linskens, 1964). Consequently, B deficiency is likely to lead to abortion and abscission of young fruits.

Calcium deficiency also leads to considerable abscission, as well as to other symptoms. It has long been recognized as a constituent of the middle lamella where Ca^{2+} acts as a bridge between the carboxyl groups of the branching chains of pectic substances. Normally, calcium levels in the middle lamella decline as abscission progresses (Sampson, 1918). An interesting aspect of the role of calcium in relation to abscission has recently come to light; that is, the ability of Ca^{2+} to retard abscission (Poovaiah and Leopold, 1973). The physiological action of calcium could be related solely to its involvement with the middle lamella. Calcium and related divalent cations also affect the permeability of cell membranes, however, and tend to reduce the permeability to water. But the role of Ca^{2+} in membrane function appears to be far reaching, certainly involving more than water permeability. It is noteworthy that Ca^{2+} has recently been found to modify plant responses to all five of the major hormones, indicating that the basic physiology of plants is intimately involved with and sensitive to the levels of Ca^{2+} (Leopold, 1977). Other divalent cations, such as Mg^{2+} and Mn^{2+}, have similar although lesser effects (see, e.g., Curtis and John, 1975).

Molybdenum deficiency induces considerable abscission (see, e.g., Evans et al., 1950). Molybdenum is an essential constituent of the enzymes of nitrogen fixation and nitrate reduction (Epstein, 1972). Thus, deficiency impairs the availability of nitrogen and in general induces the symptoms of nitrogen deficiency and the related hormonal deficiencies.

b. Toxicities

A number of mineral substances, particularly the metallic ions, can be toxic to plants (Epstein, 1972; Levitt, 1972; Childers, 1966). But abscission responses have received little attention in the literature, possibly because they were considered commonplace. Better documented are the responses to high levels of B, usually in the form of borate. Some irrigation waters contain excessive levels of B, and in other situations, toxic doses of borax have been applied to crops. These have produced a number of symptoms, including abscission of young fruit and premature leaf abscission (L. A. Richards, 1954; Oberly and Boynton, 1966; Ballinger et al., 1966). High levels of other ions, including Fe^{3+}, Zn^{2+}, Cu^{2+}, Mn^{2+}, Cl^-, and I^-, when applied to foliage, can be sufficiently toxic to induce leaf abscission (Addicott and Lynch, 1957; Herrett et al., 1962; Biggs, 1971). The mechanism of action of such ions in promoting abscission is unknown. Increased (wound) ETH accompanied abscission induced by Fe^{3+} and Cu^{2+} (Ben-Yehoshua and Biggs, 1970). No ETH was detected, however, when abscission was accelerated by Mn^{2+} (Biggs, 1971). It is likely that the action of toxic chemicals includes drastic modification of the levels of hormones and possibly other changes as well.

c. Salinity and Alkalinity

Soils that are saline (high in neutral salts) or alkaline (high in exchangeable Na) occur in arid regions where crop production is often attempted if irrigation water is available. If salinity and alkalinity are not too severe, crops can be produced by good management. Symptoms of toxicity are common, however, especially among the less tolerant crops (see Richards, 1954), and these symptoms can include considerable leaf abscission (Hayward and Wadleigh, 1949; Treshow, 1970).

One of the main effects of soil salinity is to limit the availability of water to plants. This effect of salinity serves to promote abscission in the same way as does moisture stress. Also, saline soils frequently accumulate toxic levels of Ca^{2+}, Mg^{2+}, K^+, Cl^-, HCO_3^-, or NO_3^-; but as was indicated in the previous section, there is little in the literature on the abscission physiology of these ions.

Alkaline soils have a dispersed structure that leads to poor water infiltration and poor aeration. Lack of aeration can be sufficiently severe to prevent adequate absorption of both water and nitrogen, so that the symptoms and abscission responses of plants growing in alkaline soils can resemble those of plants under flooded conditions.

7. Atmospheric Pollutants

Toxic effects of air pollutants from natural and industrial sources were actually recognized many centuries ago (see Weinstein and McCune, 1971; Middleton, 1961). Pollutants that have produced serious injury to plants include the following: illuminating gas, ETH, CO, SO_2, H_2S, fluorides, O_3, peroxyacetyl nitrates (PAN), nitrogen oxides, NH_3, Cl_2, and mercury vapor. Some of their agricultural, physiological, and biochemical effects have been investigated (Treshow, 1970, 1971; Weinstein and McCune, 1971; Middleton, 1961; Dugger and Ting, 1970; Crocker, 1948; Rich, 1964); abscission is one of the frequent symptoms (Kendall, 1918; Crocker, 1948; Treshow, 1970; Weinstein and McCune, 1971; Ormrod, 1978).

Ethylene is a common product of combustion and is recognized as the most toxic constituent of illuminating (coal) gas. It is considered responsible for much of the crop damage near metropolitan areas (Treshow, 1970). ETH is unique among the pollutants, however, in being also a plant product and in functioning as a plant hormone in some kinds of growth, fruit ripening, and abscission. The hormonal role of ETH in abscission is discussed in Chapter 4. It is of interest to note that larger than normal amounts of ETH are released by plant tissues in response to wounding and other kinds of stress and that ETH promotes adaptive responses that help to overcome the adverse effects of stress (Burg, 1968; Pratt and Goeschl, 1969). Thus, ETH the *pollutant* (in high concentrations) is a cause of stress and injury to the plant, while ETH the hormone (in lower concentrations) is an important physiological agent in the recovery of the plant from injury or stress.

Ozone is a major constituent of photochemical smog and appears to be the principal cause of the chlorotic decline that has seriously damaged *Pinus ponderosa* in the San Bernardino Mountains of southern California. Symptoms of ozone injury include chlorotic mottling of needles, abscission of needle clusters, and terminal dieback (Fig. 6:4) (P. R. Miller et al., 1963). A large number of plants have already been found to be sensitive to ozone (Rich, 1964). Leaf abscission is a frequent symptom of injury, as is young-

Fig. 6:4.
Ozone-induced needle abscission in *Pinus ponderosa*. Left, a normal shoot. Right, a shoot fumigated with 0.5 μℓ/ℓ ozone, 9 h per day for 18 days. (Courtesy of P. R. Miller, U.S. Forest Service)

fruit abscission (Weaver and Jackson, 1968; Manning and Feder, 1976). On tomatoes, long-term exposure to low levels of ozone induced abscission of 86 percent of the young fruit. In beans, sensitive cultivars exposed to low levels of ozone over an extended period retained 17 percent fewer fruit than did a resistant variety (Manning and Feder, 1976).

Fluorides produce a variety of toxic symptoms, including considerable abscission in susceptible plants (Treshow, 1971; Weinstein and McCune, 1971; Weinstein, 1977). Several varieties of *Citrus* and a number of woody ornamental plants, including *Rhododendron canescens, Bougainvillea spectabilis, Melaleuca leucadendra, Pyracantha coccinea,* and *Gardenia jasminoides,* are very susceptible to HF (MacLean et al., 1968). It has been suggested that the abortion and abscission of young fruit induced by fluorides results from their binding of Ca. The latter is one of the more important elements required for normal pollen germination and tube development (Treshow, 1971).

Fumigation of *Citrus* cultivars and woody ornamentals with NO_2 also induces considerable leaf abscission (MacLean et al., 1968). At present, however, it seems unlikely that atmospheric concentrations of oxides of nitrogen are sufficient to injure *Citrus* (C. R. Thompson et al., 1970).

Sulfur dioxide in low concentrations is not especially toxic to plants; indeed, it enters readily into the S metabolism of the plant. Much of the S of SO_2 can be incorporated into the S-containing amino acids and some accumulates as sulfates (Treshow, 1970). It is at the higher concentrations and in susceptible species that symptoms appear after toxic ions accumulate. Kondo and Sugahara (1978) found that the leaves of peanut and tomato, which are resistant to SO_2, contain relatively high levels of ABA and are able to close their stomates rapidly upon exposure to SO_2. In contrast, radish, perilla, and spinach, which are sensitive to SO_2, have low levels of ABA in their leaves and close their stomata slowly after exposure to SO_2.

Air pollution by NH_3 results from accidents in connection with its widespread industrial and agricultural use. As a result of such accidents, it was discovered that brief exposure to NH_3 can induce leaf abscission of cotton plants. Since NH_3 was relatively inexpensive, some attention was given to development of equipment for the application of NH_3 to field cotton prior to harvest (Elliott, 1964). The method has proved to be impractical, however.

Another kind of air pollution precipitates in the form of "acid rain," which appears to come from industrial pollutants discharged into the atmosphere. Most of the acidity is traceble to H_2SO_4, although sometimes HF and HNO_3 may be involved. There is little indication so far that acid rain in the field affects abscission. Simulated acid rain, however, has induced premature abscission of bean leaves (Ferenbaugh, 1976; Evans and Lewin, 1980).

8. Ecological Effects of Abscised Parts

With the abscission of leaves, twigs, branches, bark, and other plant parts many nutrients are returned to the soil, and the accumulated litter of abscission has a number of other beneficial effects (see, e.g., Kozlowski, 1973). The organic matter of forest litter has far-reaching effects on soil properties and on the biota of the soil.

One of the major roles of abscission in many environments is the return of mineral nutrients to the soil. Such recycling is of great importance in most situations, as usually one or more mineral nutrients are in short supply. In a hardwood forest community, Day and Monk (1977) found that most of the annual nutrient uptake is recycled the same season: 92% K, 79% Ca, 93% Mg, and 68% Na being returned annually. In spite of this, the long-range accumulation of nutrients is considerable in a mature forest, and on poorer soils deciduous species are at a disadvantage. For example, in a study of seven forest community types in north central Florida, Monk (1966) found a preponderance of deciduous species at the sites having the greater rainfall and the more fertile soils. In contrast, evergreen species were more numerous at the drier and less fertile sites. Monk suggested that segregation of evergreen species to the dry, less fertile sites may be related to the more gradual return of nutrients to the soil by year-round leaf fall of the evergreens. This would permit a more tightly closed mineral cycle and enable the evergreens to survive where deciduous species could not. In a subsequent investigation, the stems of deciduous and evergreen species were inoculated with Ca. In the course of a year, 85% of the Ca was lost from the deciduous trees with their abscised leaves, while only 10 to 25% of the Ca was lost from the evergreen species with their abscised leaves (Monk, 1971).

In addition to facilitating the recycling of mineral nutrients, abscised parts of plants can strongly influence the composition of plant and animal life in the immediate environment. The chemical aspects of allelopathic relations among plants has been reviewed by Evenari (1961). Leaves of some species contain toxic materials that can be tolerated by few competitive plant species, nor by animals for that matter (Janzen, 1974). For example, Went (1942) observed that the desert shrub *Encelia farinosa* has few plants growing in its shelter. He suggested that this might be due to a specific, allelopathic substance given off by the shrub. Gray and Bonner (1948a, b) found that the leaves of *Encelia* did, indeed, contain such a substance, and they were able to determine its chemical structure (3-acetyl-6-methoxybenzaldehyde). Other genera with species having similar effects include *Artemisia, Thamnosma, Eucalyptus, Ailanthus, Juglans, Picea,* and *Acer* (Evenari, 1961). The chemical structure of a number of the allelopathic substances have been determined. Phenolic substances are conspicuous among them (Evenari, 1949).

B. PHENOLOGY: SEASONAL ABSCISSION

Viewed in the Northern Hemisphere with its relatively uniform climate, the seasonal aspects of abscission present a misleadingly simple appearance. At first, autumnal leaf abscission seems to be the dominant habit, and on further examination evidence of vernal leaf abscission is found in a number of species, as well as evidence of drought abscission in others. But the more widely one becomes acquainted with the ecological habitats of the rest of the world, particularly of the tropics with its varied climatic patterns, vast numbers of species, and local environments, the more difficult it is to categorize the abscission responses. One cannot help but share with van der Pijl (1953) and Bremekamp (1936) their respect and admiration for the diversity of plant life in the tropics. Van der Pijl translates Bremekamp's comment on this point: "Botany would have developed not only in a different but also in a broader way, if its cradle had not stood in the gloomy north, but in the tropics. Even now its growth is often hampered by the ideas of botanists that have first-hand knowledge of the vegetation of temperate regions only." It is hoped that our expanding knowledge of tropical botany will continue to illuminate and extend the observations that have been made in such detail in the northern temperate regions. The discussion in this section on phenology has been organized under the subsequent headings as a matter of expediency. The headings represent the more conspicuous phenological manifestations of abscission, appropriate for discussion in this book. They are really categories of convenience, and there are many variations from the simpler and more conspicuous examples discussed here. For an introduction to the phenology of leaf fall in the tropics, the interested reader is referred to books such as those by P. W. Richards (1952), Longman and Jeník (1974), and Janzen (1975). For an introduction to some of the complexities of periodic leaf abscission in the tropics, the publications of Koriba (1958), Alvim (1964), and Alvim and Alvim (1978) are highly recommended.

1. Autumnal Defoliation

In the northern temperate regions of the world, autumnal leaf abscission is the most conspicuous manifestation of abscission. The species involved are the familiar trees and shrubs that make up the majority of useful and ornamental species and vast areas of forest.

The principal environmental signal that initiates autumnal defoliation is a shortening photoperiod; maintenance of long photoperiods is an effective way to delay or prevent autumnal defoliation.

Autumnal leaf fall does not occur at precisely the same time each year, as it would if it were controlled only by the photoperiod, but actually comes at different times in different years. The major modifying factor appears to be moisture. Drought can override the photoperiod controls and accelerate au-

Table 6:2. *Vernal Leaf Abscission*

Average daily leaf abscission from young trees in the coldhouse during late winter (l.w.) and during spring bud break (b.b.).

Species	Observation Period	Average Daily Leaf Abscission
Picea excelsa	(l.w.) 15–31 March	5.5
	(b.b.) 11–20 May	25.7
Taxus baccata	(l.w.) 9–17 April	9.3
	(b.b.) 28 April–8 May	510.0
Buxus sempervirens	(l.w.) 19–27 April	9.2
	(b.b.) 28 April–6 May	917.0

Source: Wiesner, 1904b.

tumnal leaf fall considerably. For example, in California in the drought years of 1972 and 1977, *Aesculus californica* abscised its leaves three to four weeks earlier than usual. Other factors, such as mineral nutrition, also can enter in. At the same latitudes, trees supplied with ample mineral nutrients will retain their leaves longer in the autumn than trees that are deficient.

Cold is another important factor in autumnal defoliation. In some situations it, too, can be the initiating factor (Rutland, 1888). Even when long photoperiods are maintained, as in the vicinity of street lights, the factor of increasing cold eventually brings about abscission (Wareing, 1956). Cooper and Reece (1969) observed that chilling, which induced autumnal leaf abscission of trifoliolate orange and peach, also stimulated ETH production by the leaves. The chilling did not induce leaf abscission of evergreen fruit trees (mango, avocado, lychee, and rough lemon), nor did it increase ETH production by the leaves.

2. Vernal Leaf Abscission

For many trees and shrubs, the major leaf abscission comes in the spring (Table 6:2). The vernal flush of leaf abscission is usually correlated with bud break, a fact which led Wiesner (1904b) to call vernal leaf abscission *Treiblaubfall*. Actually, there are three major patterns of leaf abscission which involve large flushes of leaf abscission in the spring: (i) abscission of marcescent leaves, (ii) abscission of a portion of the leaves of evergreen trees, and (iii) abscission of all of the leaves of "evergreen" trees.

a. Abscission of Marcescent Leaves

In several genera that are commonly deciduous in the autumn (e.g., *Quercus*, *Fagus*, *Carpinus*, and *Ostrya*) there are species that retain most of the marcescent "dead" leaves through the winter and abscise them in the spring (Fig. 3:4). The blade and most of the petiole die in the autumn, but the tissues at the very base of the petiole, including the abscission zone, remain alive. In the spring as temperatures rise and photoperiod lengthens, abscission processes go forward, and the leaf or the petiole stump, in the event that the dead leaf blade has been broken away by winter weather, are abscised. The abscission processes follow an anatomical pattern that appears identical with that of leaves that are abscised in the autumn (Berkley, 1931). The abscission of marcescent leaves is interpreted as a delayed autumnal abscission, delayed perhaps because the physiological controls in the species that are concerned do not function early enough to prevent abscission being delayed by cold weather. It is interesting that young trees, known to be higher in auxin and the growth-promoting hormones, retain a larger percentage of marcescent leaves through the winter than do older trees. For *Fagus sylvatica* the leaf retention is a juvenile character which disappears completely with age (Schaffalitzky de Muckadell, 1961). In some cases the abscission of marcescent leaves may well be correlated with bud break, although Hoshaw and Guard (1949) observed that in *Quercus coccinea* the majority of the leaves fell before bud break.

b. Abscission of a Portion of the Leaves of Evergreen Trees

In his observations of leaf abscission, Wiesner (1904b) noted that a number of evergreen trees had a major flush of leaf fall in the spring as buds were breaking. In careful observations of *Picea excelsa*, *Taxus baccata*, and *Buxus sempervirens*, among others, he observed that various species abscised from 10 percent to about 65 percent of their leaves during a very brief period in the spring. These species tend to abscise a few leaves almost every day of the year, but in the vicinity of Vienna they have a major flush of abscission in the spring at the end of April as buds are breaking. Some of Wiesner's observations are summarized in Table 6:2. It includes a representative species, *Taxus baccata*, which was abscising leaves at the rate of about 21 per day in mid-April, but as bud break came, the rate went up to 510 leaves per day for the last few days of April and the first few days of May.

Similar abscission behavior is shown by a number of tropical trees whose leaf abscission is correlated with "flushing" (bud break and branch growth). However, because tropical climates often lack well-marked seasons, the periodic abscission is correlated with variation in the weather, usually with dry spells. For example, *Theobroma cacao* in Bahia, Brazil, usually has a main flush of leaf abscission in September or October followed by two or three minor flushes between November and April. Leaf fall invariably occurs in close correlation with the flushing of shoot growth (Alvim et al., 1974).

c. Abscission of All of the Leaves of "Evergreen" Trees

Many evergreen trees abscise all of their old leaves in the spring. Typically, the abscission commences with the breaking of vegetative buds or of flower buds, or with the appearance of new leaves. With most species the flush of new leaf growth is well advanced before all the old leaves have abscised, but there is great variation among species and within species in this behavior, so that some trees may be almost completely bare for a brief period in the spring. Hence, the category "evergreen" has to be stretched a bit for such individuals. This habit is displayed by the "live" oaks such as *Quercus suber* and *Q. agrifolia*, and a wide variety of other species—e.g., *Magnolia grandiflora*, *Cinnamomum camphora*, and *Persea americana*.

d. Physiology of Vernal Leaf Abscission

The distinctive feature of vernal leaf abscission is that it comes at the beginning of the growing season in contrast to autumnal (and summer) abscission, which comes at the end of a season of growth. The ecological factors associated with these two habits are essentially opposite. Vernal abscission is correlated with rising temperature and increasing day length, while autumnal abscission is correlated with falling temperature and decreasing day length. In autumnal leaf abscission, the principal changes initiating abscission would appear to be in the leaf blade. Vigor declines, and the balance of hormones within the leaf shifts to the point that abscission is initiated. In vernal abscission, some decline in vigor of the old leaves is likewise very likely, but in these species the decline does not appear to go so far that the changes in the leaf blade are sufficient in themselves to initiate abscission. The increasing metabolic activity in the developing buds which is brought on by rising temperatures and lengthening photoperiod leads to increased synthesis of the growth-promoting hormones IAA, GA, and CK (see Wareing and Thompson, 1975). These hormones will move basipetally and serve to intensify the hormonal gradients between the leaves and the stem, thus promoting abscission. Although the hormonal relations of vernal leaf abscission have not been investigated, the foregoing explanation is well supported by experimental evidence of the action of hormones in controlling the abscission of leaves and debladed petioles from stems (see Addicott, 1970).

3. Drought (and Summer) Leaf Abscission

Drought-induced abscission is probably the most widespread of the major categories, involving far more species and larger areas of the earth's surface than either of the other categories. Such abscission is characteristic of the many regions of the world having alternating wet and dry seasons. It is a conspicuous feature of the arid and semi-arid tropics, of regions having a hot, dry summer, such as the Mediterranean, and of many deserts and semi-deserts.

The major factor in drought-induced abscission is moisture stress, which usually involves low levels of soil moisture and low atmospheric humidities. Such climates occur widely in Asia, Africa, and the New World. The dry season varies in length from almost the entire year in desert areas to perhaps only a few days or at most a week or two in the wetter tropics. In Mediterranean climates the dry season often lasts six months. In other regions two or more dry seasons per year alternate with the wet seasons; or the seasonal pattern may be quite irregular. Whatever the pattern, the stress of moisture deficiency, following the advent of dry weather, initiates leaf abscission. Typically, the trees involved remain leafless for the dry season and leaf out when rains come again. But there are many variations, some related to the species. Koriba (1958) has listed trees that fit the usual pattern and abscise all their leaves once a year. He also lists several other species that regularly abscise their leaves twice a year and one, *Ficus caulocarpa*, that abscises its leaves three times a year. Many species will not abscise all of their leaves unless the drought is especially severe. These are called *facultatively deciduous* and may retain their leaves in the wetter years and shed them in the drier years (Merrill, 1945). Bews (1925) in his observations of the thornveld of South Africa noted that most of the tree species are facultatively deciduous, but some are regularly deciduous (e.g., *Erythrina* spp., *Brachystegia* spp.). Similarly, a number of trees that are deciduous in their native habitats because of a dry season, when growing in the uniformly wet climate of Singapore, which lacks a dry season, are not deciduous. Examples include *Ficus elastica*, *Trema orientalis*, *Antidesma bunius*, *Tamarindus indica*, and *Albizia odoratissima*, which come from the mainland of southern Asia (Koriba, 1958). A final example is *Hevea brasiliensis*. In the uniform climate of Singapore it holds its leaves for 13.3 months. In north Malaysia it holds them for 12 months, while in Ceara, Brazil, and in south Malaysia it holds them for only 10 months and is bare for 2 months out of the year (Koriba, 1958; Chua, 1970).

The summer deciduous habit is also a conspicuous adaptation to the dry season of Mediterranean climates (Mooney and Dunn, 1970). This habit is shown by *Encelia californica*, *Salvia mellifera*, *S. leucophylla*, and *Artemisia californica* and several other species of the coastal sage community of California. A comparable group of species occurs in the climatically similar region of Chile (Mooney et al., 1970; Harrison et al., 1971). In the Negev of Israel there are a number of species that abscise their winter leaves in early summer and replace them by much smaller summer leaves—e.g., *Poterium spinosum*, *Ononis natrix*, *Artemisia monosperma* (Orshan, 1954). A similar adaptation is shown by *Zygophyllum dumosum*, from whose leaves the pair of leaflets is abscised from each leaf in early summer, leaving the succulent petiole to function during the dry season (Zohary and Orshan, 1954).

Another example in the wide range of summer deciduous habits is exhibited by the shrub *Fouquieria splendens* of the deserts of southwest North America. This spiny plant is leafless most of the year and puts out short leafy branchlets (short shoots) only after a rain (Figs. 6:1, 6:2) (Henrickson, 1972). The leaves seldom persist for long; they are abscised soon after the return of hot, dry weather. In years when there is more than one period of rain, *F. splendens* will leaf out and again abscise its leaves. There can be several such cycles in a year (Darrow, 1943).

A different pattern of drought leaf abscission is shown by trees that abscise only a portion of their leaves. This pattern is shown by broad-leaved evergreens, such as *Eucalyptus* spp., *Citrus* spp., and presumably many other tropical and subtropical evergreen species. Under unstressed conditions, these species may abscise an occasional leaf at any time during the year. The abscised leaves are the disadvantaged ones that, because of shading or limited supplies of nutrients, lose out in the competition with younger, more vigorous leaves. After a period of hot, dry weather these species will have a flush of leaf abscission, shedding a portion of their older, less vigorous leaves. The amount of foliage abscised is a direct reflection of the intensity of stress to which the tree was exposed. In Mediterranean climates with intermittent hot periods, these species will commonly show several flushes of leaf abscission during the summer.

4. "Hygrophobic" Leaf Abscission

A few species are known that abscise their leaves in the wet season and are in leaf during the dry season. Koriba (1958) has described this habit as hygrophobic leaf abscission. Examples include *Spondias mangifera* and *Tetramelis nudiflora* of Indonesia, *Melanorrhoea* spp. of Malaysia, and *Jacquinia pungens* of Central America (Koriba, 1958; Janzen, 1970). It is estimated that in Ceylon about 2 percent of the deciduous trees abscise their leaves in the September rainy season (Koriba, 1958). The seeming unorthodox behavior of the hygrophobic species is difficult to explain, but it is likely that the behavior gives the plant some distinct advantages in a hazardous and competitive environment. Janzen (1970) points out that a plant that is in leaf when most of its neighbors are leafless can have a great advantage in photosynthesis (and see Janzen and Wilson, 1974).

C. BIOTIC FACTORS IN ABSCISSION

1. Pathogenic Microorganisms

Abscission is a common response to infection by pathogenic microorganisms. It can serve as a defense mechanism, removing infected tissues. For example, almost complete defoliation can occur in a relatively short period, as in diseases of almond caused by *Coryneum beijerinckii* and *Cladosporium carpophilum* (Ogawa et al., 1955). Infection of fruit followed by abscission causes serious losses to farmers, of which the apple abscission from infection by *Botrytis cinerea* (Tronsmo and Raa, 1977) is but one instance.

An interesting study of fungus-induced leaf abscission with ecological overtones has been reported by Kimmey (1945). In *Ribes roezli*, one of the intermediate hosts of white-pine blister rust (*Cronartium ribicola*), normal leaf abscission began with the dry weather of summer and continued throughout the season, but rust infection caused considerable premature leaf abscission. The premature leaf abscission was greater from plants growing in the open where moisture stress was higher and in seasons when there were no summer rains. The first light frosts of autumn promoted the fall of infected leaves more than that of uninfected leaves (Kimmey, 1945). A number of similar examples are cited by Pfeiffer (1928) and Heald (1933).

Premature leaf abscission in *Quercus rubra* induced by the oak wilt fungus (*Ceratocystis fagacearum*) follows the same pattern of anatomical changes found in abscission of healthy leaves (TeBeest et al., 1973).

When the maturation of the protective layers of abscission is relatively slow, pathogenic organisms may penetrate the incompletely healed scar. Such is frequently the case in the olive knot disease caused by *Bacterium savastanoi* (Hewitt, 1938) and the bacterial canker of peach caused by *Xanthomonas pruni* (Feliciano and Daines, 1970).

Soil-borne pathogenic organisms, particularly *Verticillium* and *Fusarium* spp., can induce vascular diseases with symptoms including wilting and leaf abscission (Talboys, 1968). Little attention has been given to the mechanism of abscission induction, as other symptoms have been more important to an understanding of the disease. Many observations have been made of hormonal and biochemical responses, some of which could well be related to abscission.

Auxin (IAA) is readily synthesized by a large number of soil microorganisms, including pathogens. Higher than normal auxin level is a common (although not universal) symptom of diseased tissues (Gruen, 1959). Also, levels of scopoletin and other growth inhibitors may rise (Sequeira and Kelman, 1962; Wood, 1967). A study by Wiese and DeVay (1970) showed that stems and leaves of cotton infected with *Verticillium alboatrum* contained higher amounts of IAA than controls, and this is correlated with reduced ability of the infected tissues to decarboxylate IAA. The latter property was related to higher levels of caffeic acid and other IAA-oxidase inhibitors in the affected tissues. Levels of ABA in the affected leaves were double the levels in the controls, and ETH released by the infected plants was two to five times that by the control plants. In addition, different isolates of *V. alboatrum* varied in their ability to induce the hormonal responses. Measurements of hormone changes correlated with these diseases have not yet been made in enough tissues to enable a suggestion as to the manner in which the hormones might be inducing abscission. The increased ABA and ETH levels are consistent, however, with the usual acceleration of abscission by those hormones.

Pectolytic enzymes are produced by most of the major vascular pathogens *in situ* (Talboys, 1968). If released in or near the abscission zone, these enzymes could be major factors facilitating abscission. Also, the drastic effects of pectic enzymes on leaf tissue (Friend and Threlfall, 1976) would quite likely disrupt normal hormonal functions sufficiently to induce abscission.

Although little is yet known of the mechanisms whereby pathogenic microorganisms induce abscission, biochemical, hormonal, and enzymatic changes brought about by pathogens (Beckman, 1964; Wood, 1967; Friend and Threlfall, 1976) are likely to be involved. In addition to those changes in levels of auxin already mentioned, the rapid leaf abscission induced by infections of *Omphalia flavida* in *Coffea* is correlated with large amounts of IAA-oxidase produced by the fungus (Sequeira and Steeves, 1954). Also, increased production of ETH is a common response to infection (Williamson and Dimock, 1953). The leaf abscission of *Physalis floridana* induced by infection with tobacco mosaic virus was prevented by sprays of the synthetic auxin NAA (Dyson and Chessin, 1961). Before we can fully understand the mechanism of pathogen-induced abscission, many more investigations involving the simultaneous examination of several hormonal and physiological factors are needed (similar to the investigation of Wiese and DeVay, 1970, described above).

Toxins are involved in the pathogenicity of many microorganisms, and an impressive list has already been identified (Strobel, 1976). Understandably, plant responses to toxins have been considered almost exclusively in terms of their pathogenicity. Little is known of how toxins might affect normal plant behavior, including abscission. As research on the action of toxins progresses, it can be expected to disclose interesting relationships and pathways of action.

The fungus *Aspergillus niger* is a secondary invader and not a plant pathogen in the strict sense. Cultures of the fungus produce an unusual growth regulator, malformin, a cyclic pentapeptide. Malformin induces a wide variety of growth responses in higher plants, including accelerated leaf abscission (Curtis and Buckhout, 1977). It is especially effective at accelerating leaf abscission in the dark; malformin stimulates more rapid abscission and in lower concentration than either ABA or CEPA (Curtis, 1977).

In most situations plant diseases appear to affect shedding by modifying the physiological factors that normally influence or control abscission. But this is not always the case, and a few of the more unusual examples of shedding responses to plant pathogens are worthy of note. One example is the leaf shedding induced by plane tree anthracnose caused by *Gnomonium venata*. This disease leads to considerable necrosis of young leaves and twigs, particularly at the base of petioles near the abscission zone, and sometimes in the tender new stem tissues above a node. Thus, individual leaves, or short twigs bearing leaves, are shed as a result of petiole or stem necrosis.

The adventitious abscission induced by the shot-hole diseases is described in Chapter 3 (and see Heald, 1933).

Finally, there are diseases in which abscission is inhibited. One is the corky bark disease of grapes (*Vitis* spp.), which can rapidly and completely kill the branches of a vine. Apparently, the cells of the abscission zones are killed before they can function; the dead leaves remain on the vines indefinitely (Beukman and Goheen, 1970). Another is the leaf-roll virus of potatoes. Infected plants retain their first leaves long after healthy plants have abscised them, and the symptom is considered diagnostic for the disease (Richardson, 1976). Also, young leaves of larch infected with *Hypodermella laricus* abort, but are not abscised. Much of the needle abscission zone becomes heavily infected and unable to function normally (Cohen, 1967).

2. Insects and Mites (Acarina)

In the millions of years that plants and insects have been co-evolving, a number of interesting and important interrelations involving abscission have developed. Some of these have been mentioned in previous sections; others will be discussed below.

a. "Defoliator" Insects

The defoliators are insects, or more commonly, larvae of insects, that feed on the blades of leaves. The harm done to the plant by such feeding can be very serious. Kulman (1971) estimates that growth loss to trees infested with defoliators is directly proportional to the quantity of foliage lost—i.e., photosynthetic tissue lost. As far as abscission is concerned, most defoliators have little effect. They feed primarily on mesophyll and usually leave untouched the petioles, major veins, and almost always some of the leaf blade. From mutilation experiments (e.g., Livingston, 1948), it is clear that abscission of most leaves is still inhibited when only a small percentage of the leaf blade remains. Thus, the defoliators do not appear to be abscission inducers, as far as is now known.

The feeding of at least one defoliator appears to retard abscission if it affects it at all. This is the caterpillar that is the larva of *Stigmella argentipedella*. Its saliva contains high levels of CK. As the caterpillar moves from one feeding location on a leaf blade to another it leaves behind enough CK to stimulate the leaf to transport nutrients to the edges of the remaining leaf blade (Engelbrecht, 1971). When the caterpillar returns, the first 2 mm contain more nutrients than when the caterpillar left. Depending on the distance over which the CK is effective, it might be expected to have the effect of delaying abscission of the leaf.

b. General Abscission Effects of Insect Feeding

Often the feeding of insects has subtle and far-reaching effects. Even limited feeding on a flower bud or young fruit is often followed by abortion and abscission. Damage to seeds can be especially serious. If the seeds are injured or eaten, fruits are usually abscised, as hormones from seeds are a major source of the growth hormones essential to fruit development. There is considerable evidence that many insects as a part of their feeding inject toxins that have injurious effects well beyond the immediate physical injury of the feeding. For example, as few as four *Antestia* bugs feeding on a *Coffea* plant can induce an economically damaging abscission of green berries (Le Pelley, 1942). Sometimes the feeding by one form of an insect will be much more damaging than the feeding by another. For example, the plum curculio (*Conotrachelus nenuphar*), which infests fruits such as the apple, peach, and plum, induces considerable abscission of young fruit from feeding by larvae. However, feeding by adults or their egg cavity preparation does not induce fruit abscission (Levine and Hall, 1977).

c. Lygus

Lygus bugs are a widespread group that can do heavy damage to crops (see, e.g., Bushing et al., 1974). Symptoms of their feeding include necrosis (Addicott and Romney, 1950), abortion of flowers and young fruits, and abscission (Carter, 1939). The injury from the feeding of *Lygus hesperus* is correlated with the injection of a salivary PGU (Strong and Kruitwagen, 1968). Symptoms of injury can appear at some distance from the site of feeding (Addicott and Romney, 1950). Either the PGU or secondary necrosis- and abscission-promoting substances are transported through the plant tissues, possibly in the phloem.

d. Anthonomus

In a manner similar to that of *Lygus*, feeding by larvae of the boll weevil (*Anthonomus grandis*) induces abscission of flower buds. Since the larvae feed exclusively on the anthers in the flower bud, it was once thought that abscission would come from the destruction of that rich source of auxin. That does not seem to be the case, however, as abscission can be induced by homogenates of the larvae. Second-instar larvae about to molt produced the most active preparations (King and Lane, 1969). The principal activity was traced to a protein fraction containing an endo-polymethylgalacturonase (King, 1973). It is of considerable interest that the abscission-accelerating

Fig. 6:5.
A yellow, soon-to-abscise leaf of *Populus* with a *Pemphigus* gall. Green, uninfested leaves are shown at the sides. (Courtesy of T. G. Whitham)

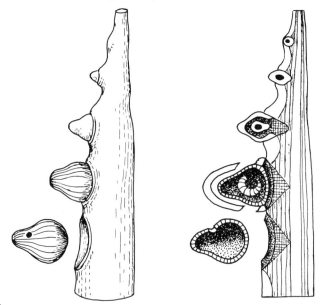

Fig. 6:6.
Diagram of the development of the gall of *Andricus sieboldi* on a twig of *Quercus* sp. Both the early covering and the empty gall are abscised. (Redrawn from Küster, 1911)

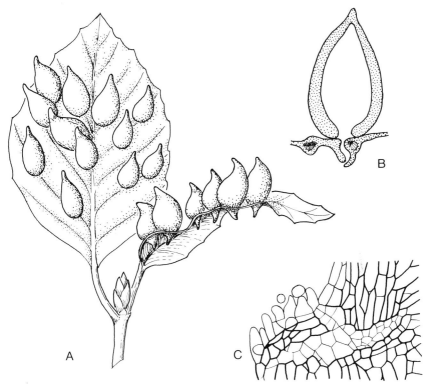

Fig. 6:7.
Pouch galls of *Mikiola fagi* on leaves of *Fagus sylvatica*. **A.** Habit sketch. **B.** Diagram of longitudinal section showing gall abscission zone. **C.** Thin-walled cells of the abscission zone. (A and B redrawn from Mani, 1964, by permission of Dr. W. Junk BV; C redrawn from Mühldorf, 1925)

factor from both *Lygus* and *Anthonomus* is a pectic enzyme that induces abscission in a zone some distance (a cm or more) from the source. It is possible that the enzyme is translocated to the abscission zone, but it is more likely, on the basis of present knowledge, that the enzyme produces sufficient local injury to modify the hormonal control mechanisms and thereby induce abscission. Pectic enzymes are known to do considerable damage to the cells with which they come in contact (Bateman and Basham, 1976).

e. Mites

As with insects, the influence on abscission of the feeding by mites varies considerably. In many cases abscission effects are not apparent; but in contrast, feeding by *Tetranychus atlanticus* on the leaves of cotton induces considerable leaf abscission while inducing few other symptoms.

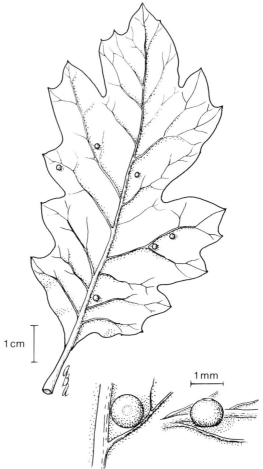

Fig. 6:8.
Jumping galls, *Neuroterus saltatorium*, still attached to leaf of *Quercus lobata*. The galls are about 1 mm in diameter.

f. Insect Galls

Among the numerous galls induced by insects on leaves there are some that accelerate and some that retard abscission of the infested leaves. In western North America some leaf galls on *Salix* and *Populus* retard abscission of the infested leaves (F. E. Strong, unpubl.). On the other hand, many gall-infested leaves abscise earlier than healthy leaves (Pfeiffer, 1928). For example, leaves of *Populus* with galls of *Pemphigus* (Fig. 6:5) abscise one to two months earlier than healthy leaves (Ziegler, 1959/60; Whitham, 1978).

Although the physiological effects of galls have not been studied in direct relation to abscission, investigations have shown increased production of IAA and CK (Yokota et al., 1974; McCalla et al., 1962; van Staden and Davey, 1978). Increased production of these hormones would be expected to delay leaf abscission. On the other hand, if the infestation weakened a leaf physiologically, it would reduce the amount of IAA, GA, or CK to reach the abscission zone, and premature leaf abscission would be expected.

When they are mature, some galls dehisce to facilitate the escape of the insect, and other galls may be abscised from the host either before or after the insect departs (Mühldorf, 1925). Such behavior can be a useful taxonomic character (see Weld, 1957). In the pouch galls it is common for a crack to develop as maturity approaches, permitting the escape of the insect—e.g., the gall of *Byrsocrypta gallarum* on the leaves of *Ulmus campestris* (Mani, 1964). Some galls dehisce by the separation of an "operculum." For example, *Cecidoses eremita* on *Schinus* separates a plug-shaped operculum, and a psyllid-induced gall on the leaf of *Newtonia insignis* has a peculiar mushroom-shaped operculum (Mani, 1964). Examples of the development and separation of two of the many galls that are abscised are shown in Figures 6:6 and 6:7.

The galls of *Neuroterus saltatorium* that form on the leaves of *Quercus*

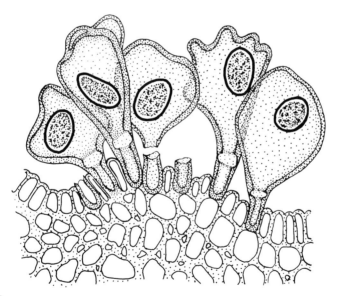

Fig. 6:9.
The single-celled galls of *Synchytrium papillatum* on the leaf epidermis of *Erodium* sp. Note abscission of the mature gall by erosion of the wall of the neck. (Redrawn from Magnus, 1893)

lobata (Fig. 6:8) and related white oaks are most unusual. The galls are small, less than 1 mm in diameter. After they abscise from the leaves in early summer, they become very active jumping galls, jumping as far as 5–7 cm (Riley, 1882).

The fungus *Synchytrium papillatum* forms single-celled galls in the epidermis of *Erodium cicutarium*. The greatly enlarged gall-cell abscises at maturity by the breakage of a thin band at the base of the cell wall (Fig. 6:9; Magnus, 1893).

7. Abscission in Lower Plants

Chapter Contents

A. FERNS, MOSSES, AND LIVERWORTS 218
   1. Ferns and Fern Allies 218
      a. Vegetative Organs 218
      b. Reproductive Structures 219
      c. Dehiscence 220
   2. Mosses and Liverworts 221

B. ALGAE 224
   1. Green Algae 224
   2. Blue-green Algae 231
   3. Brown Algae 232
   4. Red Algae 234

C. FUNGI 235
   1. Filamentous Fungi 235
      a. Somatic Fragments 235
      b. Sporangial Abscission and Dehiscence 235
      c. Conidial Abscission 236
      d. Ascus Dehiscence 242
      e. Basidiospore Discharge 244
      f. Abscission and Dispersal in the Nidulariaceae 247
   2. Yeast 250

D. LICHENS 253

E. BACTERIA 254

This chapter will survey the abscission behavior of lower plants. On first thought, one would expect to find abscission in the ferns, mosses, and possibly liverworts, but perhaps be skeptical as to whether such a process occurred in the simpler plants, such as the filamentous algae and bacteria. As the morphology of increasingly simpler forms is examined, however, a continuous series of manifestations is found, involving the separation of increasingly simple structures, until one encounters the abscission of the buds of yeast and the abscission (fragmentation) of the simpler colonial forms of algae, fungi, and bacteria. The sections of this chapter will present some of the more conspicuous examples.

A. FERNS, MOSSES, AND LIVERWORTS

1. Ferns and Fern Allies

There is abundant evidence in the fossil record of abscission of leaves, branches, and reproductive structures in ferns and related groups. In the present-day representatives abscission is not as conspicuous, but one does not need to look far to find examples.

a. Vegetative Organs

Leaf abscission occurs widely among the ferns, but is sporadic in some groups. For example, only a few of the tree ferns abscise leaves by the activity of a separation layer, exposing a smooth protective layer. Leaves of the majority of tree ferns simply droop and hang about the trunk after senescence. In some species the leaves break away, leaving a stump up to several inches long, which may persist indefinitely. In other cases, such as *Cyathea medullaris* of New Zealand, the projecting broken vascular bundles usually are abraded away by the action of hanging leaves or adjacent shrubbery, so that the lower portions of the trunk present very smooth leaf scars.

The anatomical changes in the leaf abscission zone of ferns has received some attention over the years. Separation involving dissolution of the middle lamella is fairly common, usually without previous cell division in the separation layer, but sometimes preceded by a series of two or three cell divisions (Bäsecke, 1908). In most species the protective layers become heavily lignified (Konta, 1974; Phillips and White, 1967). In species having a well-developed phyllopodium the separation layer is located at the junction of the phyllopodium and the rachis (Konta, 1974; Phillips and White, 1967).

Branch abscission is relatively rare in present-day ferns. Perhaps the most noteworthy example is the abscission of branches of the small floating aquatic ferns, *Azolla* and *Salvinia*, which enables these plants to scatter widely over bodies of water (Smith, 1938; Konar and Kapoor, 1972). Rhizomatous ferns may grow extensively, and eventually the branches separate by the death and disintegration of older portions of the plant body.

An interesting example of abscission of an entire shoot occurs in *Tmesipteris* of the Psilotales. When the sporophyte is still young, it abscises from its foot, which remains embedded in the gametophyte (Holloway, 1918).

Azolla bipinnata has the unusual ability to abscise its roots. The separation layer lies at the base of the root where it abuts on the stem and is marked by relatively large intercellular spaces that appear early in the development of the root (see Fig. 3:27).

b. Reproductive Structures

In some ferns the sporophyte can reproduce asexually by means of bulblets (adventitious buds) produced on the leaves (Bailey, 1935). They are produced by *Cystopteris bulbifera* and several species of *Asplenium*. Apparently, the bulblets of *Cystopteris* abscise from the leaf (Seward, 1931), but in other cases separation is apparently not achieved until the death of the parent leaf. The walking ferns, *Asplenium rhizophyllum*, and some other species of *Asplenium*, are capable of developing such plantlets at the tips of the leaves (Dobbie and Crookes, 1952; Bailey, 1935). In these cases, also, separation of the progeny plant is not achieved until the death of the parent leaf. Similar vegetative propagules are also found in seed plants (e.g., *Bryophyllum*) (see Chapter 3).

Smaller vegetative propagules are called gemmae. These range in size and complexity from bud-like multicellular bodies to smaller multicellular bodies showing limited differentiation, down to small groups of cells and, in some cases, even single cells. They are often produced in appreciable numbers and have the common property of being easily separated from the parent plant. The bulblets mentioned above are not usually called gemmae; however, they are really not much more complex morphologically than the bulblet gemmae of *Lycopodium*. These are flattened, heart-shaped structures composed of fleshy leaves and a small root which develop in the axils of sporophyte leaves from which they readily separate (Campbell, 1930). In *Psilotum*, gemmae can be produced in great numbers on the subterranean shoots. They are small, oval bodies of a few cells and are but one cell thick. The gemmae develop on single-celled stalks and, as far as can be determined from the literature, separate by the breakdown of the middle lamella between the stalk cell and the gemma (Fig. 7:1) (Solms Laubach, 1884).

Fern gametophytes rarely reproduce themselves. Occasionally, they may branch and fragment, owing to the death of older parts, but usually the growth of the sporophyte stops further development. A few cases have been discovered, however, in which fern gametophytes reproducing asexually by means of gemmae have established colonies that are reproducing clonally, independent of the sporophytic generation. Each clone produces gemmae characteristic of its genus (Farrar, 1967). One example is the gametophyte of *Hymenophyllum*, which produces gemmae after it has stopped producing antheridia and archeogonia. A marginal cell develops into a stalk cell and a row of cells that is the gemma proper. In this genus the gemmae often take on the form of a three-rayed star prior to detachment and development into a new gametophyte (Campbell, 1930). Gemmae of a number of related

220 *Abscission in Lower Plants*

genera have been described by Stokey (1948) and Stokey and Atkinson (1958) (Figs. 7:2, 7:3).

c. Dehiscence

Dehiscence of fern sporangia can be relatively simple, but in the higher forms there are special mechanisms to facilitate dispersal of spores. In *Ophioglossum* and *Botrychium* the sporangium "opens by a transverse slit" (Campbell, 1930), and the spores are expelled by the shrinkage of the walls. The rupture of the sporangia in *Lycopodium* and in the Marattiales is very

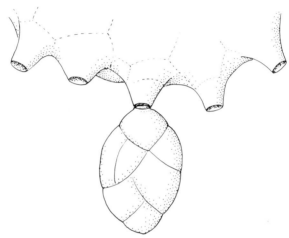

Fig. 7:1.
Gemma and scars of gemmae on a subterranean shoot of *Psilotum*. (Redrawn from Solms Laubach, 1884)

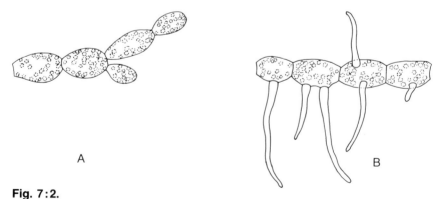

Fig. 7:2.
Reproductive (abscised) fragments of fern gametophytes. **A.** *Xiphopteris serrulata*. **B.** *Ctenopteris exornans*. (Redrawn from Stokey and Atkinson, 1958)

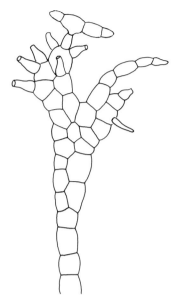

Fig. 7:3.
Portion of a gametophyte of *Trichomanes auriculatum* showing a gemma and sterigmata from which gemmae have been abscised. (Redrawn from Stokey, 1948)

much the same and does not involve an annulus. In the typical ferns, sporangium opening follows the separation of the middle lamella between the walls of specialized lip cells of the stomium. Opening of the sporangium is further facilitated by the breakage of other cells of the sporangium wall or by their separation through middle lamellae (Smith, 1938; Campbell, 1930). The wall of the typical fern sporangium includes an annulus, a row of cells which at maturity are heavily lignified on three sides (Fig. 7:4). When the sporangium dries, these cells contract with considerable tension, and upon separation of the lip cells the annulus suddenly straightens and almost reverses its position, forcibly expelling the spores. The dehiscence of the sporangium involves two factors, hydrolysis of the middle lamella between the lip cells, and a forceful, mechanical opening of the sporangium. In some ferns, such as *Osmunda* and *Schizaea*, the annulus may be a plate, usually near the base of the sporangium, which functions in a similar way to contract on drying and facilitate the opening of the sporangium (Fig. 7:5) (Smith, 1938).

2. Mosses and Liverworts

Vegetative branches of the gametophytes of mosses and liverworts are capable of indefinite growth, and in due course the branches become separated. Ordinarily, this is only by the death and disintegration of older

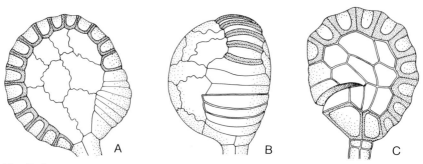

Fig. 7:4.
Fern sporangia. Note lip cells that separate at maturity. **A, B.** *Pteridium aquilinum.* **C.** *Dipteris conjugata.* (Redrawn from Smith, 1938, with permission of McGraw-Hill Book Company.)

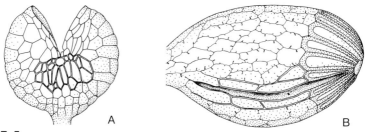

Fig. 7:5.
Dehiscence of fern sporangia. **A.** *Osmunda claytoniana.* **B.** *Schizaea bifida.* (Redrawn from Smith, 1938, with permission of McGraw-Hill Book Company.)

portions of the gametophyte. Another major method of vegetative propagation is by the separation of gemmae. Gemmae range in size from single cells in many species of the Jungermanniales to the somewhat differentiated multicellular gemmae of the Marchantiales. In species of *Marchantia* and *Lunularia* gemmae are produced in considerable numbers and contribute to the wide distribution of those genera. Their gemmae develop on a stalk one cell in diameter as more or less circular, lens-shaped structures, a few cells thick in the center (Fig. 7:6). In these genera separation of the gemmae from their stalks is facilitated by the swelling of mucilage secreted by nearby gland cells (Campbell, 1930; Smith, 1938). In the moss *Tetraphis*, somewhat similar but slightly less differentiated gemmae are produced. Again, these are formed in a kind of receptacle located at the ends of branches, and each gemma forms on the end of a single-celled filament as a lenticular disc (Fig. 7:7). In the moss *Aulacomnium*, clusters of gemmae are produced at the end of stalks that resemble gametophores. These gemmae are ovoid bodies of only a few cells on a delicate, single-celled stalk. In *Riccardia*

(*Aneura*) the production of gemmae consisting of one or at most two cells is common. They are remarkable in that they develop within the walls of pre-existing cells, and the walls of the parent cells must be weakened to liberate them. The steps in the development and liberation of these gemmae are closely similar to those of the aplanospores of the green algae (Smith, 1938). This is but another evidence of the affinity between the green algae and the higher plants.

The capsule of the typical moss sporophyte opens by the abscission of a lid, the operculum. Separation is accomplished by the drying and collapse of a ring of thin-walled cells below the operculum. Its removal is facilitated by

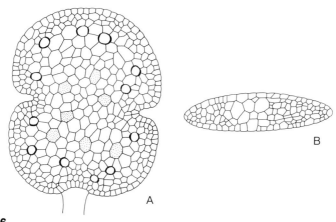

Fig. 7:6.
Gemmae of *Marchantia polymorpha*. **A.** Surface view. (Redrawn from Smith, 1938, with permission of McGraw-Hill Book Company.) **B.** Cross section. (Redrawn from Bell and Woodcock, 1968.)

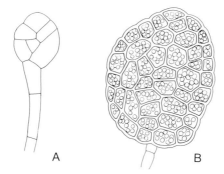

Fig. 7:7A, B.
Two stages in the development of gemmae of *Tetraphis pellucida*. (Redrawn from Campbell, 1930)

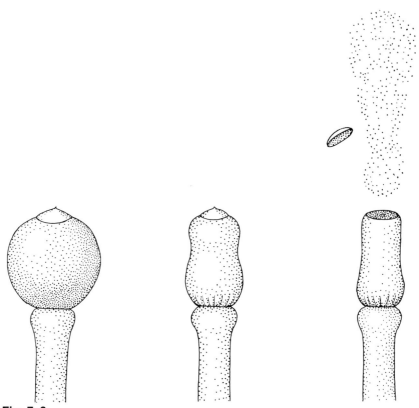

Fig. 7:8.
Stages in the drying of the capsule and explosive abscission of the operculum of the peat-moss, *Sphagnum*. The spores and operculum are sent 15 cm or more into the air. (Redrawn from Ingold, 1965)

the upward bending of the teeth of the peristome of the capsule (Smith, 1938). The capsule of the peat-moss, *Sphagnum*, is simpler in structure (Fig. 7:8), and its operculum is abscised explosively from the pressure of the drying and shrinking capsule wall (Ingold, 1965). The capsule of *Andreaea* dehisces by means of four valves (Bell and Woodcock, 1968; and see Fig. 7:9).

B. ALGAE

1. Green Algae

The wide distribution of fresh-water algae is the result of many adaptations for dispersal. Although there is considerable dispersal of resting cells and spore-like zygotes, the dissemination of vegetative cells is considered of

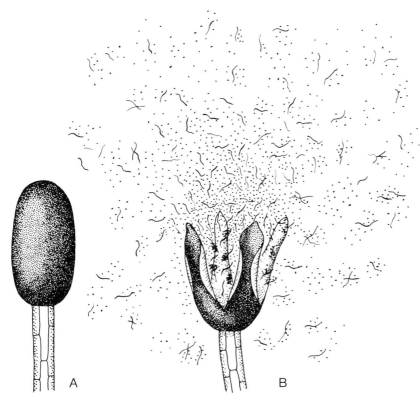

Fig. 7:9.
Sporophyte capsule of the leafy liverwort, *Cephalozia bicuspidata*, **A**, before and **B**, immediately after dehiscence. Note the scattered spores and elaters. Dehiscence takes place along the four sutures of the capsule. (Redrawn from Ingold, 1965)

much greater importance (Smith, 1950). The principal agents of dispersal are birds, wind, and water, of which the first two are of major importance in long-distance transport of vegetative cells. There is increasing evidence that the vegetative cells of many species can successfully withstand considerable desiccation (Smith, 1950).

Vegetative multiplication of colonial algae can take place in three ways. The first is accidental breakage, the result of physical factors or the feeding of animals, breaking filaments or other colonies and thereby increasing the number of colonies. This method is hardly abscission, but is somewhat analogous to the vegetative multiplication of higher plants by the mechanical separation of living branches.

A second method of multiplication is disarticulation of filaments weakened by the localized development of zoospores or aplanospores. This leads to the separation of the remaining vegetative portions and an increase in the number of colonies.

A third method involves the fragmentation of filaments into individual cells or short chains. It should be noted that the cell walls of green algae are like the primary walls of the cells of higher plants in having a distinct inner layer that is high in cellulose and an outer layer that is high in polysaccharides containing uronic acids. Chemically, the components of the outer layer are so similar to the pectic substances of higher plants that we shall call them that in this discussion. In addition, some green algae have conspicuous layers of hemicellulose, and a few have layers of chitin (Tiffany, 1924; Mackie and Preston, 1974). Both the cell-wall chemistry of the green algae and the mechanism of cell separation appear very close to those of the higher plants. Such evidence gives additional support to the general view that the higher plants evolved from ancestors similar to present-day green algae.

The fragmentation of colonial green algae is usually achieved through the softening and dissolution of the outer pectic layer. For example, in the Palmellaceae, the colonies are usually amorphous, and the cells are held together by a voluminous pectic envelope. Vegetative multiplication is achieved by fragmentation of a colony when the envelope weakens. In the Schizogoniales as the pectic envelope weakens, particularly between the cells of the filament, the filament can dissociate into a mass of spherical cells that readily separate from one another. In *Stichococcus* the filamentous cells are very loosely bound, and the chains never exceed more than a few cells in length before the cells separate.

For many of the filamentous green algae, abscission involves factors other than merely the weakening of the pectic layer between the end walls. For example, in some species of *Spirogyra* the end walls are folded. After the middle lamella has softened, osmotic pressure expands the end wall of the growing cell(s), and at the same time exerts a mechanical stress that greatly facilitates separation (Fig. 7:10) (Lloyd, 1926). In *Mougeotia* there is no fold in the end walls, but expansion of one of the cells also exerts mechanical force that promotes separation (Lloyd, 1926). In *Oedogonium* cell division is achieved in a unique way through what must be enzymatic activity within a special structure, the division ring. This brings about the separation of the entire cell wall in a narrow circular band near one end of the cell, enabling the cell to expand. The new wall is laid down in such a manner that a circular collar remains, marking the site of division. As the cell can divide repeatedly, these collars accumulate at one end of the cell (Figs. 7:11–7:13). The number of collars indicates the number of times that particular cell has divided. In a similar separation, *Oedogonium* releases the zoospore

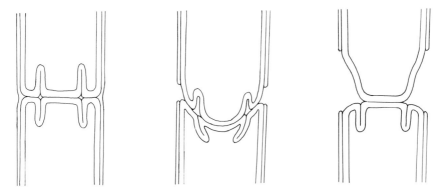

Fig. 7:10.
Cell separation in *Spirogyra weberi*. Turgor facilitates unfolding of the end wall. (Redrawn from Lloyd, 1926)

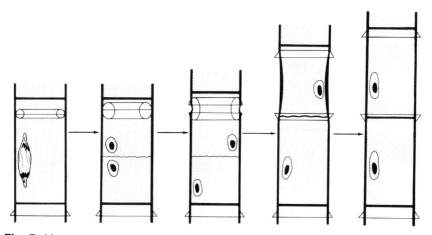

Fig. 7:11.
Diagram of cell division in *Oedogonium*. Separation of the old wall is brought about by localized hydrolysis adjacent to the division ring. (Courtesy of Pickett-Heaps, 1975)

from the filament by a circumferential rupture of the cell wall (Pickett-Heaps, 1975). Again, the separation involves a highly localized secretion of hydrolytic enzymes. In *Cephaleuros virescens* the zoosporangia are abscised in a manner suggestive of the detachment of yeast buds (Chapman, 1976; Fig. 7:14; cf. Fig. 7:40).

In the Ulvales, as well as in the Schizogoniales, vegetative multiplication by abscission of small proliferous shoots from the holdfast is common. In the Siphonales, vegetative multiplication can be by abscission of branchlets

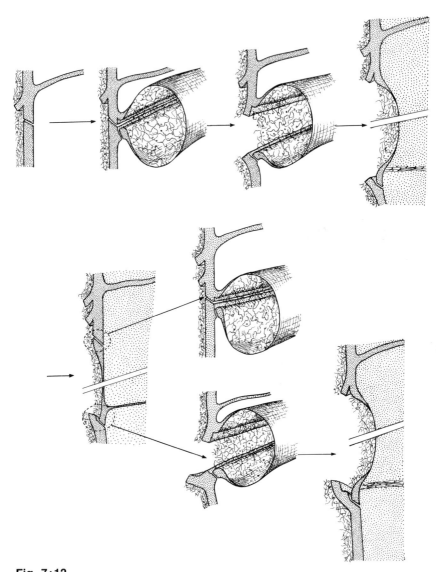

Fig. 7:12.
Diagram of the ultrastructural aspects of wall changes in cell division of *Oedogonium*. (Courtesy of Pickett-Heaps, 1975)

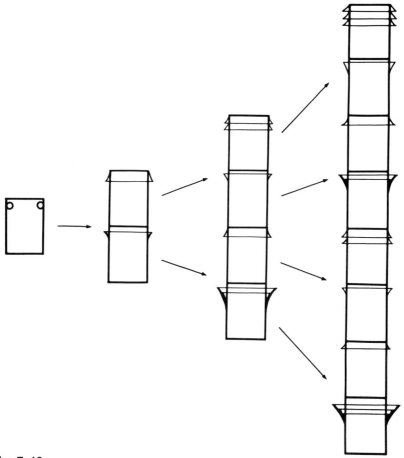

Fig. 7:13.
Diagram showing the accumulation of caps (rings of wall material) in a dividing filament of *Oedogonium*. (Courtesy of Pickett-Heaps, 1975)

(e.g., *Bryopsis*), which then form rhizoids and develop into new plants. Small buds are produced and cut off by *Protosiphon*. *Codium* reproduces vegetatively by detachable propagules (Fritsch, 1945; Smith, 1950, 1969).

Flagellar abscission occurs after the zoospore of *Oedogonium* comes to rest. There is "an unusual shivering or vibratory motion during which all the flagellae are violently shed" (Pickett-Heaps, 1975). This behavior is in contrast to that of the motile cells of most other algae, such as *Hydrodictyon*, *Stigeoclonium*, *Ulothrix*, and *Vaucheria*, whose flagellae are withdrawn and resorbed when the swimming phase is finished.

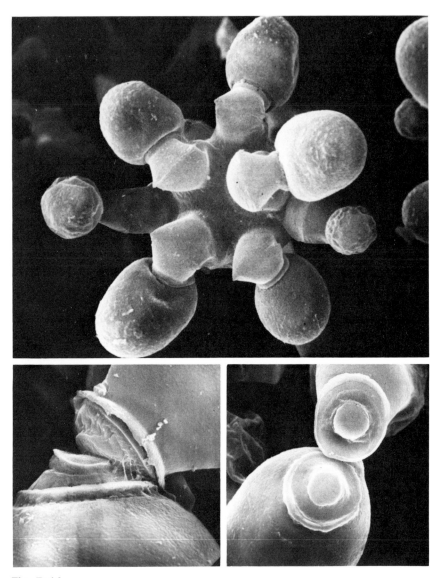

Fig. 7:14.
Scanning electron micrographs of two steps in the abscission of the sporangium of the green alga *Cephaleuros virescens*. (Courtesy of Chapman, 1976)

Ecological factors have a strong influence on the development and behavior of algae. Mineral nutrition, for example, affects abscission and fragmentation. In particular, the absence of Ca affects the strength of the cross walls; when a filament from a normal culture medium is placed in a medium lacking Ca, fragmentation is induced (Tiffany, 1924). This involves a breakdown of the "middle lamella" between the end walls, indicating that Ca is in some way necessary for the physical strength of the pectic substances in the algal middle lamella, as has been known for some time in the higher plants. Reed (1907) observed that filaments of *Spirogyra* induced to undergo cell division in media lacking Ca usually fail to deposit new cross walls. As Sampson (1918) observed in *Coleus* leaf abscission, abscission is accompanied by the exit of Ca from the middle lamellae of the abscission zone.

2. Blue-green Algae

As in the green algae, separation of colonies into subcolonies occurs in the nonfilamentous *Cyanophyta*. In some species and genera this occurs fairly readily, and the colonies remain relatively small in size. In other genera, such as *Chroococcus* and *Gloeocapsa*, the pectic envelope is more persistent and the colonies can grow to a very large size. The reasons for the difference in ease of fragmentation among the nonfilamentous forms are not clear. Obviously, in some the pectic envelope dissolves more readily (Smith, 1950), implying a somewhat more loose chemical configuration. Possibly, also, the more readily fragmenting forms produce rather more hydrolytic enzymes than the nonfragmenting ones. With the algae, as we are now coming to recognize in the higher plants, there may also be simultaneous production of both synthetases and hydrolases of pectic substances, with the balance between synthesis and hydrolysis falling at different points in different taxonomic groups.

Methods of fragmentation in the filamentous *Cyanophyta* are somewhat similar to those in the *Chlorophyta*. The cell walls between heterocysts and vegetative cells are relatively weak, and the filaments of some forms commonly fragment at those points. Another important method of multiplication is by means of hormogonia, short, delimited sections of filaments. In species of *Oscillatoria* and *Lyngbya*, the hormogonia are up to several cells long and demarked by separation discs (necridia). These are double-concave discs of largely pectic material that form by the death of cells and weaken and permit separation of the hormogonia (Bold and Wynne, 1978). In others the hormogonia develop as short chains. For example, in *Westiella* the hormogonia develop as a terminal series of several cells in pectic sheaths (Fig. 7:15) (hormospores) that break away from the parent filament (Smith, 1950). It should be noted that the hormogonia of the *Cyanophyta* and the short-chain fragments of the Chlorophyceae are remarkably similar to the

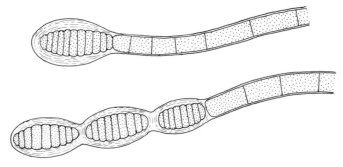

Fig. 7:15.
Hormospores of the blue-green alga *Westiella lanosa*, prior to separation. (Redrawn from Smith, 1950)

smaller gemmae of moss protonemata, which can fragment into viable propagules of a few cells each.

3. Brown Algae

The larger brown algae include some notable examples of abscission. The most famous include the pelagic species of *Sargassum*, the plants that form the extensive masses of the Sargasso Sea. These species rely entirely on vegetative fragmentation for their reproduction (Bold and Wynne, 1978). Reproduction by fragmentation is common in *Hormosira banksii* of New Zealand, and in the salt marsh fucoids of the Northern Hemisphere (V. J. Chapman, p.c.).

In the Laminariales there are a number of kelps that abscise blades at the end of the growing season. Some of the perennial species of *Laminaria* abscise the large, leathery blades in the autumn and regenerate a new blade in the spring, as does *Pterygophora* (Smith, 1969). In these algae there is a meristematic region at the junction of the stipe and blade. Separation takes place between the meristem and the blade and involves a rather extensive breakdown of cells. The next season's blade develops from the recently exposed portion of the meristem.

With *Nereocystis* (and possibly *Alaria*, *Lessoniopsis*, and some other members of the Laminariales), the portions of the blade containing the sori are abscised at maturity (Pfeiffer, 1928). Separation results from activity in several layers of cells in an irregular band through the lamina (blade), leading to dissolution of the middle lamellae (Fig. 7:16; Walker, 1980).

In a number of brown algae the blades develop by the splitting of young tissues in a manner suggestive of the development of leaf-blade openings in *Monstera*. Such splitting is characteristic of *Macrocystis* and also occurs in *Laminaria*, *Lessonia*, and *Durvillaea*. The cytology of the separation has

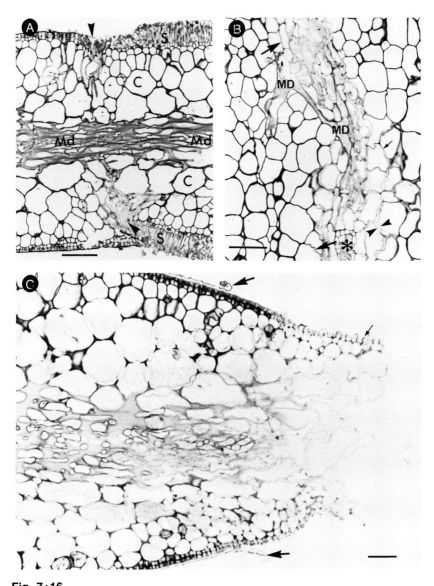

Fig. 7:16.
Sorus abscission in *Nereocystis*. **A.** An early stage of separatory activity (arrows) and sorus (S) margins. **B.** A later stage with separation in the cortex nearly complete. Note separated primary walls (arrows) after dissolution of middle lamellae. **C.** Margin of the lamina after sorus abscission. Note loss of stainable materials from cells at the margin. (Courtesy of Walker, 1980)

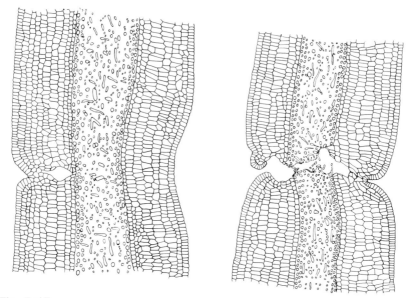

Fig. 7:17.
Cell separation in the blade-splitting of *Laminaria hyperborea*. (Redrawn from Killian, 1911)

received some attention (Fig. 7:17) and appears to involve schizolysis, separation of intact cells (Killian, 1911; Pfeiffer, 1928). In this respect, it is similar to dehiscence in the higher plants.

Numerous perforations characterize the mature lamina of the genus *Agarum*. These develop with a rapid multiplication of cells, following which the central tissue falls away (Fritsch, 1945).

4. Red Algae

Fragmentation and abscission in the red algae appears to be somewhat similar to that in the brown algae, but there is rather less information in the literature. *Constantinea* grows in a manner resembling that of *Laminaria*. The stipe terminates in one or more perfoliate horizontal circular blades. In due course, the lower blades break away, leaving a ring-shaped scar on the stipe. New blades develop on the apical portion of the stipe that projects above the next oldest blade (Fig. 7:18) (Smith, 1969).

Antithamnion cruciatum (Ceramiaceae) and some species of the Florideophyceae, at their northern limits in the Atlantic Ocean, appear to maintain themselves exclusively by fragmentation (Whittick and Hooper, 1977).

C. FUNGI

1. Filamentous Fungi

The body of the filamentous fungus is relatively simple compared with that of the seed plant. Abscission of parts of the somatic body is inconspicuous at best. In contrast, spores and spore dispersal are of great importance in the life of fungi, and they have developed a number of highly specialized mechanisms for the abscission and dispersal of spores, sporangia, and other disseminules.

a. Somatic Fragments

As with the algae, there appears to be considerable dissemination of fungi by means of hyphal fragments broken away from the parent mycelium and carried with bits of substrate to new locations by various agents such as animals.

A few fungi do produce gemmae, small bodies that can be dispersed by water and other agents in much the same way as the gemmae of bryophytes and pteridophytes. The soredia of lichens consist of a few hyphal strands and associated algal cells.

b. Sporangial Abscission and Dehiscence

A sporangium is a sac-like structure whose protoplasmic contents become converted into spores. The original wall of the sporangium remains as

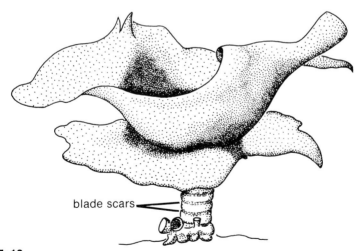

Fig. 7:18.
The red alga *Constantinea simplex*. The stipe shows scars of abscised blades. (Redrawn from Smith, 1969, with permission of Stanford University Press.)

a covering about the spores and must eventually dehisce or rupture to liberate the spores. In a few cases, a sporangium is abscised and the spores remain together until germination—e.g., *Dictyuchus*. In many cases the sporangium dehisces by the opening of an apical pore, as in *Saprolegnia* (Fig. 7:19). In other cases the spores are liberated by unoriented rupture of the sporangial wall.

One of the most interesting examples of spore dispersal is that of *Pilobolus*. In this genus the sporangium develops at the tip of a swollen sporangiophore (Fig. 7:20). During development, the two are covered by a common cell wall. At maturity, the outer wall breaks down between the sporangium and the sporangiophore, and turgor builds within the sporangiophore. Soon, the thin inner wall between the two structures ruptures with the violent ejection of the contents of the sporangiophore, which projects the sporangium up to 2.5 m. Initial velocities of the sporangium have been the subject of much interest; careful measurements have disclosed them to range from 5 to 25 m/sec. The sporangium release may be considered a kind of circumscissile dehiscence of the outer wall. Presumably, the weakening of both the outer walls and the inner walls are the result of enzymatic activity.

c. Conidial Abscission

A conidium is a nonmotile asexual spore produced at the tip or side of a special cell, the conidiophore. Many conidia are single-celled, produced in chains from the conidiophore. Some conidia are composed of adherent clusters of cells. Conidia abscise from each other or from the conidiophore by the weakening of adjoining walls in a variety of patterns. Separation can be

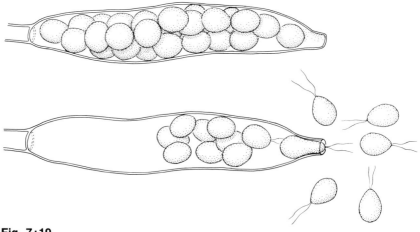

Fig. 7:19.
Sporangia of *Saprolegnia ferax* before and during zoospore release. Note the dehiscence of the tip of the sporangium.

Abscission in Lower Plants 237

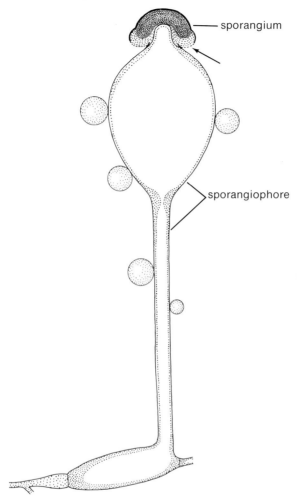

Fig. 7:20.
Sporangium and sporangiophore of *Pilobolus longipes* just before discharge. Note that the outer wall joining the sporangium and sporangiophore have already ruptured (arrow). The droplets exuded by the sporangiophore are a result of its turgor. (Redrawn from Ingold, 1971)

facilitated by turgor forces in some cases, and in other cases by partial desiccation.

Figures 7:21–7:24 include a selection from the great variety of patterns of conidial production and dehiscence that occur in the fungi. In *Coremiella* and *Geotrichum*, the middle lamella between conidia swells, rupturing the outer wall (Fig. 7:21A, B). In *Chaetosphaeria*, the conidiophore is flask-shaped and releases the loosely adherent conidia through a terminal pore (Fig. 7:21C). In *Scopulariopsis*, *Spilocaea*, and related genera, the abscission of each conidium leaves a narrow collar on the conidiophore. These can accumulate into a conspicuous series of ridges (Fig. 7:22B, C). Abscission is often facilitated by weak attachment of conidia to conidiophores, as in *Ambrosiella*, *Domingoella*, and *Drumopama* (Fig. 7:23), and especially in the genera with multicellular conidia, as in *Alternaria* and *Dactylosporium* (Fig. 7:24B, C). In some cases abscission of the conidia leaves conspicuous scars on the conidiophores (Fig. 7:24A, C) similar to the bud scars on yeast (Fig. 7:40).

Electron microscopy of wall changes in the course of conidial abscission shows patterns closely similar to those in the abscission zones of seed plants. There is cell-wall swelling, especially of the inner-wall layer (that appears similar to the middle lamellae), followed by dissolution and separation. As with the cuticle of seed plants, the outermost layer may not weaken but rup-

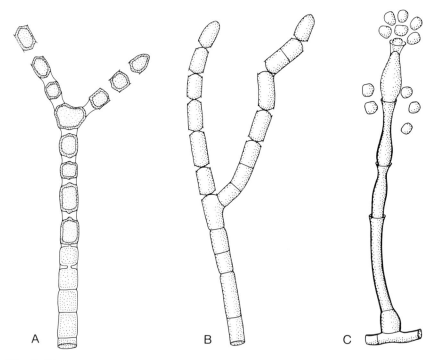

Fig. 7:21.
Conidium development and separation. **A.** *Coremiella ulmariae*. **B.** *Geotrichum candidum*. **C.** *Chaetosphaeria myriocarpa*. (Redrawn from Kendrick and Carmichael, 1973)

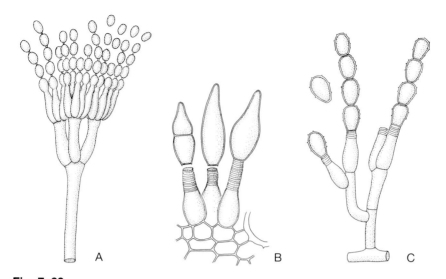

Fig. 7:22.
Conidium development and separation. **A.** *Penicillium* sp. **B.** *Spilocaea pomi*. Note collar-scars of previous conidial separations, and see Figure 7:25. **C.** *Scopulariopsis brevicaulis*. Note collar-scars of previous separations, and see Figure 7:26. (Redrawn from Kendrick and Carmichael, 1973)

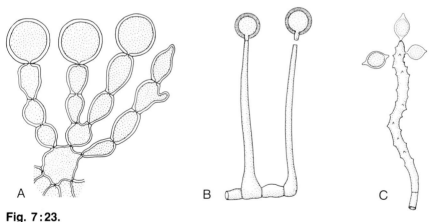

Fig. 7:23.
Conidium development and separation. **A.** *Ambrosiella xylebori*. **B.** *Domingoella asterinarum*. **C.** *Drumopama girisa*. Note conidial scar projections. (Redrawn from Kendrick and Carmichael, 1973)

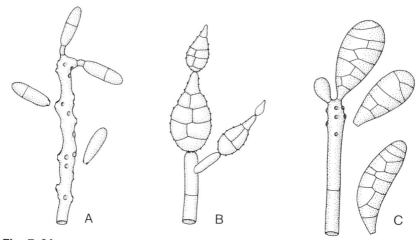

Fig. 7:24.
Conidium development and separation. Note bicellular and multicellular conidia of these examples, and the conspicuous conidial scars. **A.** *Cephaliophora nigricans.* **B.** *Alternaria alternata.* **C.** *Dactylosporium macropus.* (Redrawn from Kendrick and Carmichael, 1973)

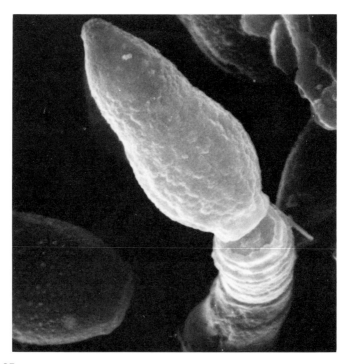

Fig. 7:25.
Scanning electron micrograph of abscising conidium of *Spilocaea pomi*. Note scar surface and collar-ring scars of previous abscissions on the conidiophore. Cf. Figure 7:22B. (Courtesy of R. Slocum)

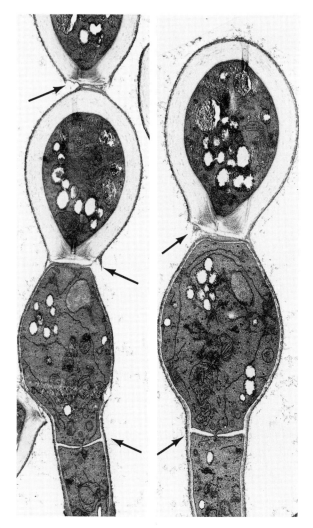

Fig. 7:26.
Electron micrographs of conidial abscission in *Scopulariopsis koningii* showing stages in conidial abscission. (Courtesy of T. M. Hammill, 1971)

ture mechanically (Figs. 7:25–7:29). The collar-like scars on *Trichurus* are similar to those on the basidiomycetous yeast *Phaffia* (Fig. 7:41).

The conidia of most species develop in air and are so small that they can be easily carried great distances and to great heights. Air currents, therefore, are the principal agency of dispersal. There are some notable examples, however, of conidial abscission and dispersal by the sudden expression of turgor. One of these is *Conidiobolus coronatus*. When the time of discharge

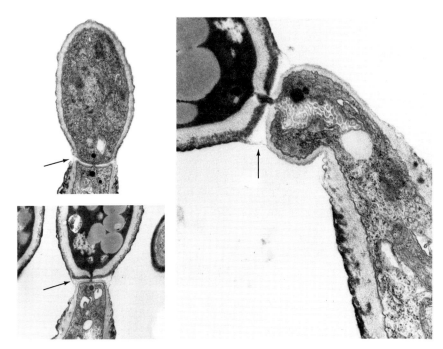

Fig. 7:27.
Electron micrographs of conidial abscission in *Trichurus spiralis*. (Courtesy of T. M. Hammill, 1977)

approaches, the common outer wall of the conidium and the conidiophore ruptures around the circular line of the transverse wall between them. Each of the adjacent inner walls bulges outward from turgor pressure, and the conidium is thrown for distances up to 4 cm (Fig. 7:30) (Ingold, 1971). In other genera, conidial abscission and dispersal are facilitated by the partial desiccation of the conidiophore. For example, the conidiophore of *Deightoniella torulosa* has a relatively thin, flat wall adjoining the mature conidium. As the conidiophore dries, tension in the cytoplasm increases until suddenly a gas phase (bubble) forms, rounding the wall adjoining the conidium (Fig. 7:31), with the result of immediate abscission and discharge of the conidium.

d. Ascus Dehiscence

In most species of the Ascomycetes, the ascus is a turgid cell which, at maturity, suddenly dehisces with the violent ejection of the ascospores. The ascus apex and mechanism of dehiscence can be characteristic of sizable taxonomic units. Some of the more common forms (after spore discharge) are shown in Figure 7:32. Dehiscence may involve an apical operculum

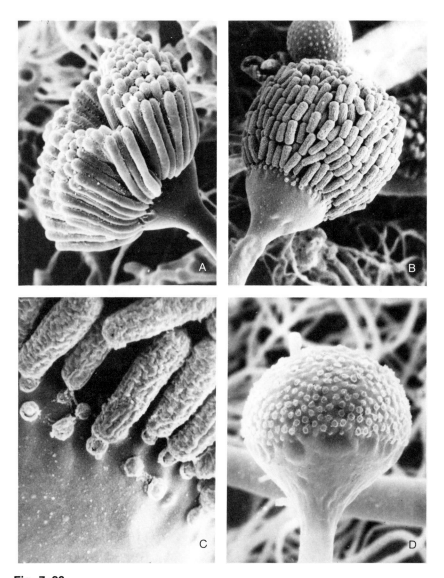

Fig. 7:28.
Scanning electron micrographs of the development and abscission of merosporangia in *Syncephalis sphaerica*. **A.** Ampulla with nearly mature merosporangia. **B.** Merospores mature and abscising. **C.** Detail of scar on ampulla after merospore abscission. **D.** Ampulla with scars after all merospores have abscised. (Courtesy of Baker et al., 1977)

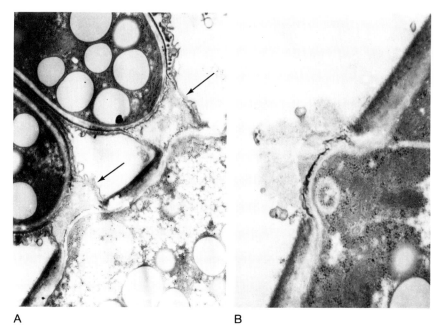

Fig. 7:29.
Electron micrographs of stages in merosporangial abscission of *Syncephalis sphaerica*. **A.** Merospores nearly detached. Note swelling of "middle lamellar" region (arrows). **B.** Fresh merospore scar on surface of ampulla. Note mass of amorphous material on surface of scar. (Courtesy of Baker et al., 1977)

(Fig. 7:32A) or a subapical operculum (Fig. 7:32B). In some taxa, dehiscence involves the opening of a vertical slit (Fig. 7:32C). In other taxa, terminal pores develop (Fig. 7:32D, E), presumably by the partial or complete hydrolysis of the wall material in the pore.

In the "bitunicate" asci the usual two wall layers are well developed. The outer primary wall, called the ectotunica, is relatively rigid, while the fibrillar structure of the inner wall, the endotunica, is more loosely organized. As turgor builds prior to spore discharge, the apex of the ectotunica dehisces, and the endotunica, with its contained ascospores, expands rapidly and discharges the spores in essentially the same manner as unitunicate asci (Fig. 7:33) (Reynolds, 1971).

e. Basidiospore Discharge

The basidiospore develops on a narrow tip of a protuberance from the basidium and is attached in an oblique manner (Fig. 7:34). Most basidiospores are forcibly discharged and are commonly referred to as ballistospores. When the ballistospore is mature, a small bubble or drop develops at

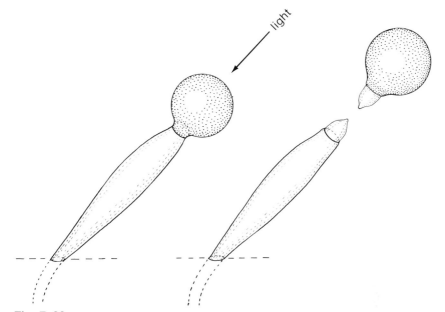

Fig. 7:30.
Turgor-propelled conidial discharge in *Conidiobolus coronatus*. Note expanded transverse walls of both conidiophore and conidiospore. (Redrawn from Ingold, 1971)

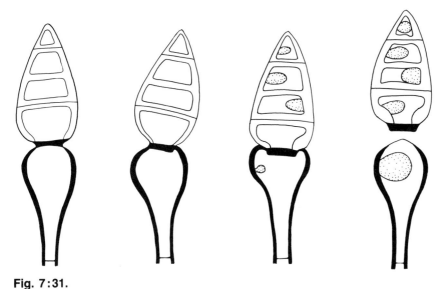

Fig. 7:31.
Conidial discharge in *Deightoniella torulosa* from partial desiccation of the conidiophore. Final rapid development of a gas phase rounds the conidiophore, separating and propelling the conidium. (Redrawn from Ingold, 1971)

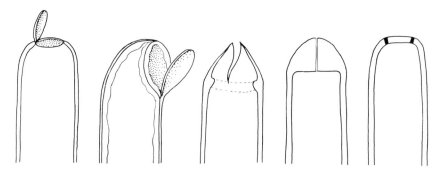

Fig. 7:32.
Ascus apices showing principal types of pore dehiscence. **A.** Apical operculum. **B.** Subapical operculum. **C.** Vertical slit. **D, E.** Terminal pores. (Redrawn from Korf, 1973, with permission of Academic Press)

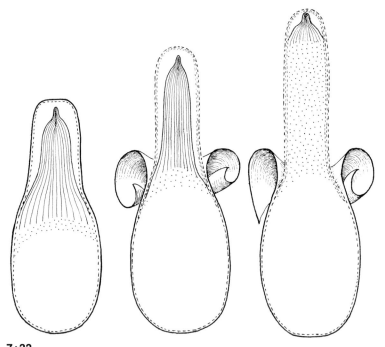

Fig. 7:33.
Dehiscence of the bitunicate ascus of *Limacinula theae*. The ectotunica (primary wall) is ruptured by the rapid expansion of the endotunica (secondary wall) and the ascus contents. (The ascospores have been omitted.) (Redrawn from Reynolds, 1971)

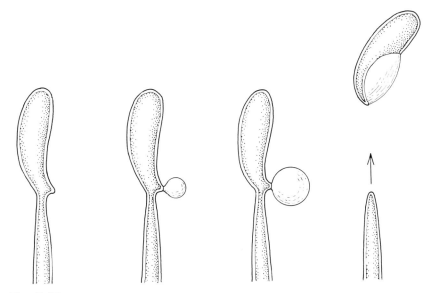

Fig. 7:34.
Discharge of the ballistospore (basidiospore) of *Calocera cornea* from its sterigma. Note the droplet that is always associated with this event. (Redrawn from Ingold, 1971)

the base of the spore and participates in abscission and discharge. The mechanism and forces involved are not yet completely understood (Alexopoulos and Mims, 1979), but the net effect is to separate the ballistospore and propel it for a short distance, usually 0.1–0.2 mm (Ingold, 1971).

f. Abscission and Dispersal in the Nidulariaceae

This group of fungi is distinguished by the production of its spores within a hyphal mass enclosed in a special structure called the peridiole, and the spores are disseminated in that unit. In the bird's nest fungi, the peridioles develop within a cup-shaped fruiting body (Fig. 7:35). In *Cyathus striatus*, the peridiole is lens-shaped and about 2 mm in diameter. It is borne on a short stalk loosely attached to supporting mycelium. The shape of the cup is such that when a drop of rain falls into it, considerable force is generated, dislodging (abscising) one or more of the peridioles a distance of 1 or 2 m. The stalk contains a remarkable device for attachment of the peridiole when it lands; namely, a coiled cord of material with a sticky tuft at the end. While the peridiole is in flight, the cord uncoils. The viscid tuft of the cord adheres to the first solid object it strikes. If that happens to be a leaf or the stalk of a plant, the cord may wind about, attaching the peridiole firmly (Fig. 7:36). Further dissemination occurs when the peridiole and its spores

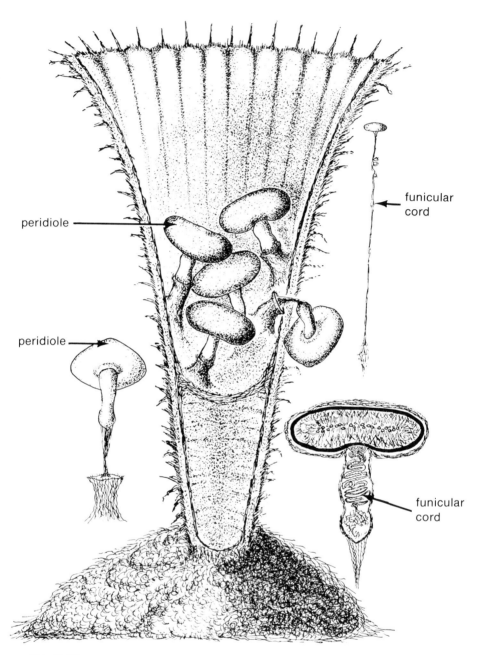

Fig. 7:35.
Fruiting body of a bird's nest fungus, *Cyathus striatus*. Note the coiled cord that extends when the peridiole is in flight. (Drawing courtesy of H. J. Brodie, 1975, *The Bird's Nest Fungi*, University of Toronto Press; and Canadian Journal of Botany)

Fig. 7:36.
Diagram of the projection of a peridiole from the fruiting body of *Cyathus striatus* by the action of a raindrop. Note the attachment of the peridiole to a plant by the viscid end of the cord. (Drawing courtesy of H. J. Brodie, 1975, *The Bird's Nest Fungi*, University of Toronto Press; and Canadian Journal of Botany)

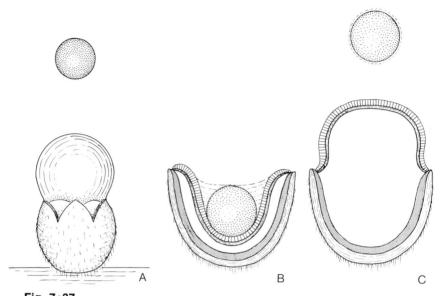

Fig. 7:37.
Peridiole discharge in *Sphaerobolus stellatus*. **A.** External view of dehisced fruiting body and discharged peridiole. **B.** Section of fruiting body after dehiscence and before discharge, showing inner and outer cups and peridiole. **C.** Section of fruiting body immediately after peridiole discharge, with inner cup turned inside out. (Redrawn from Ingold, 1971)

are ingested by a herbivorous animal and deposited some distance from the parent plant (Brodie, 1975).

Sphaerobolus is a related genus with an equally impressive dispersal mechanism. While the fruiting body of *Sphaerobolus* (Fig. 7:37) is somewhat simpler than that of *Cyathus*, the mechanism of dispersal is quite effective. The peridiole is a spherical mass of hyphae and spores about 1 mm in diameter. It develops within a spherical fruiting body about 2 mm in diameter, which, at maturity, dehisces at the apex along several lines (Fig. 7:37A). Within the fruiting body, the spore mass is cupped between two complex layers. A turgor builds in the space between the two layers to the point that the inner cup-shaped layer is violently turned inside out (Fig. 7:37B, C), discharging the spore mass a distance of up to 7 m (Ingold, 1971).

2. Yeast

Yeasts are fungi that are normally single celled and that typically reproduce by budding. Taxonomically, most yeasts are ascomycetes, but some are basidomycetes, and the affinity of a few yeasts has not yet been determined (Phaff et al., 1978).

Vegetative multiplication of yeasts is characteristically by the production of single-celled "buds" (Fig. 7:38), which are sooner or later abscised from the parent cell. Buds are produced in a variety of patterns by the different yeasts. In some forms, bud abscission is delayed and a "pseudomycelium" develops (see Phaff et al., 1978). Because of the economic importance of the yeasts and the relative ease with which they can be grown, their biochemistry and morphogenesis has been studied in some detail (Cabib, 1975). In *Saccharomyces*, shortly after the bud emerges, deposition of the primary septum commences, starting as a ring at the junction of the bud and the parent cell. There is a conspicuous accumulation of cytoplasmic vesicles near the wall of the developing bud and at the primary septum (Sentandreu and Northcote, 1969). Presumably, these are secreting both substrates and enzymes for (i) bud-wall growth, (ii) septum deposition, and (iii) eventual septum separation. In due course, the septum is completed across the entire connection between the two cells. It appears to arise and function in a manner analogous to the middle lamella of higher plants, but in *Saccharomyces* its composition is mainly chitin (polyhexosamine). Secondary septa are deposited on each side of the primary septum and become continuous with the main cell walls (Fig. 7:39). The secondary septa are composed of various glucans and mannans, as are the main cell walls (Cabib, 1975). Separation is achieved by what appears to be enzymatic hydrolysis of a portion of the primary septum. The scars that remain after separation are shown in Figure 7:40.

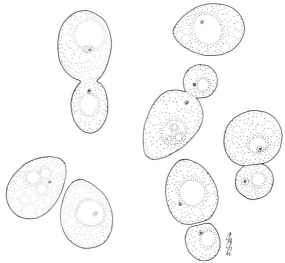

Fig. 7:38.
Budding yeast, *Saccharomyces cerevisiae*.

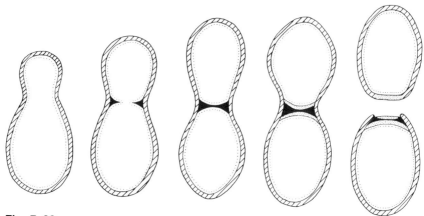

Fig. 7:39.
Diagram of septum formation and bud abscission in yeast. See text for explanation. (Redrawn with permission from Cabib, 1975. Copyright by Academic Press Inc. (London) Ltd.)

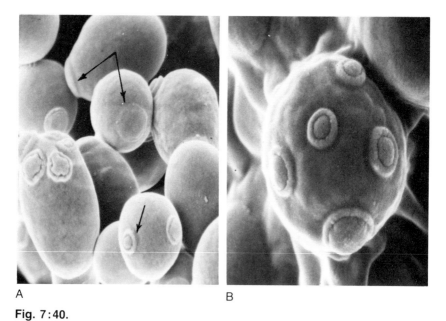

Fig. 7:40.
Scanning electron micrographs of *Saccharomyces cerevisiae*. **A.** Showing "birth" scars on young cells (upper arrows) and bud scars on an older cell (lower arrow). **B.** Older cell with several bud scars. (Courtesy of Phaff et al., 1978, reprinted by permission.)

In the basidiomycetous yeasts, budding can occur repeatedly at the same site. This gives rise to bud scars with concentric collars (Fig. 7:41) that resemble the cell-division collars in the green alga, *Oedogonium* (Fig. 7:13), and the collar-scars of certain conidiophores (Figs. 7:22, 7:25–27).

Some yeasts reproduce by a kind of fission. In *Schizosaccharomyces*, cross walls form by annular centripetal growth of the inner portion of the cell wall. After the new cross wall is complete, the central portion is hydrolyzed, separating the two daughter cells (Fig. 7:42) (Phaff et al., 1978). Characteristic scars with plugs remain on each side. Although the chemical constituents of the cell wall differ somewhat from those in higher plants, the process of separation appears cytologically similar to middle lamella separation of higher plants.

D. LICHENS

Lichens produce several kinds of "diaspores," structures that function as vegetative disseminules (*Thallusbruchstücke*). Isidia are protuberances of the thallus that develop on a number of lichens that contain both algal cells and medullary fungal filaments. These can be broken off by various mechanical agencies and moved some distance, particularly by water. Soredia

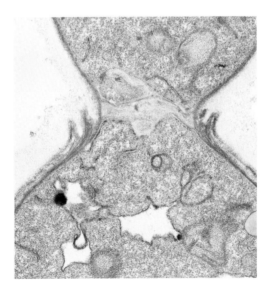

Fig. 7:41.
Electron micrograph of budding *Phaffia rhodozyma*. Note cell-wall collars from earlier budding at the same site. (Courtesy of Phaff et al., 1978, reprinted by permission.)

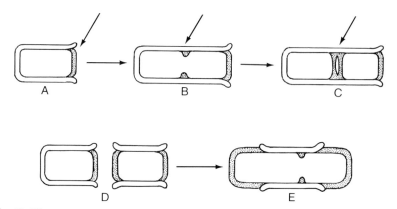

Fig. 7:42.
Diagram of cell separation in *Schizosaccharomyces*. **A.** Newly separated cell with scar-plug (arrow). **B.** Dividing cell with beginning of new cross wall (arrow). **C.** Start of cell separation in cross wall (arrow). **D.** Separation complete. **E.** Expanded cell in process of next division. (Redrawn by permission from Phaff et al., 1978)

are smaller bodies consisting of several algal cells enveloped by fungal hyphae. They are usually produced in localized structures called soralia. Because of their very small size, soredia can be dispersed by wind as well as by water (Jahns, 1973; Pyatt, 1973). Larger differentiated fragments that function in dissemination are produced by a number of foliose lichens; these include schizidia and phyllidia. The schizidia are the result of a split in the lower part of the lichen medulla that permits the upper layer to curl away and eventually separate. The phyllidia are foliose growths of the upper cortex or thallus margin, usually with a constricted base. They, too, can fold away and function as diaspores (Poelt, 1973). As far as I can determine, separation of the lichen diaspores is mechanical, involving the rupture of attaching cells.

E. BACTERIA

The abscission (separation) of bacterial cells from each other continues the trend shown in the simpler algae. As will appear from the following discussion, it is impossible to draw a sharp line of demarcation between the separatory processes of the algae and the higher plants, on the one hand, and the bacteria, on the other hand.

Fragmentation of filaments of bacterial cells for some forms is very similar, if not identical, to the blue-green algae. For example, the genus *Beggiatoa*, which is considered a colorless form of *Oscillatoria*, reproduces by breakage of the filament. This is accomplished by the formation of "sacrificial" cells (necridia) at intervals in the filament. These appear identical with the separation discs of *Oscillatoria*. Upon dying, the necridia disintegrate, separating the filament into two or more new filaments (Strohl and Larkin, 1978).

A number of bacterial genera are characterized by chain formation. The length of the chain is sometimes a species characteristic. For example, in *Streptococcus faecalis* cells normally appear as single, paired, or in very short chains. Various factors can modify this behavior and induce the formation of long chains—e.g., detergents. The short chain length is due to the activity of a lysozyme that under normal conditions promotes separation, but the lysozyme is strongly inhibited by detergents. The involvement of such an enzyme has been confirmed through the production of chain-producing mutants in which the enzyme synthesis has been blocked (Lominski et al., 1958; Lominski and Rafi Shaikh, 1968).

Although *Staphylococcus aureus* normally grows in grape-like clusters, by treatment with detergents it can be induced to form regularly arranged packets of cells. Similarly, there are mutants of the species which also form regularly arranged packets. The loss of ability to produce the normal clusters of cells is due to inactivation or failure of production of the separation (abscission) enzyme(s) (Koyama et al., 1977).

Lampropedia hyalina develops its colonies in rectangular sheets of conspicuously uniform structure. As with abscission in the higher plants, variation of environmental factors can promote separation and disorderly arrangement of cells. These factors include changes in pH, in nutrient concentration, and in temperature (Kuhn and Starr, 1965).

As might be expected, bacteriologists have been much more interested in how bacteria adhere to one another and to substrates than in how they abscise or separate from one another and substrates. There is now substantial knowledge of the outer envelope of bacteria, the glycocalyx. This consists of a tangled mat of polysaccharide fibers, which can bind to other cells or substrates either directly or through linkage with simple proteins called lectins. The binding polysaccharides are responsible for the adherence of streptococci to teeth, of *Bacteroides fragilis* to the peritoneum, and of *Vibrio cholera* to intestinal cells. A point of great significance to our discussion is the chemical similarity between the glycocalyx of bacteria, the outer cell layer of many of the algae, and the middle lamella of higher plants. They all are composed of polysaccharide chains, usually branching, and with varying degrees of fibrillar development and strength. The bacterial glycocalyx stains with ruthenium red just as do the pectic substances of higher plants (Costerton et al., 1978).

A recent review of the morphogenesis of bacterial aggregations (Hoffman, 1964) notes that there has yet been little study of the mechanism whereby the linkage of bacterial chains is broken. There is good evidence, however, of the involvement of lysozyme activity in the separation of cells, as in *Clostridium perfringens* and in cell abscission of streptococcal chains.

In consideration of separation and abscission in the bacteria one cannot fail to be deeply impressed by the close similarity between these, the morphologically simplest of present-day cells, and the cells of the most highly evolved plants in their mechanism of separation. At both of these extremes and in the innumerable intermediate forms, there are common elements of cell adhesion by means of pectin-like polysaccharides, and cell separation by the activity of secreted enzymes. Further, in all instances the processes of adhesion and abscission are subject to genetic controls and to the influence of a variety of environmental factors.

Here it may be noted that whereas the processes of separation appear closely similar throughout the plant kingdom, a diversity of structure is demonstrated among the scars and protective layers that develop as a part of the abscission process as described in this and preceding chapters.

8. Genetics of Abscission and Dehiscence

Chapter Contents

A. EVIDENCE OF GENETIC CONTROL OF ABSCISSION   257
   1. In Cultivated Plants   257
   2. In Taxonomy   259

B. EVIDENCE OF GENETIC CONTROL OF DEHISCENCE   262
   1. In Cultivated Plants   262
   2. In Taxonomy   262

C. IDENTIFIED GENES AFFECTING ABSCISSION AND DEHISCENCE   264

The ease with which small changes in ecological and physiological factors can affect abscission and dehiscence indicates the great plasticity and sensitivity of the control mechanisms. Nevertheless, genetic factors certainly underlie and determine these patterns of behavior. For each plant, the genetic code dictates the limits and mechanisms within which ecological and physiological factors can function. Evidence of genetic control of abscission is widespread, although it is not yet well documented. The frequent use of the various kinds of abscission and dehiscence behavior as taxonomic characters indicates the firm genetic basis of such behavior. Further, when horticultural varieties are compared, particularly varieties of the tree fruits, many differences in leaf, flower, and fruit abscission become apparent. In recent years a number of genes controlling these processes have been identified.

A. EVIDENCE OF GENETIC CONTROL OF ABSCISSION

1. In Cultivated Plants

Among the varieties of fruits are conspicuous examples of differences in abscission behavior. For example, most varieties of apples tend to abscise their fruit shortly before it is fully mature. There is great variation among

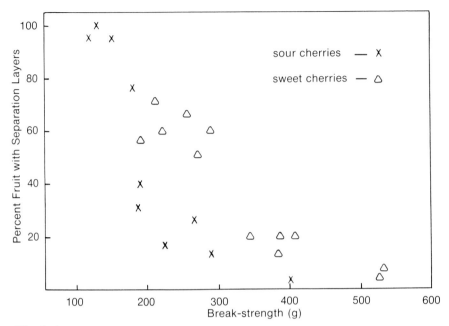

Fig. 8:1.
Inverse correlation between degree of separation-layer development and the force required to separate the mature fruit of different cherry varieties. (Redrawn from Stösser, 1971)

apple varieties in this respect. McIntosh apples are especially prone to "preharvest drop." At the other extreme is the variety Cortland, which does not abscise its fruit until long past maturity (C. G. Forshey, p.c.). Similar differences have been found among peach varieties. For example, the varieties Gaume, Halford, and Pullar regularly abscise 30 percent or more of their fruit shortly before maturity, while the varieties Golden Queen and Phillips abscise 10 percent or less (Jerie, 1976). In an extensive study of fruit abscission of cherries, Stösser (1971) found a close correlation between the degree of development of the fruit abscission layer and the ease with which fruit are abscised. The force required to remove fruit is much greater for varieties that have poorly developed separation layers (Fig. 8:1). The foregoing differences among fruit varieties, while undoubtedly related to anatomical and underlying biochemical differences, reside ultimately in the genetic composition of the varieties.

Young-fruit abscission is another aspect of cultivated fruit trees that shows considerable inherited variation. Trees of many varieties tend to retain more of their young fruit than they can develop to marketable size.

Such varieties must be thinned by hand or by chemical agents to reduce the number of developing fruit to a desirable load. In peaches the variety Redhaven characteristically abscises very few of its young fruit, while the variety Elberta abscises a moderate number. In apples the variety Wealthy abscises few young fruit, but the variety McIntosh abscises a large number of its young fruit (Batjer, 1954). Similar varieties are well known in plums, pears, and other tree fruits. The plants of field crops show similar differences. In cotton, most varieties (e.g., Acala, Deltapine) abscise about two-thirds of their young fruit, but genetic strains have been selected that retain almost 100 percent of the young fruit. In cowpeas the variety Adzuki abscised only 36 percent of its young fruit, while the variety Sikira abscised 53 percent (Adedipe and Ormrod, 1975). The potato variety Russet Burbank abscises many more of its flowers and young fruit than does the variety Menominee (Weinheimer and Woodbury, 1967). In tomatoes, some fruit abscission behavior is genetically controlled and related to morphological expression of single genes (Fig. 8:2) (Rick, 1967).

Similarly, there are varieties of trees, ornamental plants, and crops which tend to retain their leaves much longer than others. For example, the soybean variety Ruse abscised most of its leaves by 105 days from sowing, while the variety Bragg had lost only a few lower leaves (Constable and Hearn, 1978). Similar differences have been observed among varieties of dry beans (*Phaseolus vulgaris*) (Fig. 8:3). In both cases, the leaf abscission behavior appears to be under rather simple genetic control (Hardwick, 1979).

In his extensive study of the clones of the aspens *Populus tremuloides* and *P. grandidentata*, Barnes (1969) found that the clones of *P. tremuloides* retained their leaves longer in the autumn than clones of *P. grandidentata* did. Such variations among clones of ornamental trees have long been known (e.g., Koriba, 1958). In the lemon, partial defoliation of the tree results from an inherited disorder, "lemon sieve-tube decline" (Schneider et al., 1961, 1978).

From his observations of branch abscission in *Populus tremuloides*, M. D. Morgan (1975) recognized a number of clones and identified three as being unusually efficient at self-pruning (branch abscission).

2. In Taxonomy

Another body of evidence that indicates the firm genetic basis of abscission is demonstrated in taxonomy. Within many taxonomic entities, abscission behavior is uniform and is a valuable taxonomic character. For example, in *Quercus*, *Magnolia*, and *Taxodium* certain species are deciduous; others are evergreen. In *Lupinus*, two groups of species are separated by the abscission or persistence of their floral bracts (Jepson, 1925).

In *Pinus*, the sheaths of the leaf-clusters are abscised by the end of the

Fig. 8:2.
Expression of abscission genes of tomato. **A.** Inflorescence of "jointless" (j2). **B.** A normal pedicel. **C.** Pedicel showing a modification of "jointless" (j2). **D.** Pedicel showing "arthritic articulation" (j2^{in}). (Photographs courtesy of Rick, 1967, with permission of the New York Botanical Garden.)

Fig. 8:3.
Specimens of two varieties of dry beans ready for seed harvest. The variety at the right has abscised all of its leaves. (Courtesy of Hardwick, 1979, with permission of the Journal of Experimental Botany)

first year by some species, but are persistent in other species. Also in *Pinus*, the cones of most species abscise, but are persistent in a few species (Munz and Keck, 1959). In the conifers, cone scales are abscised by some genera (e.g., *Agathis, Araucaria, Cedrus*), but are persistent in others (e.g., *Picea, Pinus, Tsuga*). Abscission behavior is often consistent within larger categories, such as genera, families, and orders. For example, the conifer genus *Larix* is deciduous, abscising all its branchlets annually, while many other conifer genera, such as *Cedrus, Picea,* and *Tsuga*, are evergreen.

A notable example is that of the Ginkgoales and Czekanowskiales. Harris et al. (1974) have separated these two related orders by their leaf abscission habits; the Ginkgoales abscise their leaves singly, while the Czekanowskiales abscised their leaves in clusters that resemble the leaf fascicles of present-day *Pinus* (Fig. 9:12).

The pattern of bark abscission can be characteristic of a genus or vary widely among the species within a genus. For example, the genus *Arbutus* is distinctive for its smooth red bark that is abscised annually in thin sheets. Conspicuous bark-abscission differences among species are found in genera such as *Platanus, Pinus,* and *Eucalyptus,* and many others. In *Eucalyptus,* a number of species are smooth barked as a result of the annual abscission of a uniform layer of bark (Fig. 3:22). Among other species in the genus, the patterns of bark abscission vary greatly. Some shed long stringy strips, and others, flakes or patches. On other species bark accumulates, sometimes in deep furrows, and shows little or no abscission (Penfold and Willis, 1961).

B. EVIDENCE OF GENETIC CONTROL OF DEHISCENCE

1. In Cultivated Plants

The agricultural production of many seed crops is made difficult by the premature "shattering" of seeds. This results from the combination of early dehiscence and seed abscission. In most cases, plant breeders have been able to select genetic forms in which seed shattering is minimal. For example, older varieties of sesame were very susceptible to shattering, but a nonshattering variety has now been developed (D. M. Yermanos, p.c.). In flax the varieties show varying degrees of dehiscence. The variety Punjab 47 is indehiscent, Marine is semidehiscent, and Coimbricum is dehiscent (Holden, 1956).

Shattering is a frequent problem among the many kinds of small-grained cereals (e.g., wild rice, *Zizania aquatica*; Woods and Clark, 1976), but this is the result of ease of caryopsis abscission rather than of dehiscence.

2. In Taxonomy

Habits of dehiscence, like abscission, can be quite uniform within a taxonomic group and, therefore, a useful taxonomic character. Many examples could be cited; a few are given below.

In the Liliaceae the fruits of some genera are indehiscent berries (e.g., *Clintonia* and *Trillium*), but most genera have fruits that are dry capsules. A number of these show loculicidal dehiscence (e.g., *Chlorogalum, Hesperocallis, Zigadenus, Fritillaria,* and *Lilium*; Fig. 8:4A). Other genera are characterized by capsules that show septicidal dehiscence (e.g., *Veratrum* and *Calochortus*; Fig. 8:4B). In the Primulaceae, some genera are dehiscent by valves (e.g., *Primula, Androsace,* and *Glaux*), while others show circumscissile dehiscence (e.g., *Anagallis,* Fig. 2:10; and *Centunculus*). Some spe-

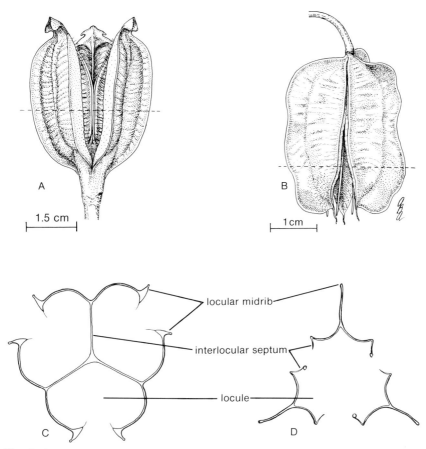

Fig. 8:4.
Two patterns of capsule dehiscence in the Liliaceae. **A.** Loculicidal dehiscence in *Tulipa* sp. **B.** Septicidal dehiscence in *Calochortus* sp. **C, D.** Sections through the dehisced capsules at the indicated levels.

cies of *Dodecatheon* have valvate dehiscence, others are circumscissile (Munz and Keck, 1959).

In the Papaveraceae most genera have fruits that are dehiscent by apical pores or valves. However, *Dendromecon* and *Eschscholzia* (Fig. 3:13) have capsules that are bivalvate and dehisce from the base toward the apex.

The fruit of the Scrophulariaceae is a capsule having two locules. The genera within the family vary greatly, however, with respect to their patterns of dehiscence. There are genera that dehisce septicidally and those that dehisce loculicidally. Most dehiscence is two-valved, but the fruits of at least one genus open with four valves. A few genera dehisce by means of pores near the apex of the capsule, and at least one does not dehisce at all, but bursts irregularly (Munz and Keck, 1959).

C. IDENTIFIED GENES AFFECTING ABSCISSION AND DEHISCENCE

Scattered through the literature on plant breeding and genetics is evidence on the nature of the genetic control of various abscission phenomena. In tomatoes, a gene named "defoliator" has the not surprising phenotypic manifestation of greatly accelerated leaf abscission (Chmielewski, 1968). Another gene in tomatoes called "jointless" results in the unusually firm attachment of mature fruit and prevents normal fruit abscission at the base of the pedicel (Fig. 8:2). This gene promotes maturation and heavy cell-wall deposition in the pedicel abscission zone; in the normal plants the abscission zone remains relatively undifferentiated, physiologically active, and hence capable of abscission (Joubert, 1965).

In jute (*Corchorus olitorius*), mutants have been identified that show earlier leaf abscission than normal plants (Sen, 1968). The leaves of such mutants show increased sensitivity to abscission promotion by GA and ETH, and show acceleration of the amino acid changes correlated with senescence (Chatterjee and Chatterjee, 1971).

Capsule dehiscence in sesame is controlled by a single gene. The dehiscent (normal) phenotype shows extensive development of the sclerenchymatous endocarp and early breakdown of parenchyma at the sutures (Fig. 8:5). Both characteristics facilitate opening (dehiscence) of the capsules at maturity as the mesophyll dries and shrinks (Ashri and Ladijinski, 1964).

The opening, "dehiscence," of cones at maturity is typical of most species of *Pinus*. In a few species (e.g., *P. banksiana*, *P. attenuata*) the cones are serotinous—i.e., they open late, usually not for many years or until heated by a fire (see Chapter 6, sec. A.3.c.). Analysis of natural populations indicates that serotiny is controlled by a single gene with two alleles (Teich, 1970).

Large, foliaceous floral bracts are characteristic of the cultivated species of cotton; they contribute significantly to the growth and development of the fruit. The fimbria at the edges of the bracts also contribute highly noxious particles to the dust of cotton mills. Two species of wild cotton have caducous bracts, *Gossypium armourianum* and *G. harknessii*. Their genetics is now under investigation (Muramoto, 1977).

There is much additional evidence that abscission and dehiscence are genetically controlled, even though definitive studies have not yet been conducted. For example, in soybeans, genotypes differing in rate of leaf senes-

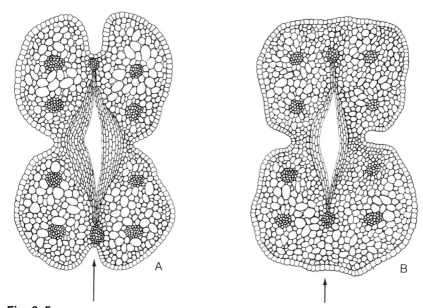

Fig. 8:5.
Diagrammatic cross sections of capsules of sesame. **A.** Dehiscent variety (No. 45). **B.** Indehiscent variety (Renner 15). Note the dehiscence zone in A (arrow) and its absence in B. (Redrawn from Ashri and Ladijinsky, 1964)

cence (and time of abscission) have been recognized (Abu-Shakra et al., 1978). In *Eucalyptus*, species characterized by bark abscission may hybridize with species having persistent bark. Analysis of hybrid populations indicates that bark abscission is under multiple-factor control (Pryor et al., 1956).

9. Paleontology and Evolution of Abscission

Chapter Contents

A. PALEONTOLOGY 266
    1. The Precambrian Era (more than 600 million years ago) 267
    2. The Paleozoic Era (250–600 million years ago) 268
    3. The Mesozoic Era (100–250 million years ago) 277
    4. Abscission in the Cenozoic Era (from 70 million years ago) 279

B. EVOLUTION 281
    1. Evidence of Recent Evolution and "Fine Tuning" of Abscission Behavior 281
    2. Some General Comments on Abscission Evolution 283
    3. A Theory of the Evolution of Abscission 283
        a. Some General Considerations 283
        b. A Brief Statement of the Abscission Evolution Theory 285
    4. Abscission in the Animal Kingdom 285

A. PALEONTOLOGY

Consideration of the diversity of abscission manifestations in present-day plants can hardly fail to arouse curiosity as to the nature of their origins and evolution. Somewhere in the long history of the plant kingdom, abscission had its beginnings. On first thought, the precise origins seemed likely to be obscure. The fossil record is far from complete; it still shows only tantalizing fragments of the early floras. Also, paleontologists seldom give more than passing attention to the evidences of abscission. Their comments, widely dispersed in a voluminous literature, can be discovered only with difficulty. For the evidence that I have found and present here, I am indebted to James A. Doyle and several other colleagues whose pertinent publications are cited in this section. The record does provide appreciable evidence of abscission, particularly of differentiated structures such as leaves, branches, seeds, and fruits. Collectively, the evidence outlines a remarkably consistent picture of the evolutionary development and diversification of abscission phenomena.

Undoubtedly, there was abscission and fragmentation in the colonial algae and fungi of the Precambrian era; the now substantial fossil evidence shows forms much like those of the present day. After the evolution of vascular plants with lignified tissues, the variety in the fossil record expanded. The mid-Paleozoic era saw the great development and diversification of lower vascular plants and, toward its end, the emergence of gymnosperms. Evolution of discrete organs appears to have been soon followed by the evolution of the ability to abscise those organs. In the Mesozoic era, the seed plants diversified greatly, and many present-day families and some genera were well established by the end of that era. As the variety of climates increased during the Cenozoic era the number of species and the variety of their adaptations increased.

1. The Precambrian Era (more than 600 million years ago)

This era saw the development of prokaryotic organisms, the bacteria and blue-green algae, and the appearance of the first simple eukaryotic organisms similar to the green algae. It seems obvious that mechanisms of cell division and cell separation must have evolved quite early in the era. The organisms could hardly have become as widely distributed as they were without such mechanisms.

The earliest known fossil record is of bacteria-like organisms that existed in Western Australia 3.5 billion years ago (Awramik et al., 1981). They show a variety of structural and colonial forms indicative of several patterns of separation (or adherence). Spheroidal microfossils in various stages of cell division (separation) are found in the Swaziland System of South Africa, which is considered to be at least 3 billion years old (Knoll and Barghoorn, 1977). Well-preserved fossils of filamentous blue-green algae occur in dolomitic limestone stromatolites estimated to be 2.2 billion years old; the alga closely resembles the present-day genus *Raphidiopsis* (Nagy, 1974). The Precambrian form, named *Petraphera*, is so similar to *Raphidiopsis* in structure that it seems certain that it must have grown and reproduced in a similar manner by cell division and spores (cf. Chapter 7). Other fossils from the Precambrian appear to represent various members of the green as well as the blue-green algae (Fig. 9:1; Schopf and Oehler, 1976; Schopf, 1978).

From the fossil record, it is clear that the mechanism of cell division was well established early in the Precambrian. Further, the existence of colonies and filaments of various shapes indicates that there had already been evolution to different patterns of separation that enabled the establishment of new colonies. Considering the more than 2-billion-year duration of the era, it appears likely that most, and possibly all, of the patterns of colony growth and fragmentation (abscission) that exist today had evolved at least once during the Precambrian.

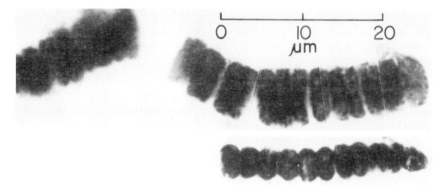

Fig. 9:1.
Photomicrographs of fossil blue-green algae of the late Precambrian Bitter Springs formation of central Australia, about 850 million years old. Pattern of cell separation appears identical with that of present-day forms. (Courtesy of Schopf, 1972)

2. The Paleozoic Era (250–600 million years ago)

The middle and later parts of this era saw the emergence of land plants with the appearance of woody and vascular tissue, the development of various vegetative and reproductive appendages and organs, and extensive diversification of vascular plants, in general.

One of the more primitive plants whose fossils suggest abscission is *Palaeostigma sewardi* from the Devonian. Its fossils contain no evidence of lignification, but bear a number of regular marks that have been interpreted as scars of vegetative buds or bulbils (Fig. 9:2; Plumstead, 1967); or the marks may possibly represent scars of some other vegetative structure.

At least two patterns of sporangial dehiscence evolved in the Lower Devonian: transverse, as exemplified by *Zosterophyllum*; and longitudinal, as exemplified by *Dawsonites* (Chaloner and Sheerin, 1979). Also at that time, *Rhynia gwynne-vaughanii* was abscising its sporangia after the spores were shed. A tissue suggestive of an abscission protective layer developed across the sporangial scar. Typically, in this species an adventitious branch developed from immediately below the sporangial scar (Edwards, 1980). The branching pattern is remarkably suggestive of the sympodial branching habit of angiosperms that is correlated with abscission of terminal buds and fruits (cf. Chapter 3).

It seems something of a paradox that soon after plants evolved leaves, they also evolved the ability to abscise them. (Yet, many of the reasons for which leaves are abscised (see Chapter 1) are the result of their morphology and must have existed from the earliest times.) By the end of the Devonian, leaf abscission was apparent in the Lycophyta; the earliest unambiguous example appears to be that of *Cyclostigma* (Figs. 9:3, 9:4; Chaloner, 1968).

Fig. 9:2.
Photograph of the Devonian fossil *Palaeostigma sewardi*. The markings are considered scars of detached organs. (Courtesy of Plumstead, 1967)

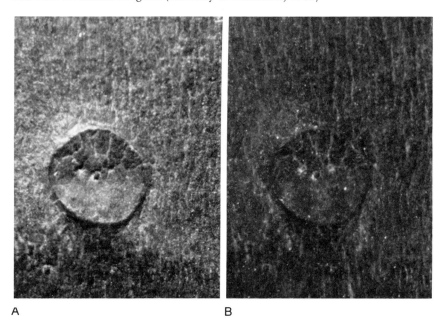

A B

Fig. 9:3.
Leaf scar of *Cyclostigma kiltorkense* from the Upper Devonian, X23. **A.** Photograph of the dry specimen. **B.** Photograph under xylene. Note the single, central vascular scar and the pair of parichnos scars. (Courtesy of Chaloner, 1968, by permission of the Council of the Linnaean Society of London.)

270 Paleontology and Evolution of Abscission

Of the numerous genera of Lycophyta that are known from the Devonian and Carboniferous, at least twelve abscised their leaves (e.g., *Lepidodendron*, Fig. 9:5; Chaloner, 1967).

A remarkable group of leaf-deciduous plants thrived in the late Carboniferous and Permian of the Southern Hemisphere in the supercontinent known as Gondwanaland. At that time, the Northern Hemisphere had a mild tropical climate, uninterrupted by seasons, and the coal-forming plants grew luxuriantly in extensive swamps. In contrast, Gondwanaland had a strongly seasonal climate that was cold enough to support active glaciation. There the coal-forming plants grew in and about shallow, glacier-carved lakes and swamps. In this environment the autumn-deciduous shrubs and trees of the Glossopteridae (Fig. 9:6) were the dominant plants and contributed greatly to the coal deposits (Plumstead, 1958, 1966; Gould and Delevoryas, 1977).

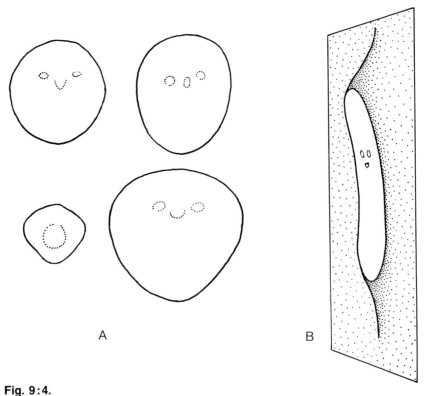

Fig. 9:4.
Diagrams of leaf scars of *Cyclostigma kiltorkense*. **A.** Surface views showing variations in size and shape. **B.** Oblique view of a compression fossil. (Redrawn from Chaloner, 1968, by permission of the Council of the Linnaean Society of London.)

Paleontology and Evolution of Abscission 271

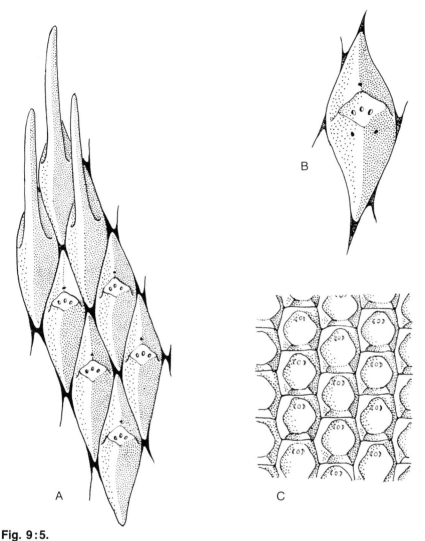

Fig. 9:5.
Leaf scars from the Carboniferous. **A, B.** *Lepidodendron* sp. **C.** *Sigillaria* sp. (Redrawn from Foster and Gifford, 1974)

Branch abscission was fairly common in the larger plants of the Paleozoic era. Some of the more notable examples include *Lepidodendron*, *Bothrodendron*, *Lepidophloios*, as well as *Calamites* (Fig. 9:7) and members of the Cordaitales. Analysis of the branch scars of the Paleozoic lycopods shows them to be remarkably similar to the branch scars of present-day gymnosperms and angiosperms, although differing in some details (Figs.

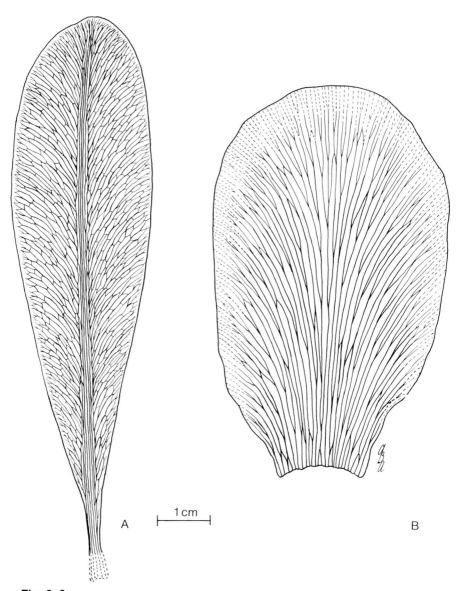

Fig. 9:6.
Leaves of deciduous Carboniferous Glossopteridae. **A.** *Glossopteris* (composite drawing based on specimens from South Africa, India, and Zimbabwe). **B.** *Gangamopteris* (drawing from photographs in Plumstead, 1975).

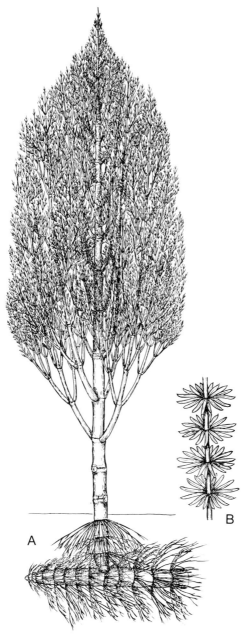

Fig. 9:7.
A. Reconstruction of a tree of *Calamites carinatus*. Note branch scars. **B.** Detail of leaves of a related species. (Drawing courtesy of Weier et al., 1974)

Fig. 9:8.
Branch scar of *Lepidodendron aculeatum* of the Carboniferous. (Photograph courtesy of Jonker, 1976)

9:8, 9:9; and cf. Fig. 3:7) (Jonker, 1976). Strobili were often abscised in much the same manner as vegetative branches (Jonker, 1976).

Other interesting evidence of early abscission relates to the axillary tubercles of *Crenaticaulis* and *Gosslingia* (Psilopsida) of the Lower Devonian. Investigation of their vascular anatomy has indicated that the tubercles are scars of lateral, root-like organs (rhizophores) similar to those of present-day *Selaginella* (Banks and Davis, 1969).

The root-like structures of *Lepidodendron* and related genera were regularly abscised. Since they were discovered, and commonly occur, separated from the stem and leaves of the parent plants, they were at first assigned to the genus *Stigmaria* and are still commonly referred to as "stigmarian appendages." Their anatomy has recently been described in some detail by Frankenberg and Eggert (1969). Their abscission zones are remarkably similar to those of the leaves of present-day deciduous trees. Three layers are readily identifiable in the abscission zone of the stigmarian appendage: (i) The distal layer at the very base of the appendage corresponds to the lignified layer that occurs at the base of the petiole in many angiosperm trees (see Chapter 2). (ii) A central layer corresponds to the separation layer proper, and appears to join a separation layer beneath the outer cortex of the main root. (iii) The proximal layer corresponds to the protective layer of the leaf scars of angiosperms (Figs. 9:10, 9:11).

Seeds form an important part of the fossil record and are of particular interest, as a distinctive feature of seeds is their abscission from the parent plant. Their typically hard coats favor preservation, but because they are detached, identification of seeds with other fossils such as leaves and stems presents a formidable and continuing task. The principal seed-producing

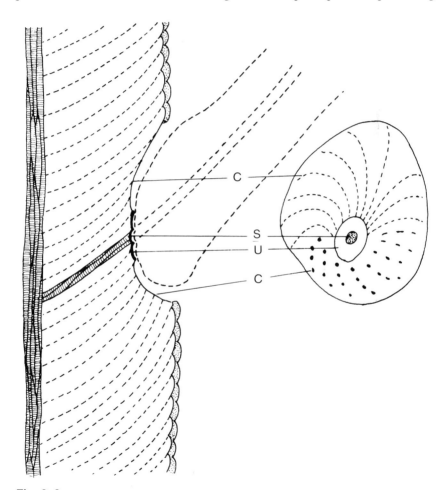

Fig. 9:9.
Diagrammatic interpretation of the *Lepidodendron* branch abscission zone and scar. (Courtesy of Jonker, 1976)

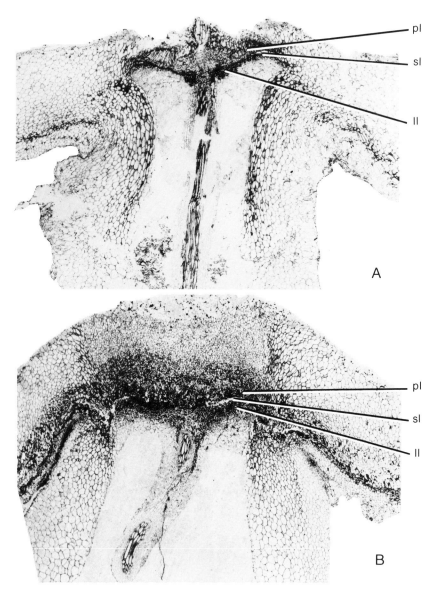

Fig. 9:10.
Photomicrographs of peel preparations of sections of the abscission zone of the stigmarian appendages [roots of Lepidodendraceae] at an early (A) and later (B) stage of development. There is a proximal protective layer (pl), a separation layer (sl), and a distal layer that corresponds to the distal lignified layer of the abscission zone of the angiosperm leaf (ll) (cf. Fig. 2:3A). (Courtesy of Frankenberg and Eggert, 1969)

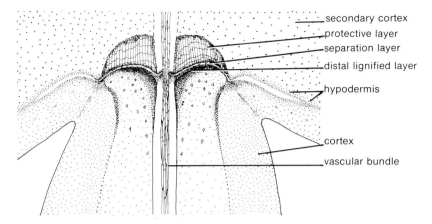

Fig. 9:11.
Diagram of the abscission zone and adjacent tissues of the stigmarian appendage. (Redrawn from Frankenberg and Eggert, 1969)

groups were the Pteridospermeae and Cordaitales, whose seeds resemble those of recent Ginkgoales and Cycadales. Certain of the Lepidodendrales also evolved to the point of seed production; *Lepidocarpon* abscised its megasporangia enclosed in a primitive "integument" formed by growth of the sporophyll upward around the megasporangium (Arnold, 1938).

3. **The Mesozoic Era (100–250 million years ago)**

This era saw the decline of the lycopods, pteridosperms, and the previously dominant fern-like plants, and a great expansion and diversification among the seed plants. The fossil record, of course, contains many new examples of abscission, especially of leaves and branches of the newly dominant groups.

In addition to the deciduous Gondwana flora that persisted well into the Era, the record of the Mesozoic includes a remarkable series of floras in the northern latitudes of Siberia and Canada, often dominated by deciduous Gingkoales and Czekanowskiales (Fig. 9:12) and by various conifers (Barnard, 1973). Here, too, there was a strongly seasonal climate, as well as great annual fluctuation in photoperiod.

The seasonal climates spread southward, and by the Cretaceous they had affected the floras throughout much of the Northern Hemisphere. These climates often included wide annual fluctuations in moisture and temperature. The earlier, relatively equable temperate and tropical climates had seen the development of many families and genera. From these, some representatives adapted to the new seasonal climates by the simple expedient of complete defoliation under adverse conditions (Fig. 9:13). A common signal for de-

278 Paleontology and Evolution of Abscission

foliation appears to have been the annual advent of drought (Axelrod, 1966), but as was already noted, temperature (especially cold) and photoperiod are also significant triggers to defoliation and probably were already functioning in the Mesozoic.

The foregoing comments are an example of the great flexibility and lability of abscission-control mechanisms (cf. Chapter 8). It is obvious that the deciduous habit has evolved independently in numerous families, and in many families it was lost from at least some genera and species as the need disappeared.

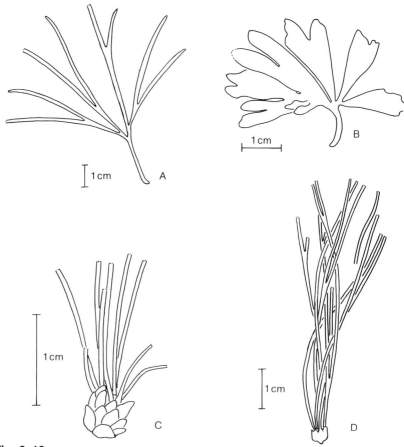

Fig. 9:12.
Abscised leaves of the Jurassic. Gingkoales and Czekanowskiales. The two orders are related, but the leaves of the Gingkoales are shed singly, whereas the leaves of the Czekanowskiales were always shed in clusters. **A, B.** *Baiera furcata, Gingko huttoni* of the Gingkoales. **C, D.** *Solenites vimineus, Czekanowskia blacki* of the Czekanowskiales. (Redrawn from Harris et al., 1974)

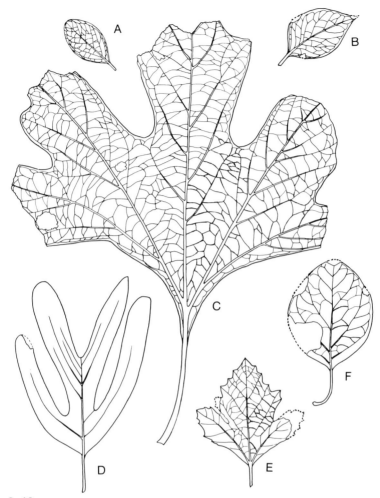

Fig. 9:13.
Abscised leaves from the Lower Cretaceous. Present-day species of these genera are all autumn deciduous. **A.** *Diospyros rotundifolia*. **B.** *Nyssa snowiana*. **C.** *Sassafras dissectum*. **D.** *Liriodendron wellingtonia*. **E.** *Crataegus tenuinervis*. **F.** *Populus hyperborea*. (Redrawn from Lesquereux, 1891)

4. Abscission in the Cenozoic Era (from 70 million years ago)

This was an era of diversification of families and genera, many of which were well established at the beginning of the era. We can assume that the basic morphological patterns and structures were established by the start of the era and that changes involved the diversifications and adaptations to emerging ecological conditions. Undoubtedly, this was also a time of much coevolution. Abscission of leaves, flowers, seeds, and fruits was one of the factors that plants utilized in developing their strategies and counter-strategies for survival in a biologically dynamic environment.

By the mid-Cenozoic, regions of arid climate had emerged in the Northern Hemisphere. These climates were further conducive to the expansion of the deciduous habit, and in many places the record of fossil leaves provides excellent evidence of the composition of the flora and of the nature of the climate (see, e.g., Axelrod, 1966). The expansion and recession of genera and species can in some cases be followed with considerable accuracy. For example, fossil branchlets of *Metasequoia* indicate that the genus was common at middle latitudes throughout the Northern Hemisphere, especially during the early Cenozoic era, but during the Miocene it disappeared from North America and now survives only as relict stands in upland valleys of the upper Yangtze River (Smiley, p.c.; Chaney, 1950). In a similar way, abscised leaves, fruits, and seeds have been used to reconstruct the rise and fall of the species and floras.

The annual habit of growth had a major expansion during the Oligocene and Miocene. The habit has doubtless evolved independently in various groups of plants, but it is most notably an adaptation to climates having a pronounced dry season. Families such as the Compositae, Cruciferae, and Leguminosae have given rise to many genera and species of annual plants. In each of these the early members of the family were trees and shrubs of tropical climates. All seem to have abscised their leaves and to have had many representatives that were deciduous. This was especially true in the Leguminosae. But typically, as an entity adopted the annual habit it lost the ability to abscise all or most of its leaves. Even a cursory survey of annual dicotyledonous plants shows leaf abscission to be poorly expressed.

Lee (1911) gave some consideration to the possible evolution of the anatomical features of leaf abscission. He reasoned that the separation layer was probably the first to appear in the course of evolution. Other layers, such as the protective layers and the lignified layer in the proximal portion of the abscission zone, probably evolved later, in his view. His point was that a separation layer, and its activity, is essential to abscission, whereas other layers appear to be of secondary importance, and their evolution could have come much later. It appears unlikely, however, that a species could survive abscission more than briefly without some provision for the protection of exposed tissues. It is more reasonable to assume that the evolution of separation layers and their associated protective layers must have evolved concurrently. This view is supported by what has been learned of the most ancient abscission zone yet to be investigated anatomically, the stigmarian appendages of *Lepidodendron* (Figs. 9:10, 9:11). Here the separation layer and a protective layer had already evolved and were functioning together.

B. EVOLUTION

1. Evidence of Recent Evolution and "Fine Tuning" of Abscission Behavior

As the local environments became more sharply delimited, specialized adaptations evolved. One of the most sensitive of these adaptations is the leaf abscission of *Fouquieria splendens* of the southwest desert of North America. This spiny shrub is leafless most of the year, but develops leaves promptly after a rain and abscises them within a few weeks. It is not at all limited to one such cycle a year, but has been observed to go through the cycle up to five times in a year when rainfall came at appropriate intervals (see Chapter 6).

Cold now appears to be one of the regular signals for the induction of autumnal defoliation of trees. For the floras of the middle latitudes of the Northern Hemisphere, the involvement of this signal appears to be of relatively late origin. There is little indication that cold was a factor in the widespread, mild climates of the region during the Paleozoic era. (It does appear that cold was a factor in the autumnal leaf abscission in Gondwanaland, as well as in the arctic Siberian-Canadian region in the late Paleozoic and early Mesozoic.) The present-day floras of the temperate Northern Hemisphere adapted to seasonal climates after an extended period of evolution in mild, uniform climates. From comparison with present-day tropical climates, it seems probable that drought was the first aspect of seasonal climates to which leaf abscission (of at least the angiosperm floras) became responsive. As climates became colder the factor of cold itself became a kind of stress. As was noted in Chapter 4, stress-inducing factors of several kinds are capable of inducing abscission. For example, cotton is very sensitive to drought and can be almost completely defoliated by a severe drought. Yet if a period of frost comes before the drought, the cold will induce defoliation. It seems reasonable that in regions of summer rainfall and in years where there is little or no drought toward the end of the growing season, it is advantageous for deciduous species to be able to respond to cold.

Similarly, sensitivity to light seems to have been refined in the course of evolution. We can assume that there has been a long-standing sensitivity that originated in tropical climates. Shaded leaves, no longer able to contribute to the carbohydrate economy of the plant, have probably been abscised from the earliest times. As plants migrated to higher latitudes, both lower light intensities and particularly the shortening photoperiods of autumn became of increasing importance as triggers of defoliation (see Chapter 6).

Recent ecological investigations have drawn attention to the importance of coevolution among plants and between plants and various animals. In such evolution abscission has a significant part. Plants have modified their behavior patterns, including patterns of leaf, fruit, and seed abscission as their strategies for survival have changed. The usual pattern for trees is to flower and fruit when they are in full leaf with a maximum capacity for photosynthesis. However, in regions of the tropics having a dry season a number of trees have developed the strategy of flowering and fruiting after leaf abscission. This habit appears to have coevolved with pollinator species, particularly species of birds and mammals whose activity would be facilitated by the greater visibility in defoliated trees. Similarly, this habit of flowering and fruiting facilitates the work of many animal species involved in seed dispersal (Janzen, 1967).

If it is assumed that the primitive condition of seed abscission involved easy detachment from the parent plant, abscission behavior today in relation to seed dispersal is very finely tuned indeed. In many species coevolution has led to complete inhibition of fruit and seed abscission, usually to accommodate flying or climbing seed-dispersal agents. In other species the process of abscission commences, but stops short of complete detachment. Such plants present disseminules that are loosened and easily detached by dispersal agents. This pattern is associated with seed disseminules adapted to be attached to the animal agents or to be carried by the wind. In still other species coevolution has reinforced prompt and complete separation of seeds and fruits at maturity. One example, from many that might be cited, is the relation between *Acacia tortilis*, large herbivores, and bruchid seed beetles of East Africa. The abscised fruits are readily available to both the beetles and the herbivores, and a high percentage of seeds have been entered by the beetles before the fruits are ingested by the herbivores. A small but appreciable number of seeds survive both the insect attack and passage through the digestive system of the large herbivores. Penetration of the seed coat by the beetles facilitates germination, and the digestive fluids of the herbivores further hastens the germination of those seeds which are sufficiently complete to be viable (Lamprey, 1967; Halevy 1974; Lamprey et al., 1974). In addition to being carried some distance from the parent tree by the herbivores, viable seeds are deposited with an appreciable amount of "fertilizer." Since the herbivores involved in this seed dispersal include the African elephant (as well as antelopes), there is little that can interfere with the dissemination of *Acacia tortilis*. This must be an example of one of the most eminently successful instances of coevolution in the annals of ecology.

Another interesting example is in the palms. In this family, senescent leaves are abscised from most species. However, species with spiny leaves rarely abscise their leaves; instead, the dead leaves droop and are retained. The hanging foliage provides a protective mesh of spines that serves as a strong deterrent to climbing herbivores such as squirrels and coati. By suppressing leaf abscission, these palms are better able to protect flowers and developing fruits (Williamson, 1976).

2. Some General Comments on Abscission Evolution

From consideration of the fossil record, it is clear that abscission is a character of very ancient origin. It is difficult to escape the conclusion that abscission of organs must be almost as ancient as the organs themselves, and that as discrete structures and organs evolved, the mechanisms for their abscission and the control of this abscission evolved, if not simultaneously, then very soon thereafter. Once it is recognized that the phenomena of cell separation and fragmentation of single-celled and colonial organisms are legitimate manifestations of abscission, it becomes apparent that abscission not only is a primitive character, it may indeed be one of *the* most ancient of morphological characters.

Plants seem to have been able to utilize abscission wherever and whenever it was advantageous in a strategy of survival. For example, consider abscission and dissemination of seeds. In many species, simple seed abscission appears adequate for dissemination and reproduction. Other species, however, have found it advantageous to include other parts of the plant in the disseminule, parts which by their structure or food value would facilitate their transportation. Seed dissemination then may involve the abscission of fruit containing seeds. In other cases, flower parts remain with the disseminule, and, in still other species, the bracts are included. In some, such as *Salicornia pusilla*, flowers are produced at each node, and each disseminule consists of a small fruit together with the attached node and internode. Perhaps the most spectacular example of strategic selection of the site of abscission to facilitate dissemination is that of the tumbleweeds, in which the entire aerial part of the plant is abscised at ground level to function as the disseminule.

Another noteworthy aspect of abscission is that it is a character which is far from fixed. As was noted in Chapter 8, there are innumerable examples of variations of abscission behavior of leaves, of flowers, and of fruits within a single species. These examples indicate the plasticity of abscission in present-day plants. Such variations are fertile ground for natural and artificial selection. The plant breeder attempting to select for a particular abscission or dehiscence pattern is reasonably likely to succeed. Thus, it has been possible to select flowers having excellent petal retention and fruits that abscise readily to facilitate harvest.

3. A Theory of the Evolution of Abscission

a. Some General Considerations

The materials considered in this chapter were gathered in a desire to trace the antecedents of present-day abscission back as far as possible in the geological record. As the search progressed, I became increasingly fascinated with this question: Where, when, and how did it all begin?

As was summarized in the section on paleontology, examples of abscission are surprisingly conspicuous back to the mid-Paleozoic. Then they become sparse. But the evidence gives us no reason to doubt that the earliest plants—back as far as the record goes into the Precambrian—were abscising (fragmenting and separating) in the same manner as their present-day counterparts (e.g., the blue-green algae).

Perhaps manifestations of abscission were at their lowest ebb in the early Paleozoic, particularly in the larger aquatic thalloid plants similar to the present-day marine kelps. Abscission offers relatively little that is advantageous to such plants. Nevertheless, these plants—some present-day members at least—do have the ability for abscission and display interesting examples (see Chapter 7). In contrast, the processes of abscission are utilized in an extensive variety of ways by the angiospermous land plants, particularly in situations where the environment presents a variety of hazards. Here the many options of abscission often contribute impressively to the homeostasis of the individual, to its survival, and to the survival and extension of the species.

In tracing the antecedents of abscission back to the simpler manifestations in the primitive lower plants, one cannot escape the conclusion that he is viewing samples of what must be a continuous array. At no place in the spectrum—from the abscission of the magnificent leaves of present-day palms, such as *Cocos* and *Roystonea*, to the separation of the cells of the prokaryotes in the Precambrian—can a significant break be identified. Abscission in the simpler, earlier forms appears to be but an anatomically simpler version of what takes place in the more complex recent forms.

In considering the evolutionary spectrum of abscission, perhaps the most difficult region is at the emergence of land plants. Here the fossil record is meager indeed, certainly as far as abscission is concerned. But one can get some idea of what probably happened by comparing abscission manifestations in gametophytes of present-day ferns and mosses with abscission in the filamentous green algae. The gametophytes include moss protonemata and fern prothallia that are very slender, even filamentous. Among the gemmae which these gametophytes produce are "string-of-beads" types and spindle-shaped types (Farrar, 1967), and filaments which in appearance are much like the hormogonia that are separated by filamentous algae in their vegetative reproduction (Stokey, 1948; Stokey and Atkinson, 1958).

The view that the manifestations of abscission, separation, and fragmentation represent a continuous spectrum is strongly supported by recognition that the chemistry and biochemistry of cell separation is essentially identical in all present-day plant forms from the lowest to the highest. While there are some differences in composition among the cementing polysaccharides, these differences are relatively small when we consider the great diversity in other aspects of evolution. In all cases separation appears to involve the secretion of polysaccharide-degrading enzymes that break the bonds of branching chains holding cells together.

Recognition that some types of abscission involve little or no obvious hydrolytic activity does not detract from the importance of biochemical degradation as the central aspect of abscission. In the higher plants, at least, abscission in which mechanical factors are to a greater or lesser extent responsible for separation can be viewed as modifications of the main body of manifestations.

b. A Brief Statement of the Abscission Evolution Theory

From consideration of the great spectrum of abscission manifestations in the higher and lower forms of present-day plants and in the fossil record, one can reach no other conclusion than that the process of abscission had its beginnings in the simple plants of the Precambrian. Further, it is clear that the biochemical machinery must have originated then and developed essentially to its present form in order to enable the separation of the cells of a colony.

One can say, therefore, that *the evolution of abscission began when two cells separated for the very first time*. Separation of previously attached cells is the distinctive characteristic of abscission.

In various ways the early plants used this simple form of abscission, cell separation, to facilitate reproduction and dissemination. As plants became more complex, they developed the ability to localize abscission to particular regions within the plant body, which we now call abscission zones. In them, cells remain physiologically responsive and capable of the biochemical changes of cell separation. One can recognize two major thrusts of abscission evolution. The first is the localization of abscission to particular places in the plant. This finds its greatest diversity in the angiosperms with the localization of abscission zones for leaflets, leaves, branches, petals, seeds, fruits, and the other discrete organs. The second evolutionary thrust has involved the development of control systems that regulate the basic mechanism of secretion of abscission enzymes. The machinery of these systems is still largely unknown. What is now known is a surprisingly great variety of ecological and physiological factors that can affect abscission (as was discussed in Chapters 4 and 6), and whose influences are channeled and coordinated by the control systems.

4. Abscission in the Animal Kingdom

A cursory survey of the animal kingdom indicates that a biologist with a reasonable knowledge of animals would have little difficulty in assembling a book on abscission in animals at least as sizable as this one. Numerous examples of fragmentation, detachment, and shedding of discrete structures come readily to mind. A few examples follow.

In the lower invertebrates, asexual reproduction by fragmentation and detachment of buds is common. For example, many of the lower forms can reproduce by fragmentation, and hydroids commonly reproduce by the detachment of buds.

In the many species of arthropods the testa is shed and molted at intervals in the growth of the individual. Reproductive individuals of ants and termites shed (abscise?) their wings after their brief period of flight.

Amphibians and most reptiles regularly shed their skin. A more spectacular abscission phenomenon shown by some species of lizards is the ability to abscise their tails to distract a predator.

In birds and mammals the outer layers of skin tend to flake away in a kind of continual abscission, somewhat resembling the continual abscission of the cells of the root cap. Feathers and fur are often shed seasonally. Some household pets seem to shed hair continually. A number of animals shed their horns each year and, of course, the decidua of mammals is shed at the time of birth.

Perhaps the most striking parallel between the abscission phenomena of plants and of animals is the evidence of involvement of hydrolytic enzymes, suggesting that the fundamental mechanisms of abscission and separation of cells and tissues may be much more similar than they might seem at first glance. The similarity may turn out to be much closer than would have been imagined a few years ago. There is increasing evidence that the adherence of animal cells involves the functioning of glycoprotein in which the polysaccharide moieties consist of branching chains very much like the branching chains of pectic substances that cement plant cells. Such considerations point to abscission as but another phenomenon in which plants and animals are far more similar than earlier generations had imagined.

10. Basics of Agricultural Abscission

Chapter Contents

A. CULTURAL PRACTICES THAT AFFECT ABSCISSION 288

B. LEAF ABSCISSION 290
   1. Acceleration 290
   2. Retardation 293

C. FLOWER ABSCISSION 294

D. YOUNG-FRUIT ABSCISSION 296
   1. Fruit Thinning (Promotion of Young-Fruit Abscission) 296
   2. Prevention of Young-Fruit Abscission 299

E. FRUIT DEHISCENCE AND SEED ABSCISSION 300

F. MATURE-FRUIT ABSCISSION 301
   1. Prevention 301
   2. Promotion 303

G. BRANCH ABSCISSION 305

H. BARK REMOVAL 306

The abscission habits of cultivated plants are rarely expressed to the complete satisfaction of the farmer. Almost every species under one condition or another displays either more or less abscission than the farmer finds desirable. There are several approaches to the practical solution of such agricultural problems. Given sufficient time, plant breeding and selection can result in the establishment of improved cultivars, relatively free of objectionable abscission behavior. This has been accomplished in numerous instances, as was noted in Chapter 8. For many crop plants, though, finding and developing cultivars with desired abscission behavior has not yet been successful, and the farmer must rely on cultural practices in order to regulate abscission.

Until recently, methods for the manipulation of abscission have been limited. The farmer had to thin fruit by hand when natural abscission was insufficient. He had to remove leaves by hand to reduce foliage, and traditionally, he picked or knocked fruit from trees when it was ripe but did not abscise.

Beginning in the 1930s, observations and experimental results appeared that indicated abscission might be capable of regulation by agricultural chemicals. Laibach's demonstration that auxin could delay abscission led to field tests and eventually to the regular use of synthetic auxins to delay abscission of leaves and fruits. At about the same time, the ability of other chemicals to thin young fruit was discovered; the early 1940s saw the development of chemical defoliation of cotton.

In the following sections, mention will be made of agricultural chemicals that have been effective in the regulation of abscission. The discussion will not include specific recommendations of either chemicals or methods of application. In the first place, new chemicals and methods are tested each year, and improvements are brought forth regularly. Thus, any specific comments could well be out of date by the time this book is published. Further, the effectiveness of both chemicals and methods varies from cultivar to cultivar. Effectiveness also varies with the physiological condition of the plants, as well as with local conditions of weather and soil (see, e.g., Edgerton, 1973). Although most agricultural chemicals are relatively harmless to humans, a number show some degree of toxicity, and some diluents and impurities can also be harmful. No one should handle an agricultural chemical without ascertaining its hazards and taking due precautions for personal safety. The reader who is interested in a particular application should obtain the most recent information on the chemical and its use in his area from the local Agricultural Extension Service or Department of Agriculture.

The following discussion is intended to introduce the general reader to the problems and possibilities of the agricultural regulation of abscission. A more technical discussion of the role of agricultural chemicals in abscission regulation was recently published (Addicott, 1976). Also, the two recent Symposia on Growth Regulators in Fruit Production provide a very good overview of current research on fruit abscission (Wellensiek, 1973; Luckwill et al., 1978). The more widely used regulator chemicals were recently described by Nickell (1978).

A. CULTURAL PRACTICES THAT AFFECT ABSCISSION

A considerable part of the work of a farmer involves manipulation of the environment in the ways within his means to facilitate the production of a crop. Just how much a farmer can do depends on the particular environment within which he must work and the economic costs of manipulations he would like to make. Ideally, he would like to control all of the factors discussed in the chapter on ecology in order to obtain the maximum productivity of his crop.

Maintenance of optimal mineral nutrition is one of the most powerful tools a farmer has to ensure high productivity and to prevent premature abscission of leaves, flowers, or fruits. As was discussed in Chapter 6, nitrogen is the essential mineral element that is most frequently deficient and one of the easiest to manipulate. Even when chemical forms of nitrogen are not readily available, amounts available to the plants can be kept high by organic fertilizers or crop rotation. As with other cultural practices, application of nitrogen to the soil must be made judiciously. Excessive amounts, while guaranteeing a minimum of abscission, can induce crops such as cotton and tomatoes to grow in a strongly vegetative manner that precludes flowering and fruiting. Other mineral nutrients, such as Zn^{2+}, when deficient, induce considerable abscission. Most such deficiencies can be readily diagnosed and corrected with little difficulty or expense.

Where irrigation is practiced or water is available to supplement uncertain rainfall, a steady supply of water is imperative in the prevention of undesired abscission of leaves, flowers, and fruits. As was discussed in previous chapters, the stress of water deficiency (drought) is probably the most common single cause of leaf abscission. With few exceptions, cultivated trees are sensitive to drought; the common deciduous fruit and nut trees of the Northern Hemisphere are especially so. Consequently, when a farmer desires to accelerate abscission, he can sometimes accomplish it by withholding irrigation water.

When crops are grown in greenhouses, the "farmer" can control a number of other factors that are important in the prevention of abscission. In greenhouse culture, one of the most important factors is light, which is often deficient because of extended periods of overcast during the winter months. Light is important, both for photosynthesis and for an appropriate photoperiod for the crop being cultivated. Similarly, greenhouse culture permits some control of temperature, which must be adjusted to the requirements of the crop if the crop is to develop normally.

Other factors that can be utilized by the farmer are pruning and thinning. Judicious pruning of crop trees can do much to keep subsequent growth in balance and productive, with a minimum of abscission. In spite of the best cultural practices, a number of valuable crop trees regularly produce far more young fruit than they can mature into marketable fruit. With these crops, thinning is necessary, and the required abscission is accomplished by hand or by chemical agents.

Other important aspects of abscission over which the farmer has some

290 Basics of Agricultural Abscission

control are the biotic factors, such as injurious insects and microorganisms. The symptoms induced by many of these insects and diseases include premature leaf, flower, and fruit abscission (see Chapter 6).

B. LEAF ABSCISSION

1. Acceleration

Chemical defoliation as an agricultural practice had its beginnings with the accidental discovery that the fertilizer calcium cyanamid could defoliate cotton when applied to the foliage as a light dust. In many situations defoliation at harvest time greatly facilitated picking of the cotton by hand, and especially picking by machine (Fig. 10:1). While cotton can often be machine picked while some leaves are present, leaves add stain to machine-picked cotton fibers, and where the foliage is heavy, it handicaps the picking.

Chemicals used to defoliate cotton include ammonium nitrate, endothall, paraquat, sodium cacodylate, sodium chlorate, tributyl phosphorotrithioate, and tributyl phosphorotrithioite (Elliott et al., 1968). In years when rainfall has been deficient, toward the end of the cotton growing season

Fig. 10:1.
Demonstration plot of chemical-defoliated cotton ready for harvest. Adjacent plots were untreated. (Photograph courtesy of V. T. Walhood)

there is sufficient leaf abscission to make chemical defoliation unnecessary. From such observations, farmers growing cotton under irrigation have learned that by careful withholding of irrigation water at the end of the growing season, they can induce an adequate amount of leaf abscission without affecting the yield of cotton.

Another practice that can be a satisfactory substitute for chemical defoliation prior to cotton harvest is the use of a chemical "wiltant" such as neodecanoic acid. A few hours after application of neodecanoic acid, cotton can be machine harvested; by that time, the wilted leaves offer little resistance to the picking machine and contribute little stain to the fiber (Miller et al., 1971).

Defoliation of young nursery trees is often required prior to digging and shipment. Such defoliation enables the young trees to be dug prior to adverse weather, and more important, it prevents both moisture loss and development of foliage diseases during shipment. Experiments to date have not resulted in the establishment of fully satisfactory practices. At the time when defoliation of nursery stock is desired, temperatures are cool, but trees often have not approached dormancy to the extent that leaf abscission is readily accomplished. Compounds that have shown the most promise include KI, Bromodine, ethephon, Harvade, Dupont-WK surfactant, and ABA (Larsen, 1973; Larsen and Lowell, 1977).

Chemical defoliation as an aid in the control of insects or disease has been explored in a number of situations, although no widely important practice appears to have developed. An interesting example is the disease South American leaf blight of *Hevea*. The spread of the disease can be prevented by chemical defoliation of nearby healthy trees (Osborne, 1968). While trees may suffer from lack of foliage, such defoliation offers a practical method of containing and eradicating a serious disease. Control of such phyllophagous insects as the pine needle borer appears feasible if defoliation can be achieved while the insects are still within the needles. Incidence of tsetse flies in the vicinity of human habitations could be reduced if practical means could be found to defoliate the wide variety of species which support the flies (Osborne, 1968). The potential value of defoliation for the control of insects and diseases remains to be developed.

Chemical defoliants are used to accelerate autumnal leaf abscission of orchard trees. This enables the trees to be pruned earlier in the winter season and enables the farmer to prune more efficiently (Gerdts et al., 1977). Use of ethephon to induce mulberry leaf abscission is being investigated; if it is successful, the practice will facilitate the harvest of leaves for feeding to silkworms (de Wilde, 1971).

In forest areas, defoliation has found application as a military tactic of some value, and is credited with saving many lives from ambush during campaigns in the tropics. Such "defoliation" is more an herbicidal action than defoliation proper. Tropical vegetation tends to be resistant to the relatively mild chemicals exemplified by the cotton defoliants. To obtain effective improvement in visibility, it is usually necessary to apply an herbicidal chemical that often kills vegetation while incidentally promoting leaf abscission (see Way and Chancellor, 1977; Kozlowski, 1973).

The physiological action of chemical defoliants is slowly becoming understood, although the information is far from complete. The more mild defoliants seem to do little more than accelerate the normal preabscission changes of senescent leaves (see Chapter 4). Each defoliant produces characteristic symptoms, usually affecting the degree of chlorophyll change, portion of the leaf affected, and development of anthocyanins. The stronger defoliants may have a fairly rapid and drastic action often involving blade desiccation. There has been some investigation of biochemical modifications of metabolism by defoliants, but these have been relatively few and difficult to evaluate. The normal biochemical events involved in the early stages of abscission control must be defined before biochemical studies will be useful.

A number of changes in hormone levels and metabolism have been noted among the responses to chemical defoliants; these changes appear to be important early steps in the induction of abscission. Auxin levels in leaves drop rapidly after defoliant application (Swets and Addicott, 1955). The lowered levels of auxin could be the result of either impaired biosynthesis or accelerated inactivation or quite likely both. Also, there is considerable evidence for increased activity of IAA-oxidase when abscission is accelerated, as by phenolic substances (Schwertner and Morgan, 1966). The involvement of lowered levels of blade auxin in abscission has been recognized for some time (Shoji et al., 1951). The major effects of chemical defoliants, as well as ETH when it is used as a defoliant, are on the leaf blade. Treatment of the petioles does relatively little to accelerate abscission (Walhood, 1955; Beyer, 1975). Thus, one of the primary effects of a defoliant is to reduce the supply of auxin coming from the leaf blade. In addition, chemical treatment can reduce the ability of the petiole to transport auxin. Thus, auxin-transport inhibitors, of which ETH is one, are promotive of abscission (Beyer, 1972, 1973).

It has been known for some time that applications of chemical defoliants increase ETH production by leaves (Jackson, 1952); however, from experiments with explants, it appears unlikely that wound ETH alone is sufficient to bring about leaf abscission (Marynick, 1977). The amounts released appear too small and too transitory to be effective under field conditions. Nevertheless, wound ETH probably contributes to defoliation, augmenting to a limited degree the other abscission-promotive changes initiated by chemical defoliants. It appears likely that there is an increase in ABA as a result of treatment with chemical defoliants; it increases rapidly in response to other stresses, particularly water stress (see Chapter 4).

When ETH is used as a chemical defoliant, it rapidly affects auxin metabolism, as was noted above, and reduces the amounts reaching the abscission zone. Its effectiveness is increased by combination with other defoliating treatments and substances, such as auxin transport inhibitors (Morgan and Durham, 1973; Morgan et al., 1977).

2. Retardation

The need to retard or to prevent leaf abscission arises in various circumstances, particularly in the florist and nursery trades. Probably the most important example is the application of NAA to the leaves and berries of holly to prevent abscission during shipment (Roberts and Ticknor, 1970). Such treatments are usually effective. The cost of applications can be avoided in some instances by careful culture of the plants. It is especially important that mineral nutrition be adequate, particularly nitrogen, and that water stress be avoided at every stage. Also, the atmosphere around the plants during storage and shipment should be controlled to prevent the accumulation of ETH (Kays et al., 1976) and to keep respiration at a low level. When exposure to ETH is unavoidable, pretreatment with Ag^+ shows promise of preventing leaf abscission and other undesirable ETH effects (Beyer, 1976).

Needle abscission from conifers used as Christmas trees is often excessive, particularly when the trees have become somewhat desiccated during shipment. The auxin regulators such as NAA and 2,4-D have not been effective in preventing needle abscission. Indeed, they have often accelerated abscission (Worley and Grogan, 1941; Addicott, unpubl.). To date, the most effective way to retard needle abscission of Christmas trees is to maintain high levels of moisture within the tree. For this purpose, antitranspirants can be helpful (Dingle, p.c.). To reduce the leaf abscission that sometimes follows the application of insecticide oil sprays to *Citrus*, 2,4-D has been found useful (Hield et al., 1964).

Ample moisture in leaves and needles avoids the stress that induces rapid synthesis of ABA. Thus, the retardation of leaf abscission involves augmenting the natural supply of auxin and avoiding practices that would lead to the build up of either ABA or ETH. Also, early experiments showed that applications of GA would retard senescence and abscission of a number of species (Brian et al., 1959). The increased GA in the leaves appears to increase the sink strength and delay preabscission changes, as do the auxin regulators.

In susceptible crops, such as coffee, leaf diseases (e.g., coffee leaf rust) and insects (e.g., leaf miners) can cause severe defoliation and reduced yields. Sprays containing copper or other fungicides have a markedly beneficial effect on leaf retention and crop production (Huxley, 1970). Such sprays have a "tonic" effect, delaying leaf fall by two to three months, and increasing production even when the coffee trees are not noticeably diseased (van der Vossen and Browning, 1978).

C. FLOWER ABSCISSION

Many species produce large numbers of flowers, but have a capacity to carry only a limited number through to full fruit maturity. In the normal course of events, excess flowers and potential fruits are abscised at one or more stages in development. Stages when abscission commonly occurs include young flower buds, older flower buds, young fruit (a few days after anthesis), partially developed fruit, and mature fruit.

Flower-bud drop is characteristic of species that flower over an extended period, and usually does not occur until after a plant has already set a number of fruit. In these circumstances the plant has the morphogenetic ability to produce flower buds, but lacks the physiologic capacity to provide all of them with adequate nutrients (see Farrington and Pate, 1981b). Here, abscission is again functioning to facilitate homeostasis. Plants such as cotton, *Lupinus*, and others (especially those that have their flowers in racemes) have a strong tendency to abscise the later developing, more terminal flower buds. Such abscission, when it occurs on plants growing in favorable conditions, is difficult to modify. However, when it is brought on by adverse weather, such as heat or cold, the bud abscission can sometimes be prevented by applications of auxin regulators. As is noted in the next section, abscission at the young-fruit stage is controllable in many circumstances.

Practices that would delay the abscission of ornamental flowers or their petals would find wide use. However, trials with auxin regulators have met with only limited success. Of fifteen species of flowering trees and shrubs sprayed with NAA, NOXA, and 4-CPA, only the petals of Japanese flowering cherries and the bracts of white flowering dogwood showed any delay of abscission (Wester and Marth, 1950). Flower abscission of *Lupinus* and *Begonia* was delayed for several days by application of NAA (Warne, 1947; Wasscher, 1947). *Bougainvillea* can be grown as an attractive pot plant; however, the flowers with their showy bracts abscise soon after the plants are removed from the greenhouse. This abscission can be delayed for two or three weeks by a spray of NAA (Fig. 10:2) (Hackett et al., 1972). Similarly, abscission of the showy bracts (as well as leaves) of potted *Poinsettia* was delayed by application of 2,4,5-trichlorophenoxyacetamide (Carpenter, 1956). Bud and petal abscission of *Lupinus hartwegii* were delayed and vase-life was prolonged by adding citric acid, aluminum sulfate, MH, or sucrose to the water in which cut stalks were placed (Mohan Ram and Ramanuja Rao, 1977). Floriculturalists have given considerable attention to practices that will delay senescence and prolong the keeping of cut flowers.

Fig. 10:2.
Flower and bract abscission of potted *Bougainvillea* can be delayed two to three weeks by sprays of NAA. (Photograph courtesy of Hackett et al., 1972)

Chemicals that are effective when added to the water in which flowers are kept include sugars; various salts, especially of K, Al, and Ag; germicides; plant hormones; and growth regulators. The extensive literature was recently reviewed by Halevy and Mayak (1981). Although most varieties used for cut flowers have been selected because they show little or no petal and flower abscission, treatments that delay senescence can be expected to delay abscission.

In the hybridization of snapdragons (*Antirrhinum*), emasculation can be accomplished by sprays of ethephon before flower opening. The sprays induce abscission of the corollas, which take the anthers with them (de Wilde, 1971).

The persistent styles of Valleria bananas (*Musa*) lower fruit quality; they will abscise in response to ethephon (de Wilde, 1971).

There are numerous situations in which flower parts or fruit falling from ornamental trees are a nuisance. When the flowers of such trees are not attractive or desirable, sprays that lead to the early abortion and abscission of flowers can be applied. If the flowers are desired but the fruit are not, sprays of NAA applied toward the end of the flowering period can prevent fruit development, and the aborted fruit will be abscised (Batjer, 1954). Ethephon has been effective on ornamental trees of *Olea*, *Pittosporum*, *Prunus*, and *Malus*, as well as on *Cocos nucifera* (de Wilde, 1971).

A number of species abscise their flower petals promptly after physical or chemical stimulation (see Chapter 3). Flowers of the Linaceae, Geraniaceae, and Papaveraceae are especially susceptible. This behavior is a source of considerable difficulty to student and professional botanists when they attempt to press collected specimens. To reduce the amount of abscission, specimens are sometimes dipped in boiling water, but that can be damaging to delicate material. A more satisfactory method is a dip or soak in "Odourless Carrier," a clear petroleum fraction with a boiling range of 176°–206°C (Guillarmod, 1976). The treatment largely prevents abscission and many undesirable color changes as well.

D. YOUNG-FRUIT ABSCISSION
1. Fruit Thinning (Promotion of Young-Fruit Abscission)

Many varieties of the tree fruits, such as apples, apricots, peaches, and *Citrus*, tend to retain a large number of young fruits, and thinning has been practiced for centuries. The principal advantage of thinning is an increase in the size of the remaining fruit. Another desirable result is the maintenance of regular annual bearing, in contrast to biennial bearing of some varieties. The latter tend to bear a heavy crop in alternate "on" years and a light crop in the "off" years. A further advantage of fruit thinning is the prevention of the breakage that can come to overloaded fruiting branches.

It was learned, somewhat fortuitously, that various insecticides, fungicides, and other agricultural chemicals would often reduce fruit set by inducing abortion and abscission of flowers or young fruit. The first of these chemicals to gain acceptance were the dinitrophenols. Soon thereafter, in the early 1940s, it was discovered that auxin regulators, particularly NAA and related substances, could also be effective fruit-thinning agents (Burkholder and McCown, 1941; Thompson, 1957; Hield et al., 1966). Somewhat later, carbaryl (Batjer and Thomson, 1961) was found to be effective. A large number of other compounds have been tested as well, including thiourea, 3-chlorophenoxypropionamide, naptalam (Fig. 10:3), ethephon, and other ETH-releasing chemicals (see Addicott, 1976).

The response of *Coffea arabica* to ethephon provides an interesting example of the potential benefits, as well as limitations, of chemical induction of young fruit abscission. *Coffea* tends to flower at irregular intervals, but farmers want to eliminate the young fruit from the "off-season" flowering so that the productivity of the trees can be concentrated on fruit (berry) production during the most favorable season for development. Sprays of ethephon induced abscission of a substantial number of unwanted young fruits. The sprays also induced abscission of about 10 percent of the leaves, but this degree of leaf abscission is considered tolerable. As a second beneficial effect, the sprays accelerated maturation of main-crop fruit by two to four weeks, giving the farmer an additional option of extending the harvest season (Browning and Cannell, 1970).

None of the regulator-chemicals has proved completely reliable. Even the most satisfactory among them occasionally promotes an excessive amount of young-fruit abscission, a most undesirable result. At times there have

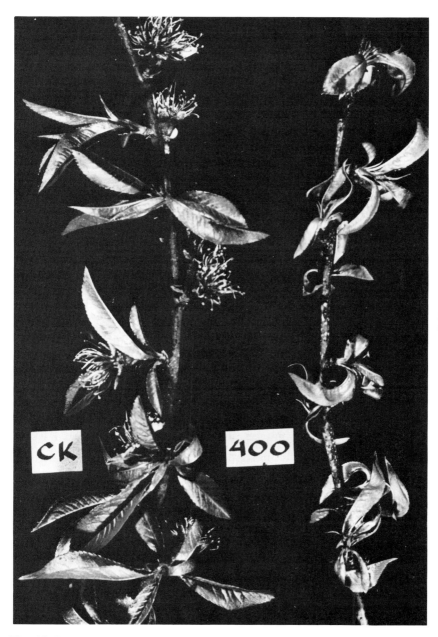

Fig. 10:3.
Fruit thinning of peach with NPA (400 mg/ℓ) applied two to four days after anthesis. At the time of the photograph, there were still numerous flowers on the control (ck), whereas most of the flowers on the treated branch had aborted and abscised. (Photograph courtesy of J. A. Beutel)

been injurious side effects to the trees (e.g., Bondad, 1976; Unrath, 1978). Variations in weather appear to strongly affect the absorption and response to the chemicals.

Some progress has been made toward understanding the physiological action of some of the fruit-thinning regulators. For example, the dinitrophenols act as pollenicides, destroying pollen either before or after pollination (Hildebrand, 1944). Responses to NAA and related regulators have received considerable attention. Their ability to induce abortion and abscission of young fruit was at first puzzling because in some situations NAA acts much like IAA, the endogenous auxin, and shows few deleterious effects. In fruit thinning, however, NAA acts more like the auxin herbicides. There is usually an initial cessation of abscission that lasts about two weeks (Luckwill, 1953); this is a "direct" effect of NAA on abscission processes. After that period, abscission of the treated fruit increases well above the rates for the untreated fruit. In this action of NAA, several intermediate effects have been observed. When applied at full bloom, NAA affected the styles and stigmas of apples so that most of the germinated pollen tubes seemed incompatible—i.e., they grew slowly and had characteristically swollen tips (Luckwill, 1953). Fruit growth ceased soon after application of NAA (Crowe, 1965; Teubner and Murneek, 1955), and levels of diffusible auxin from the fruit declined sharply during the first 24 h (Crowe, 1965). Considerable embryo abortion resulted from NAA applications (Luckwill, 1953; Teubner and Murneek, 1955; but see Leuty and Bukovac, 1968), and translocation of nutrients to the affected young fruit appeared blocked soon after spraying (see Crowe, 1965; Martin, 1973). NAA sprays almost certainly lead to an increased release of ETH, but whether this ETH persists long enough to affect abortion and abscission of the fruit has not been ascertained. Abscission-inducing sprays of carbaryl on Red Rome apples did not affect ETH evolution of the young fruits or pedicels (Schneider, 1975). However, after a fruit-thinning spray of NAA, the responding young fruit of kaki persimmon showed a remarkable increase in ABA (Yamamura and Naito, 1975). Reduced nutrient transport continues to be identified as an important effect of NAA, carbaryl, and other regulators in fruit thinning (Williams and Batjer, 1964; Schneider, 1977). Almost certainly, the regulators weaken the sink strength of the young fruits and possibly damage the phloem of the pedicels as well.

From the foregoing, it is evident that the action of NAA on young fruit involves, first, a temporary inhibition of abscission that is presumably the result of the sprays directly augmenting the levels of endogenous auxin at the fruit abscission zone. Second, NAA causes injuries of several kinds, which result in impaired translocation and embryo abortion. A major symptom of these injuries is greatly reduced export of auxin (and presumably other growth hormones) from the young fruit to their abscission zones. The reduced auxin alone is sufficient to account for the increased abscission, but its effect is almost certainly aided by increased ETH and ABA from the injured, wilting fruit.

The physiological action of other fruit-thinning chemicals has received little attention. The few studies that have been conducted also detected changes in auxin metabolism and impaired nutrient transport (see Addicott, 1976).

2. Prevention of Young-Fruit Abscission

In a number of crops, abscission of young fruit can be excessive. For example, the first flowers produced by tomatoes growing in the field are almost invariably abscised. In other crops, such as green beans, high temperatures will induce considerable young-fruit abscission. In still others, such as pears, a late frost in the spring will induce abscission. Further, when a plant breeder attempts to pollinate, manipulation of flowers often results in the abortion and abscission of the fruit which the breeder is most interested in seeing develop to maturity. Most of these difficulties can be at least partially overcome by the use of auxin-regulators, GA, or CK. In at least one situation, foliar sprays of zinc [sulfate ?] greatly increased fruit set of litchi (Hoda and Syamal, 1975). The sprays probably increased auxin levels within the trees and vigor, generally. When days are short and light intensity is low, greenhouse-grown tomatoes commonly abscise most of their young fruit. Similarly, field-grown tomatoes usually abscise the young fruit of the first flower clusters that develop when weather is still cool (see Wittwer, 1954). Under such circumstances, auxin-regulators are used commercially to increase the number of fruit set. The present practice in southern California is to spray 25 mg/l 4-CPA on the first flower cluster, four or five times, at ten-day intervals (B. J. Hall, p.c.).

Green beans (and related varieties of *Phaseolus vulgaris*) abscise most of the flowers and young fruit during periods of hot, dry weather. Under these conditions, also, the auxin regulators can reduce losses from young fruit abscission. 4-CPA at 0.4–0.8 g/ha has given very good results; NAA, NOXA, and α-(2-chlorophenoxy)propionic acid have also been effective (Wittwer, 1954).

GA and CK have also increased fruit set in various experiments. One or both of these are effective on apples, Calimyrna figs, *Citrus*, grapes, tomatoes, muskmelons, and *Pinus* (e.g., Rappaport, 1957; Hield et al., 1958; Crane, 1969; Krezdorn, 1969; Dennis, 1973; Sweet, 1973; Weaver, 1973; Wertheim, 1973). On pears, GA has greatly improved fruit set of frost-damaged flowers (Gyuró et al., 1978; Flick and Hermann, 1978). On the apple cultivar Cox's Orange Pippin, a spray containing GA, DPU, and NOXA greatly increased retention of young fruit from unpollinated flowers, and reduced the abscission of young fruit from pollinated flowers (Goldwin,

1978). On a self-sterile sour cherry, sprays containing GA and 2,4,5-TP stimulated fruit growth (and prevented abscission) so successfully that a profitable crop was produced (Kökéndyné-Inántsy et al., 1978).

Regulator chemicals have proven of great value where pollination and fertilization are essential for the success of a research investigation. It is common knowledge that flowers manipulated by emasculation and pollination are much more likely than untouched flowers to be abscised. Also, in the case of some kinds of pollen incompatibilities, the ovary may be abscised before the slowly growing pollen tube can reach the embryo sac. Such unwanted abscission can be prevented or at least greatly reduced by application of regulator chemicals at the time of pollination (Emsweller, 1954). For example, NAA (0.0025% in lanolin) applied to tomato ovaries at the time of hand pollination reduced young-fruit abscission and enabled double the number of seeds to be harvested from a cross than would otherwise be possible (C. M. Rick, p.c.). With cotton, a spray of GA (50–100 mg/l) is applied to the flower at the time of hand pollination. The treatment is highly effective in assuring the development of young fruit until they are past the period of abscission (V. T. Walhood, p.c.).

In preventing abscission, the hormone regulators appear to act by making the fruit a better sink, better able to attract nutrients from the rest of the plant (see Crafts and Crisp, 1971; Rodgers, 1980). Conversely, if vigorous vegetative growth is prevented at the time of young-fruit abscission, more young fruit are retained. That can be accomplished by pruning or by the use of chemical growth retardants (Farrington and Pate, 1981a). In either case, the vigorous competition of vegetative tissues is reduced to the benefit of fruit growth and retention (Luckwill, 1970). As a result, the treated fruit are more vigorous and better able to inhibit their own abscission. Presumably, inhibition is accomplished by greater production of growth hormones, particularly auxin (see Addicott, 1970). However, it is important to note that when all sinks of a plant are treated simultaneously, little is gained. For example, with both cotton and peaches, when the entire plant was sprayed with GA, there was some increase in number of fruit set, but the total yield of the plant was not different from the controls (e.g., Stembridge and Gambrell, 1972).

E. FRUIT DEHISCENCE AND SEED ABSCISSION

In nature, many plants shed their seeds as a result of the dehiscence of pods and abscission of seeds from the pods. Commonly, this dehiscence and abscission occur gradually during the dry season as the plants slowly become desiccated. Such a strategy of seed abscission has the advantage of releasing seeds for dissemination over an extended period of time. In contrast, in cultivated plants it is desirable that there be a minimum of seed abscission prior to harvest, but that dehiscence and abscission occur promptly after harvest. Most of the established cultivars of seed crops have been selected for this property and abscise relatively few seeds prematurely. However, from time to time, premature seed "shattering" is encountered. Attempts to prevent such shattering by the use of regulators have met with limited success; e.g., a spray of NAA increased seed retention of a new cultivar of *Phalaris tuberosa* from 50 percent to 76 percent (Mullett, 1966). Unfortunately, subsequent large-scale trials gave such variable results that no commercial practice could be established. In that case, and in most similar cases—e.g., veldtgrass (*Ehrharta calycina*)—non-shattering forms are eventually found and established within the cultivar (Love, 1963; R. N. Oram, p.c.).

Many legumes are strongly inclined to shatter seed with the explosive dehiscence of the pods. In some instances—e.g., alfalfa grown for seed—the careful coordination of desiccant chemical application and harvest operations can greatly reduce losses from shattering. Harvest can be accomplished successfully by combine after the foliage is dry but before the pods have opened. Desiccants used for this purpose include diquat, dinoseb, endothall, hexachloroacetone, and petroleum solvents (Hoover and Walhood, 1974; Pedersen et al., 1972).

With walnuts (*Juglans*) and pecans (*Carya*), preparation of harvested fruit for market can often be expedited by accelerating the dehiscence and abscission of the exocarp (fruit hull or shuck). ETH gas, ethephon, and dikegulac have been used for this purpose (Sorber and Kimball, 1950; Olson et al., 1977).

In cotton, dehiscence of the fruits is often retarded when the plants are leafy and moisture is retained in the vicinity of the fruits. Dehiscence and opening of the fruits is facilitated by the use of chemical defoliants. These remove the leaves, thereby reducing sources of moisture in the vicinity of the fruits; and in addition, the defoliant chemicals speed the drying out and curling back of the dehiscing fruit walls. The drying out also reduces the incidence of fungi that damage fiber in the partly opened fruits, and enables simultaneous harvest of the cotton from all the fruits.

F. MATURE-FRUIT ABSCISSION

1. Prevention

Some cultivars of apples, pears, prunes, oranges, grapefruit, lemons, and other fruits have a tendency to abscise full-sized fruits shortly before they are ready for harvest. This affliction, preharvest drop, can cause heavy losses of the marketable crop (Fig. 10:4). Following the discovery that auxin could retard abscission of debladed petioles, Gardner et al. (1939)

302 *Basics of Agricultural Abscission*

Fig. 10:4.
Eight-year-old apple tree (cv. McIntosh), which has lost 50 percent of its crop in preharvest drop. This abscission can be prevented by sprays of appropriate growth-regulator chemicals. (Photograph courtesy of C. G. Forshey)

showed that the auxin-regulator NAA and its derivatives could delay the abscission of apples long enough for them to fully mature and be harvested normally. These and related regulators now find wide usage on apples and pears, as well as a number of other fruit crops. These regulators include 2,4-D, MCPA, and 2,4,5-TP. With some *Citrus* varieties, GA is used in conjunction with 2,4-D because of the ability of GA to retard *Citrus* fruit maturation. The treatment both delays abscission and prolongs storage life (Coggins, 1973). In the oil palm (*Elaeis guineensis*), fruit abscission was delayed by application of GA, NAA, 2,4,5-TP, or ethephon (Chan et al., 1972). Retardation of abscission by ethephon is an unexpected result, as the ETH released by ethephon almost invariably promotes abscission. The dosage in this case may have been somewhat phytotoxic. With peaches, ETH production and fruit abscission were retarded by holding the fruit in an atmosphere of 15% CO_2 (Jerie, 1976).

It has recently been found that the growth retardant daminozide applied early in the season can substantially reduce the preharvest drop of apples

(Batjer and Williams, 1966) and of pears. On pears, effective applications can be made any time from shortly after flowering until a month before harvest (Martin and Griggs, 1970), although application late in the season can lead to the undesirable side effect of increased fruit set the next season (Edgerton, 1971). The responses to daminozide appear to be due, at least in part, to retardation of vegetative growth and consequent higher levels of mineral and other nutrients for the developing fruit. The more vigorous fruit would also be higher in hormones. A promising recent development is the delay of abscission of mature sour cherries by application of Ca^{2+} (as $CaCl_2$, Stösser, 1975).

2. Promotion

Man has been harvesting fruits and nuts since his advent on the face of the earth. From time to time, he has attempted to develop methods and devices to facilitate and expedite fruit harvest. In spite of these efforts, which must extend back over a period of at least a few million years, there was relatively little progress until quite recently. Now, considerable research is in progress in an attempt to develop new chemical abscission agents and to coordinate their use with mechanical devices. One can hope that this current research activity will be more successful than past efforts.

Rising costs of fruit harvest have stimulated the intensive research directed to increasing harvest efficiency. Mechanical devices are now available that are capable of shaking fruit from the trees if the abscission processes are fairly well advanced. However, with many cultivars, the fruit are not yet abscising at the optimum time for harvest and cannot be removed by the limited degree of shaking that trees can tolerate.

The requirements of a successful promotor of fruit abscission are so rigorous that only a few of the many chemicals tested have shown any indication of becoming commercially acceptable. Such a regulator must meet the obvious requirements of reasonable cost and practicability of application, and must not affect the quality of the fruit nor leave a toxic residue. But perhaps the most stringent requirement is that the regulator must be capable of inducing the abscission of a substantial number of fruit with a minimum of other adverse side effects, such as leaf abscission (Fig. 10:5). None of the regulators examined to date fully meet the requirements except in restricted circumstances. What is known of their action is discussed in the following paragraphs.

For some time, ETH has been used to promote abscission in various circumstances where plant materials can be enclosed for the day or more required for treatment. Such practices cannot be employed in orchards, but they have been used to speed dehiscence of outer layers of the fruit (shucks) from walnuts and pecans (Sorber and Kimball, 1950). The development of agricultural chemicals that break down readily and release ETH (e.g., eth-

Fig. 10:5.
Mature navel oranges induced to abscise by spray of erythorbic acid. (Photograph courtesy of R. M. Burns)

ephon, Alsol) was a major step forward. In effect, these substances bathe the treated plants with ETH and release some within the plant tissues as well. They have now been widely tested, particularly where responses similar to those induced by ETH are desired (de Wilde, 1971). The chemicals have been notably effective in promoting the abscission of mature apples, cherries, plums, prunes, olives, oranges, and coffee berries (de Wilde, 1971; Oyebade, 1976). Farmers growing such crops can harvest them efficiently by carefully timing their work with the advance of the induced abscission in the fruits. The induced abscission is especially helpful when the fruits are to be harvested with the use of mechanical tree-shaking devices (Al-Jaru and Stösser, 1973a, b; Wilson, 1978; Edgerton, 1973). Unfortunately, ethephon and Alsol have a number of adverse side effects, such as induction of leaf abscission and abnormal vegetative growth. When used to accelerate the coloring of fruit, ethephon often has the undesirable side effect of increasing preharvest (premature) fruit abscission (e.g., Child, 1978). If ways could be found to prevent the side effects, the compounds would find extensive application. Encouraging results have recently come from the combination of Ca^{2+} with Alsol applied to olives. In preliminary tests the Ca^{2+} completely eliminated the side reactions (G. C. Martin, p.c.).

CHI has also been tested extensively for the promotion of mature fruit abscission. It is an antibiotic that inhibits protein synthesis, but when it is applied as an agricultural chemical, effective dosages are somewhat toxic. It has a strong tendency to induce leaf abscission and to injure the surface of fruits. Such injury is tolerable on oranges produced for juice, and CHI has found considerable use on that crop. A number of chemicals have been tested with CHI in order to induce a fully satisfactory abscission response and reduce the side effects. The injury to fruit from CHI results in a prolonged release of wound ETH. The amount of such ETH produced is proportional to the abscission response, and it is generally accepted that the abscission induction from CHI is through ETH production (Cooper and Henry, 1974; Holm and Wilson, 1977; Wilson, 1978). For example, the addition of GA to CHI applied to *Citrus* inhibits the abscission response, whereas ABA promotes it (Cooper and Henry, 1974; and see Addicott, 1976).

The available chemicals are still almost completely ineffective on some species. High dosages of ethephon and of CHI failed to increase cone abscission of *Pinus taeda* or *P. elliottii*, yet caused considerable needle abscission and branch injury (McLemore, 1973).

G. BRANCH ABSCISSION

Many forest trees do not abscise their branches readily, especially when young (see Chapter 3). If methods for the abscission of lower branches could be developed, they would be valuable tools in timber production. There is some reason to hope that self-pruning (branch-abscising) varieties of forest trees can be selected. Differences with respect to branch-abscission habits within and among species have already been noted, but utilization of these tendencies will be many years in development. For most timber species, branch abscission is the result of death of the shaded and otherwise disadvantaged lower branches, followed by their decay and eventual falling away from the tree (Kozlowski, 1973). The fewer the number of persistent lower branches, the more valuable the tree is for lumber. Pruning of lower branches can be effected by hand tools, by careful use of fire, and by herbicidal chemicals (Millington and Chaney, 1973). Lower branches of oak and other hardwood trees can be readily killed by sprays of 2,4-D and 2,4,5-T. After such sprays, branches up to 5 cm in diameter decayed and fell away in four to seven years (MacConnell and Kenerson, 1964). An interesting experiment on twig abscission of *Quercus alba* was performed by Chaney and

Leopold (1972). They applied ethephon in lanolin to the base of twigs on a mature tree, and the treatment doubled the number of twigs abscised by the tree. Twigs of juvenile trees did not respond to the ethephon treatment. Also, *Juglans nigra* did not respond to such treatments (W. R. Chaney, p.c.).

H. BARK REMOVAL

Bark abscission occurs widely among tree species (see Chapter 3), but as it occurs in nature it is only the outer layers of dead tissues that are abscised. In the harvest of trees for wood or timber, it is often advantageous to remove the entire bark, including the living tissues of phloem, right down to the cambium. Such removal eliminates tissues that foster decay and insect attack and enables the wood to dry, lowering the weight of a log by as much as one-third (Kozlowski, 1973).

In chemical bark removal, a herbicide is applied to the sapwood in a band around the base of the tree after cutting away a ring of bark. The herbicide moves upward in the trunk, initially at least, in the sapwood. Ultimately, the cambium and adjacent thin-walled cells collapse and permit the bark to separate from the more mature sapwood (Kozlowski, 1960). The intensity of the response varies with many factors, such as season of application and tree vigor, and not all tree species are capable of a commercially acceptable response. The most satisfactory herbicides for bark removal have been arsenic compounds, especially sodium arsenite (Wilcox et al., 1956). Other herbicides, such as 2,4-D and 2,4,5-T, may be satisfactory in some circumstances (Raphael et al., 1954).

Epilogue

In this book I have surveyed the phenomena of abscission and attempted to give the reader an appreciation of the scope and variety of abscission manifestations, as well as to outline the mechanisms of the process as they are presently understood. The survey has disclosed some remarkable aspects of abscission that could hardly have been appreciated without some systematic attempt to gather and relate innumerable small pieces of information.

One of the first things that may strike the reader is how widespread the phenomena of abscission are in the living world. In higher plants themselves, there is ample evidence that each discrete organ and structure can be abscised by one or more, usually many more, species. Further, abscission is also widespread in the lower plants; its manifestations vary with the structural complexity of the plant groups; nevertheless, it occurs extensively, particularly in relation to the functions of propagation. In passing, it may be noted that abscission is an important aspect of the life of animals and is utilized in a variety of ways by both higher and lower animals; significantly, abscission in animals appears to involve much the same kind of biochemical change as in plants.

Another remarkable aspect of abscission is the variety of benefits for which plants utilize abscission. These range from its conspicuous involvement in propagation and dissemination of the species to what might be called sanitation, including the removal of organs and structures whose presence is no longer needed or whose retention may actually be hazardous to the plant.

The paleontological evidence of abscission is much less accessible than one would like, but the conclusion is still inescapable that abscission has been a part of plant life from the earliest geologic times. There is no question but that leaf and branch abscission, and abscission of propagules of various sorts, were well established by the Carboniferous. What is most impressive is the knowledge that the earliest evidence of the existence of cellular organisms in the Precambrian also includes evidence that these cells were dividing and separating. Study of the entire paleontological record, as well as abscission in the diverse present-day forms, indicates that the same basic biochemical changes are acting today to separate the cells in a leaf abscission zone as were acting over 3 billion years ago to separate the cells of a chain of microorganisms!

To a plant physiologist, it is indeed impressive that a basically simple process like abscission is sensitive to so many internal and external factors. It would be difficult, if not impossible, to name a physiological or ecological factor that has not been observed to influence abscission in one way or another. The ability of plants to respond to changes in such factors in some effective way undoubtedly accounts for the survival both of individuals and of species. Abscission is one of the important mechanisms that plants have utilized to maintain homeostasis in the face of a changing environment.

Abscission appears to be an unusually favorable subject for research, particularly physiological and biochemical research. The process of abscission can be easily regulated and manipulated. For precise experiments, abscission zones can be excised and studied under controlled conditions. One of the most attractive aspects of research on abscission is the relative simplicity of the end point, involving the production and activity of one, or at most a few, cell-wall-degrading enzymes.

While the overview of the process of abscission and its control which emerges from this book is in many ways very satisfying, there are still some aspects that are far from clear and others that call for reinvestigation by modern methods and techniques. Among these topics, certainly the biochemistry and cytology of enzyme production and secretion need much wider investigation to confirm the strong indications now in hand. Corollary investigations of the "retightening" of separation layers offer an opportunity to study the balance between degradative and synthetic activities in the cell walls of plants. Retightening of loosened *Citrus* fruits and cotton leaves is now well known and would provide excellent research material.

Other biochemical aspects of abscission are much in need of further study and clarification. What is the role of non-pectic, non-cellulosic polysaccharides—i.e., those components of the primary cell wall attacked by various "cellulases"? Are true cellulases and cellulose involved at all in abscission? Is lignase involved in abscission of woody tissues and structures? On this question, scattered but provocative evidence suggests that lignases *are* involved, but the evidence is still fragmentary.

While the numerous observations of branch abscission suggest that it can be much like leaf abscission, the various abscission-related changes at the base of woody twigs and branches call for thorough investigation by modern methods.

The process of separation of vascular tissues of leaves, petals, and similar structures has never been carefully investigated. Recent workers often repeat uncritically the unsupported assertions of earlier workers. A careful and critical investigation of this aspect of abscission is needed.

The most fundamental problem of abscission has as yet received almost no attention. The question "What physiological and morphogenetic factors determine the location and sensitivity of abscission zones?" awaits serious investigation utilizing the new tools of physiology, biochemistry, and cytology.

Finally, we can hope that physiological investigations will someday clarify the mechanism whereby the many stimuli that can affect abscission are interpreted and integrated, signaling to the abscission zone either promotion or retardation of abscission activity.

In conclusion, it is gratifying to note that our understanding of abscission has advanced substantially in recent decades and that because of widening research interest in the problems of abscission we can look forward to substantial progress in coming years.

APPENDIX
Check List of Literature on Abscisic Acid in Abscission

I. CORRELATIVE OCCURRENCE, INDICATING A HORMONAL ROLE

A. In leaves:

Acer pseudoplatanus	Eagles and Wareing, 1964
	Wareing et al., 1964
	Dörffling et al., 1978
Acer rubrum	Perry and Hellmers, 1973
Betula pubescens	Eagles and Wareing, 1964
	Wareing et al., 1964
Citrus sinensis	Monselise et al., 1967
Cochorus olitorius	Sen, 1968
Coleus blumei	Chang and Jacobs, 1973
Coleus rehneltianus	Böttger, 1970b
Euonymus alatus	Hemphill and Tukey, 1973
Euonymus japonica	Osborne et al., 1972
Glycine max	Samet and Sinclair, 1980
Malus hupehensis	Weinbaum and Powell, 1975
Phaseolus vulgaris	Osborne et al., 1972
	Dörffling et al., 1978
Populus tremula	Eliasson, 1969
Rosa rugosa	Tomaszewska and Tomaszewski, 1970
Streptocarpus molweniensis	Bornman, 1969

B. In branches:

Spirodela	Witztum and Keren, 1978b

C. In flowers:

Begonia hybrids	Hänisch ten Cate et al., 1975b
Citrus sinensis	Goldschmidt, 1980

312 Appendix

Hibiscus rosa-sinensis	Swanson et al., 1975
Lupinus luteus	Porter, 1977
Rosa sp.	Mayak et al., 1972

D. In young fruit:

Citrus unshiu	Takahashi et al., 1975
Diospyros kaki	Yamamura and Naito, 1975
Gossypium hirsutum	Addicott et al., 1964
	Carns, 1958
	Cognée, 1975
	Davis and Addicott, 1972
	Rodgers, 1980
	Varma, 1977
	Vaughan and Bate, 1977
Lupinus luteus	Rothwell and Wain, 1964
	van Steveninck, 1959a, b
Phaseolus vulgaris	Tamas et al., 1979
Prunus persica	Yamaguchi and Takahashi, 1976
Pyrus communis	Martin et al., 1980

E. In mature fruit:

Carya illinoensis	Lipe et al., 1969
Citrus sinensis	Goldschmidt et al., 1972, 1973
	Rasmussen, 1975
Fragaria sp.	Rudnicki, Pieniążek, and Pieniążek, 1968
Lycopersicon esculentum	V. Sjut and F. Bangerth (unpubl.)
Malus sylvestris	Dörffling et al., 1978
	Rudnicki and Pieniążek, 1970
Prunus domestica	Shaybany et al., 1977
Prunus persica	Yamaguchi and Takahashi, 1976
Pyrus communis	Rudnicki, Machnik, and Pieniążek, 1968
Vitis vinifera	Coombe, 1973

F. In fruit dehiscence:

Carya illinoensis	Lipe et al., 1969
Gossypium hirsutum	Davis and Addicott, 1972
	Rodgers, 1980

Note: Stress-induced increases in ABA are correlated with accelerated leaf and fruit abscission.

II. ABSCISSION IN RESPONSE TO APPLIED ABA

A. By leaves, debladed petioles, or explants:

Acer rubrum	Perry and Hellmers, 1973
Ailanthes glandulosa	El-Antably et al., 1967
Betula pubescens	Eagles and Wareing, 1964
Citrus sinensis	Cooper and Henry, 1967
	Sagee et al., 1980
Coleus rehneltianus	Böttger, 1970a
Glycine max	Sloger and Caldwell, 1970
Gossypium hirsutum	Addicott et al., 1964
	Smith et al., 1968
Hevea brasiliensis	Sethuraj (unpubl.)
Malus sylvestris	Larsen, 1969
	Edgerton, 1971
	Pieniążek, 1971
Olea europaea	Hartmann et al., 1968
Perilla ocymoides	El-Antably et al., 1967
Phaseolus vulgaris	Craker and Abeles, 1969
	Osborne et al., 1972
Prunus mazzard	Larsen, 1969
Prunus persica	Lipe, 1966
Pyrus communis	Larsen, 1969
Ribes nigrum	El-Antably et al., 1967
Salix viminalis	Saunders et al., 1974

B. By branches:

Lemna minor	Schwebel, 1973
Spirodela oligorhiza	Witztum and Keren, 1978b

C. By flowers:

Begonia hybrid	Hänisch ten Cate and Bruinsma, 1973a, b
Gossypium hirsutum	Varma, 1976
Linum lewisii	Addicott, 1977
	Wiatr, 1978

      Lupinus luteus   Porter, 1977
      Rosa sp.   Mayak and Halevy, 1972
      Vitis vinifera   Weaver and Pool, 1969

D. By fruit:
      Citrus sinensis   Cooper and Henry, 1974
      Gossypium hirsutum   Addicott et al., 1964
                                           Varma, 1976
      Linum lewisii   Addicott, 1977
      Malus sylvestris   Edgerton, 1971
      Prunus cerasus   Zucconi et al., 1969
      Prunus persica   Ga'ash and Lavee, 1973
      Vitis vinifera   Weaver and Pool, 1969

E. Under hypobaric conditions; i.e., minimal ETH:
      Citrus fruit   Cooper and Horanic, 1973
                                           Rasmussen, 1974
      Gossypium explants   Marynick, 1977

III. BASIC ABA RESPONSES PROBABLY INVOLVED IN PROMOTION OF ABSCISSION

A. Increased cellulase activity in abscission zone:
                                      Craker and Abeles, 1969

B. Increased permeability to water and cations:
                                      Reed and Bonner, 1974
                                      van Steveninck, 1972

C. In general: Promotion of the "senescence pattern" of enzyme secretion:
                                      Addicott, 1972

Literature Cited

ABELES, F. B. 1967. Mechanism of action of abscission accelerators. Physiol. Plant. 20:442–54.
———. 1969. Abscission: role of cellulase. Plant Physiol. 44:447–52.
———. 1973. Ethylene in Plant Biology. Academic Press, New York.
Abeles, F. B., and R. E. HOLM. 1967. Abscission: role of protein synthesis. Ann. N.Y. Acad. Sci. 144:367–73.
Abeles, F. B., G. R. LEATHER, L. E. FORRENCE, and L. E. CRAKER. 1971. Abscission: regulation of senescence, protein synthesis, and enzyme secretion by ethylene. HortScience 6:371–76.
Abu-SHAKRA, S. S., D. A. PHILLIPS, and R. C. HUFFAKER. 1978. Nitrogen fixation and delayed leaf senescence in soybeans. Science 199:973–75.
ADATO, I., and S. GAZIT. 1977. Role of ethylene in avocado fruit development and ripening. I. Fruit drop. J. Exp. Bot. 28:636–43.
ADDICOTT, F. T. 1945. The anatomy of leaf abscission and experimental defoliation in guayule. Am. J. Bot. 32:250–56.
———. 1965. Physiology of abscission. Pp. 1094–1126 in W. Ruhland (ed.), Encyclopedia of Plant Physiology 15/2. Springer-Verlag, Berlin.
———. 1970. Plant hormones in the control of abscission. Biol. Rev. 45:485–524.
———. 1972. Biochemical aspects of the action of abscisic acid. Pp. 272–80 in D. J. Carr (ed.), Plant Growth Substances 1970. Springer-Verlag, Berlin.
———. 1976. Actions on abscission, defoliation and related responses. Pp. 191–217 in L. J. Audus (ed.), Herbicides. Physiology, Biochemistry, Ecology. Vol. 1. 2nd ed. Academic Press, London.
———. 1977. Flower behavior in *Linum lewisii*: Some ecological and physiological factors in opening and abscission of petals. Am. Midl. Nat. 97:321–32.
Addicott, F. T., and R. S. LYNCH. 1951. Acceleration and retardation of abscission by indoleacetic acid. Science 114:688–89.
———. 1957. Defoliation and desiccation: Harvest-aid practices. Adv. Agron. 9:67–93.
Addicott, F. T., and J. L. LYON. 1972. Current status of abscisin I and other abscission substances in cotton burs. Pp. 32–36. Proc. 1972 Beltwide Cotton Prod. Res. Conf. (Nat. Cotton Council, Memphis).
Addicott, F. T., and V. E. ROMNEY. 1950. Anatomical effects of *Lygus* injury to guayule. Bot. Gaz. 112:133–34.
Addicott, F. T., and S. M. WIATR. 1977. Hormonal controls of abscission: biochemical and ultrastructural aspects. Pp. 249–57 in P. E. Pilet (ed.), Plant Growth Regulation. Springer-Verlag, Berlin.

Addicott, F. T., R. S. LYNCH, and H. R. CARNS. 1955. Auxin gradient theory of abscission regulation. Science 121:644–45.
Addicott, F. T., R. S. LYNCH, G. A. LIVINGSTON, and J. K. HUNTER. 1949. A method for the study of foliar abscission *in vitro*. Plant Physiol. 24:537–39.
Addicott, F. T., H. R. CARNS, J. L. LYON, O. E. SMITH, and J. L. McMEANS. 1964. On the physiology of abscisins. Pp. 687–703 in J. P. Nitsch (ed.), Régulateurs Naturels de la Croissance Végétale. Cent. Natl. Rech. Sci., Paris.
ADEDIPE, N. O., and D. P. ORMROD. 1975. Absorption of foliar-applied ^{32}P by successive leaves, and distribution patterns in relation to early fruiting and abscission in the cowpea (*Vigna unguiculata* L.). Ann. Bot. 39:639–46.
AHARONI, N. 1978. Relationship between leaf water status and endogenous ethylene in detached leaves. Plant Physiol. 61:658–62.
AINSWORTH, G. C. 1971. Dictionary of the Fungi. 6th ed. Commonwealth Agricultural Bureaux, Farnham Royal, England.
ALBERSHEIM, P., and U. KILLIAS. 1963. Histochemical localization at the electron microscope level. Am. J. Bot. 50:732–45.
Albersheim, P., M. McNEIL, and J. M. LABAVITCH. 1977. The molecular structure of the primary cell wall and elongation growth. Pp. 1–12 in P. E. Pilet (ed.), Plant Growth Regulation. Springer-Verlag, Berlin.
ALEXOPOULOS, C. J., and C. W. MIMS. 1979. Introductory Mycology. 3rd ed. Wiley, New York.
Al-JARU, S., and R. STÖSSER. 1973a. Möglichkeiten der Erleichterung der maschinellen Ernte von Süss- und Sauerkirschen auf chemischem Weg. I. Der Erwerbsobstbau 15:53–55.
———. 1973b. Der Einfluss von Ethrel und Cycloheximid auf die Trenngewebeausbildung und die Reduktion der Haltekräfte bei Zwetschen und Mirabellen. Der Erwerbsobstbau 15:171–74.
ALVIM, P. de T. 1964. Tree growth and periodicity in tropical climates. Pp. 479–95 in M. H. Zimmermann (ed.), The Formation of Wood in Forest Trees. Academic Press, New York.
Alvim, P. de T., and R. ALVIM. 1978. Relation of climate to growth periodicity in tropical trees. Pp. 445–64 in P. B. Tomlinson and M. H. Zimmermann (eds.), Tropical Trees as Living Systems. Cambridge Univ. Press, Cambridge.
ALVIM, R., P. de T. ALVIM, R. LORENZI, and P. F. SAUNDERS. 1974. The possible role of abscisic acid and cytokinins in growth rhythms of *Theobroma cacao* L. Rev. Theobroma 4:3–12.
Alvim, R., E. W. HEWETT, and P. F. SAUNDERS. 1976. Seasonal variation in the hormone content of willow. I. Changes in abscisic acid content and cytokinin activity in the xylem sap. Plant Physiol. 57:474–76.
ARNOLD, C. A. 1938. Paleozoic seeds. Bot. Rev. 4:205–34.
ASHRI, A., and G. LADIJINSKI. 1964. Anatomical effects of the capsule dehiscence alleles in sesame. Crop Sci. 4:136–38.
AUDUS, L. J. 1972. Plant Growth Substances. Vol. 1. Chemistry and Physiology. 3rd ed., Barnes & Noble, New York.
AVERY, G. S., JR., and L. POTTORF. 1945. Auxin and nitrogen relationships in green plants. Am. J. Bot. 32:666–69.

Avery, G. S., Jr., P. R. BURKHOLDER, and H. B. CREIGHTON. 1937. Nutrient deficiencies and growth hormone concentration in *Helianthus* and *Nicotiana*. Am. J. Bot. 24:553–57.
AWAD, M., and R. E. YOUNG. 1979. Postharvest variation in cellulase, polygalacturonase, and pectinmethylesterase in avocado (*Persea americana* Mill, cv. Fuerte) fruits in relation to respiration and ethylene production. Plant Physiol. 64:306–8.
AWRAMIK, S. M., J. W. SCHOPF, M. R. WALTER, and R. BUICK, 1981. Filamentous fossil bacteria 3.5×10^{-9} years-old from the Archean of Western Australia. (unpubl.)
AXELROD, D. I. 1966. Origin of deciduous and evergreen habits in temperate forests. Evolution 20:1–15.
BAILEY, D. K. 1970. Phytogeography and taxonomy of *Pinus* subsection *Balfourianae*. Ann. Mo. Bot. Gard. 57:210–49.
BAILEY, L. H. 1935. The Standard Cyclopedia of Horticulture. Macmillan, New York.
BAKER, K. L., G. R. HOOPER, and E. S. BENEKE. 1977. Ultrastructural development of merosporangia in the mycoparasite *Syncephalis sphaerica* (Mucorales). Can. J. Bot. 55:2207–15.
BALLINGER, W. E., H. K. BELL, and N. F. CHILDERS. 1966. Peach nutrition. Pp. 276–390 in N. F. Childers (ed.), Temperate to Tropical Fruit Nutrition. Somerset Press, Somerville, N.J.
BANKS, H. P., and M. R. DAVIS. 1969. *Crenaticaulis*, a new genus of Devonian plants allied to *Zosterophyllum*, and its bearing on the classification of early land plants. Am. J. Bot. 56:436–49.
BARLOW, H. W. B. 1950. Studies in abscission. I. Factors affecting the inhibition of abscission of synthetic growth-substances. J. Exp. Bot. 1:264–81.
BARNARD, C. 1926. Preliminary note on branch fall in the Coniferales. Proc. Linn. Soc. N.S.W. 51:114–28.
BARNARD, P. D. W. 1973. Mesozoic floras. Spec. Pap. Palaeontol. 12:175–87.
BARNES, B. V. 1969. Natural variation and delineation of clones of *Populus tremuloides* and *P. grandidentata* in northern Lower Michigan. Silvae Genet. 18:130–42.
BÄSECKE, P. 1908. Beiträge zur Kenntnis der physiologischen Scheiden der Achsen und Wedel der Filicinen. Bot. Zeitg. 66:25–87.
BATCHELDER, D. G., J. G. PORTERFIELD, and G. McLAUGHLIN. 1971. Thermal defoliation of cotton. P. 36, Proc. 25th Cotton Defoliation-Physiol. Conf. (Nat. Cotton Council, Memphis).
BATEMAN, D. F., and H. G. BASHAM. 1976. Degradation of plant cell walls and membranes by microbial enzymes. Pp. 316–55 in R. Heitefuss and P. H. Williams (eds.), Encyclopedia of Plant Physiology N.S. 4. Springer-Verlag, Berlin.
BATJER, L. P. 1954. Plant regulators to prevent preharvest fruit drop, delay foliation and blossoming, and thin blossoms and young fruits. Pp. 117–31 in H. B. Tukey (ed.), Plant Regulators in Agriculture. Wiley, New York.
Batjer, L. P., and B. J. THOMSON. 1961. Effect of 1-naphthyl N-methylcarbamate (Sevin) on thinning apples. Proc. Am. Soc. Hortic. Sci. 77:1–8.

Batjer, L. P., and M. W. WILLIAMS. 1966. Effects of N-dimethyl amino succinamic acid (Alar) on watercore and harvest drop of apples. Proc. Am. Soc. Hortic. Sci. 88:76–79.

BEAL, J. M., and A. G. WHITING. 1945. Effect of indoleacetic acid in inhibiting stem abscission in *Mirabilis jalapa*. Bot. Gaz. 106:420–31.

BECKER, D. A. 1968. Stem abscission in the tumbleweed, *Psoralea*. Am. J. Bot. 55:753–56.

———. 1978. Stem abscission in tumbleweeds of the Chenopodiaceae: *Kochia*. Am. J. Bot. 65:375–83.

BECKMAN, C. H. 1964. Host responses to vascular infection. Annu. Rev. Phytopathol. 2:231–52.

BELL, P. R., and C. L. F. WOODCOCK. 1968. The Diversity of Green Plants. Addison-Wesley Pub. Co., Reading, Mass.

BEN-YEHOSHUA, S., and R. H. BIGGS. 1970. Effects of iron and copper ions in promotion of selective abscission and ethylene production by *Citrus* fruit and the inactivation of indoleacetic acid. Plant Physiol. 45:604–7.

BERG, C. C. 1977. Abscission of anthers in *Cecropia* Loefl. Acta Bot. Neerl. 26:417–19.

BERGER, R. K., and P. D. REID. 1979. Role of polygalacturonase in bean leaf abscission. Plant Physiol. 63:1133–37.

BERJAK, P., and T. A. VILLIERS. 1970. Ageing in plant embryos. I. The establishment of the sequence of development and senescence in the root cap during germination. New Phytol. 69:929–38.

BERKLEY, E. E. 1931. Marcescent leaves of certain species of *Quercus*. Bot. Gaz. 92:85–93.

BEUKMAN, E. F., and A. C. GOHEEN. 1970. Grape corky bark. Pp. 207–9 in N. W. Frazier (ed.), Virus Diseases of Small Fruits and Grapevines. Univ. Calif. Div. Agric. Sci., Berkeley.

BEWS, J. W. 1925. Plant Forms and Their Evolution in South Africa. Longmans, Green, London.

BEYER, E. M., JR. 1972. Auxin transport: A new synthetic inhibitor. Plant Physiol. 50:322–27.

———. 1973. Abscission. Support for a role of ethylene modification of auxin transport. Plant Physiol. 52:1–5.

———. 1975. Abscission: The initial effect of ethylene is in the leaf blade. Plant Physiol. 55:322–27.

———. 1976. Silver ion: A potent antiethylene agent in cucumber and tomato. HortScience 11:195–96.

Beyer, E. M., Jr., and P. W. MORGAN. 1971. Abscission. The role of ethylene modification of auxin transport. Plant Physiol. 48:208–12.

BIALE, J. B. 1960. Respiration of fruits. Pp. 536–92 in W. Ruhland (ed.), Encyclopedia of Plant Physiology 12/2. Springer-Verlag, Berlin.

BIGGS, R. H. 1971. *Citrus* abscission. HortScience 6:388–92.

Biggs, R. H., and A. C. LEOPOLD. 1957. Factors influencing abscission. Plant Physiol. 32:626–32.

———. 1958. The two-phase action of auxin on abscission. Am. J. Bot. 45:547–51.

BLYDENSTEIN, J. 1966. Root systems of four desert grassland species on grazed and protected sites. J. Range Manage. 19:93–95.
BÖCHER, T. W. 1964. Morphology of the vegetative body of *Metasequoia glyptostroboides*. Dtsch. Bot. Ark. 24:1–70.
BÖHLMANN, D. 1970. Anatomisch-histologische Untersuchungen im Bereich der Astabzweigung bei Nadel- und Laubbäumen. Allg. Forst. Jagdztg. 141: 224–30.
BOLD, H. C., and M. J. WYNNE. 1978. Introduction to the Algae. Prentice-Hall, Englewood Cliffs, N.J.
BONDAD, N. D. 1976. Response of some tropical and subtropical fruits to pre- and post-harvest applications of ethephon. Econ. Bot. 30:67–80.
BORGER, G. A. 1973. Development and shedding of bark. Pp. 205–36 in T. T. Kozlowski (ed.), Shedding of Plant Parts. Academic Press, New York.
BORNMAN, C. H. 1967a. The relation between tylosis and abscission in cotton (*Gossypium hirsutum*) explants. S. Afr. J. Agric. Sci. 10:143–53.
———. 1967b. Some ultrastructural aspects of abscission of *Coleus* and *Gossypium*. S. Afr. J. Sci. 63:325–31.
———. 1969. Laminal abscission in *Streptocarpus*. P. 18, Abstracts. XI Intl. Botanical Congress. Seattle, Wash.
Bornman, C. H., F. T. ADDICOTT, and A. R SPURR. 1966. Auxin and gibberellin effects on cell growth and starch during abscission in cotton. Plant Physiol. 41:871–76.
Bornman, C. H., A. R. SPURR, and F. T. ADDICOTT. 1967. Abscisin, auxin, and gibberellin effects on the developmental aspects of abscission in cotton (*Gossypium hirsutum*). Am. J. Bot. 54:125–35.
Bornman, C. H., F. T. ADDICOTT, J. L. LYON, and O. E. SMITH. 1968. Anatomy of gibberellin-induced stem abscission in cotton. Am. J. Bot. 55:369–75.
Bornman, C. H., A. R. SPURR, and F. T. ADDICOTT. 1969. Histochemical localisation by electron microscopy of pectic substances in abscising tissue. J. S. Afr. Bot. 35:253–63.
BÖTTGER, M. 1970a. Die hormonale Regulation des Blattfalls bei *Coleus rehneltianus* Berger. I. Die Wechselwirkung von Indol-3-essigsaure, Gibberellin- und Abscisinsaure auf Explantate. Planta 93:190–204.
———. 1970b. Die hormonale Regulation des Blattfalls bei *Coleus rehneltianus* Berger. II. Die naturaliche Rolle von Abscisinsaure im Blattfallprozess. Planta 93:205–13.
BRADFORD, K. J., and D. R. DILLEY. 1978. Effects of root anaerobiosis on ethylene production, epinasty, and growth of tomato plants. Plant Physiol. 61:506–9.
BREMEKAMP, C. E. B. 1926. On the opening mechanism of the acanthaceous fruit. S. Afr. J. Sci. 23:488–91.
———. 1936. Over Boomen en wat daarop lijkt. De Tropische Natuur (Batavia) 25:1–11.
BRENNAN, J. R. 1966. Anatomical structure of the two abscission zones in the petiole of *Parthenocissus tricuspidata*. Am. J. Bot. 53:620.
BREWSTER, SIR DAVID. 1855. Memoirs of the Life, Writings, and Discoveries of

Sir Isaac Newton. Vol. 1. Reprinted from the Edinburgh Edition of 1855. Johnson Reprint Corp., New York (1965).
BRIAN, P. W., H. G. HEMMING, and M. RADLEY. 1955. A physiological comparison of gibberellic acid with some auxins. Physiol. Plant. 8:899–912.
Brian, P. W., J. H. P. PETTY, and P. T. RICHMOND. 1959. Effects of gibberellic acid on development of autumn colour and leaf-fall of deciduous woody plants. Nature 183:58–59.
BRODIE, H. J. 1955. Springboard plant dispersal mechanisms operated by rain. Can. J. Bot. 33:156–67.
———. 1975. The Bird's Nest Fungi. Univ. Toronto Press, Toronto.
BROWN, H. S., and F. T. ADDICOTT. 1950. The anatomy of experimental leaflet abscission in *Phaseolus vulgaris*. Am. J. Bot. 37:650–56.
BROWNING, G., and M. G. R. CANNELL. 1970. Use of 2-chloroethane phosphonic acid to promote the abscission and ripening of fruit of *Coffea arabica* L. J. Hortic. Sci. 45:223–32.
BURG, S. P. 1962. The physiology of ethylene formation. Annu. Rev. Plant Physiol. 13:265–302.
———. 1968. Ethylene, plant senescence and abscission. Plant Physiol. 43:1503–11.
BURKHOLDER, C. L., and M. McCOWN. 1941. Effect of scoring and of α-naphthyl acetic acid and amide spray upon fruit set and of the spray upon preharvest fruit drop. Proc. Am. Soc. Hortic. Sci. 38:117–20.
BURROWS, W. J., and D. J. CARR. 1969. Effects of flooding the root system of sunflower plants on the cytokinin content in the xylem sap. Physiol. Plant. 22:1105–12.
BUSHING, R. W., V. E. BURTON, and C. L. TUCKER. 1974. Dry large lima beans benefit from *Lygus* bug control. Calif. Agric. 28(5):14–15.
CABIB, E. 1975. Molecular aspects of yeast morphogenesis. Annu. Rev. Microbiol. 29:191–214.
CAMPBELL, D. H. 1930. The Structure and Development of Mosses and Ferns. 3rd ed. Macmillan, New York.
CARNS, H. R. 1951. Oxygen, respiration and other critical factors in abscission. Ph.D. dissertation. Univ. Calif., Los Angeles.
———. 1958. Present status of the abscission accelerator from young cotton bolls. P. 39, Proc. 13th Cotton Defoliation Conf. (Nat. Cotton Council, Memphis).
Carns, H. R., and M. N. CHRISTIANSEN. 1975. Influence of UV-B radiation on abscission. Climatic Impact Assessment Program Monograph 5. Part 1. Pp. 4-93 to 4-95.
Carns, H. R., F. T. ADDICOTT, and R. S. LYNCH. 1951. Some effects of water and oxygen on abscission *in vitro*. Plant Physiol. 26:629–30.
Carns, H. R., F. T. ADDICOTT, K. C. BAKER, and R. K. WILSON. 1961. Acceleration and retardation of abscission by gibberellic acid. Pp. 559–65 in R. M. Klein (ed.), Plant Growth Regulation. Iowa St. Univ. Press, Ames.
CARPENTER, W. J. 1956. The influence of plant hormones on the abscission of *Poinsettia* leaves and bracts. Proc. Am. Soc. Hortic. Sci. 67:539–44.
CARR, D. J., and W. J. BURROWS. 1967. Studies on leaflet abscission of blue lupin leaves. I. Interaction of leaf age, kinetin and light. Planta 73:357–68.

CARTER, W. 1939. Injuries to plants caused by insect toxins. Bot. Rev. 5:273–326.
CHALMERS, D. J., and B. VAN DEN ENDE. 1975. A reappraisal of the growth and development of peach fruit. Aust. J. Plant Physiol. 2:623–34.
CHALONER, W. G. 1967. Lycophyta. Pp. 434–804 in E. Boureau (ed.), Traité de Paléobotanique. Vol. II. Bryophyta, Psilophyta, Lycophyta. Masson, Paris.
———. 1968. The cone of *Cyclostigma kiltorkense* Haughton, from the Upper Devonian of Ireland. J. Linn. Soc. Bot. 61:25–36.
Chaloner, W. G. and A. SHEERIN. 1979. Devonian macrofloras. Spec. Pap. Palaeontol. 23:145–61.
CHAMBERLAIN, C. J. 1935. Gymnosperms, Structure and Evolution. Univ. Chicago Press, Chicago.
CHAN, K. W., R. H. V. CORLEY, and A. K. SETH. 1972. Effects of growth regulators on fruit abscission in oil palm, *Elaeis guineensis*. Ann. Appl. Biol. 71:243–49.
CHANDLER, W. H. 1950. Evergreen Orchards. Lea & Febiger, Philadelphia.
———. 1951. Deciduous Orchards. Lea & Febiger, Philadelphia.
CHANEY, R. W. 1950. A revision of fossil *Sequoia* and *Taxodium* in western North America based on the recent discovery of *Metasequoia*. Trans. Am. Philos. Soc. 40:171–263.
CHANEY, W. R., and A. C. LEOPOLD. 1972. Enhancement of twig abscission in white oak by ethephon. Can. J. For. Res. 2:492–95.
CHANG, Y. P., and W. P. JACOBS. 1973. The regulation of abscission and IAA by senescence factor and abscisic acid. Am. J. Bot. 60:10–16.
CHAPMAN, R. L. 1976. Ultrastructure of *Cephaleuros virescens* (Chroolepidaceae; Chlorophyta). I. Scanning electron microscopy of zoosporangia. Am. J. Bot. 63:1060–70.
CHATTAWAY, M. M. 1953. The anatomy of bark. I. The genus *Eucalyptus*. Aust. J. Bot. 1:402–33.
———. 1955. The anatomy of bark. VI. Peppermints, boxes, ironbarks, and other eucalypts with cracked and furrowed barks. Aust. J. Bot. 3:170–76.
CHATTERJEE, S., and S. K. CHATTERJEE. 1971. Analysis of abscission process in cotton (*Gossypium barbadense* L.) leaves of different maturity. Indian J. Exp. Biol. 9:485–88.
Chatterjee, S., and A. C. LEOPOLD. 1963. Auxin structure and abscission activity. Plant Physiol. 38:268–73.
CHATTERJEE, S. K., and A. C. LEOPOLD. 1964. Kinetin and gibberellin actions on abscission processes. Plant Physiol. 39:334–37.
CHILD, R. D. 1978. The interaction of growth regulators in the control of ripening and abscission of apples. Pp. 49–57 in The Effect of Interactions Between Growth Regulators on Plant Growth and Yield. British Plant Growth Regulator Group, Monograph 2, Wantage (U.K.).
CHILDERS, N. F., ed. 1966. Temperate to Tropical Fruit Nutrition. Somerset Press, Somerville, N.J.
CHIN, T.-Y., and L. BEEVERS. 1970. Changes in endogenous growth regulators in nasturtium leaves during senescence. Planta 92:178–88.
CHMIELEWSKI, T. 1968. New dominant factor with recessive lethal effect in tomato. Genet. Pol. 9(1–2):39–48.

CHOUDHURI, M. A., and S. K. CHATTERJEE. 1970. Seasonal changes in abscission responses to auxin in debladed petioles. J. Exp. Bot. 21:693–701.
CHRISPEELS, M. J. 1976. Biosynthesis, intracellular transport, and secretion of extracellular macromolecules. Annu. Rev. Plant Physiol. 27:19–38.
Chrispeels, M. J., and J. E. VARNER. 1967. Hormonal control of enzyme synthesis: on the mode of action of gibberellic acid and abscisin in aleurone layers of barley. Plant Physiol. 42:1008–16.
CHUA, S. E. 1970. The physiology of foliar senescence and abscission in *Hevea brasiliensis* Muell. Arg. Res. Arch. Rubber Res. Inst. Malaya, Doc. 63.
CLEMENS, J., and C. J. PEARSON. 1977. The effect of waterlogging on the growth and ethylene content of *Eucalyptus robusta* Sm. (swamp mahogany). Oecologia (Berl.) 29:249–55.
CLOWES, F. A. L., and B. E. JUNIPER. 1968. Plant Cells. Blackwell Scientific Publications, Oxford.
COGGINS, C. W., JR. 1973. Use of growth regulators to delay maturity and prolong shelf life of *Citrus*. Acta Hortic. 34(1):469–72.
COGNÉE, M. 1975. Variation in the Physiological and Hormonal States of Cotton Fruit and Their Relationship with the Initiation of Abscission. D.Sc. dissertation. Univ. Paris. Translated from French by the Al Ahram Center for Scientific Translations and E. E. King, 1978.
COHEN, L. I. 1967. The pathology of *Hypodermella laricis* on larch, *Larix occidentalis*. Am. J. Bot. 54:118–24.
COKE, L., and J. WHITTINGTON. 1968. The role of boron in plant growth. IV. Interrelationships between boron and indol-3yl-acetic acid in the metabolism of bean radicles. J. Exp. Bot. 19:295–308.
CONSTABLE, G. A., and A. B. HEARN. 1978. Agronomic and physiological responses of soybean and sorghum crops to water deficits. I. Growth, development and yield. Aust. J. Plant Physiol. 5:159–67.
COOK, A. A. 1975. Diseases of Tropical and Subtropical Fruits and Nuts. Hafner Press, New York.
COOK, J. A., C. D. LYNN, and J. J. KISSLER. 1960. Boron deficiency in California vineyards. Am. J. Enol Vitic. 11:185–94.
COOMBE, B. G. 1973. The regulation of set and development of the grape berry. Acta Hortic. 34(1):261–73.
COOPER, W. C., and W. H. HENRY. 1967. The acceleration of abscission and coloring of citrus fruit. Proc. Fla. State Hortic. Soc. 80:7–14.
———. 1968. Effect of growth regulators on the coloring and abscission of *Citrus* fruit. Isr. J. Agric. Res. 18:161–74.
———. 1974. Abscisic acid, auxin, and gibberellic acid in wound-ethylene production and abscission of fruit of *Citrus sinensis* (L.) Osbeck on two rootstocks. Pp. 1052–61 in Y. Sumiki (ed.), Plant Growth Substances 1973. Hirokawa Publishing, Tokyo.
Cooper, W. C., and G. HORANIC. 1973. Induction of abscission at hypobaric pressures. Plant Physiol. 51:1002–4.
Cooper, W. C., G. K. RASMUSSEN, B. J. ROGERS, P. C. REECE, and W. H. HENRY. 1968. Control of abscission in agricultural crops and its physiological basis. Plant Physiol. 43:1560–76.
Cooper, W. C., and P. C. REECE. 1969. Effect of chilling on ethylene production,

senescence, and abscission in leaves of evergreen and deciduous fruit trees. Proc. Fla. State Hortic. Soc. 82:270–73.
COSTERTON, J. W., G. G. GEESEY, and K.-J. CHENG. 1978. How bacteria stick. Sci. Am. 238:86–95.
CRAFTS, A. S. 1968. Water deficits and physiological processes. Pp. 85–133 in T. T. Kozlowski (ed.), Water Deficits and Plant Growth. Vol. II. Academic Press, New York.
Crafts, A. S., and C. E. CRISP. 1971. Phloem Transport in Plants. Freeman, San Francisco.
CRAKER, L. E., and F. B. ABELES. 1969. Abscission: role of abscisic acid. Plant Physiol. 44:1144–49.
CRANE, J. C. 1964. Growth substances in fruit setting and development. Annu. Rev. Plant Physiol. 15:303–26.
———. 1965. The chemical induction of parthenocarpy in the Calimyrna fig and its physiological significance. Plant Physiol. 40:606–10.
———. 1969. The role of hormones in fruit set and development. HortScience 4:8–11.
Crane, J. C., I. Al-SHALAN, and R. M. CARLSON. 1973. Abscission of pistachio inflorescence buds as affected by leaf area and number of nuts. J. Am. Soc. Hortic. Sci. 98:591–92.
CROCKER, W. 1948. Growth of Plants. Reinhold, New York.
CROWE, A. D. 1965. Effect of thinning sprays on metabolism of growth substances in the apple. Proc. Am. Soc. Hortic. Sci. 86:23–27.
CURTIS, R. W. 1977. Studies on the stimulation of abscission by malformin on cuttings of *Phaseolus aureus* Roxb. Plant Cell Physiol. 18:1331–41.
———. 1978. Induction of resistance to dark abscission by malformin in white light. Plant Physiol. 62:264–66.
Curtis, R. W., and T. J. BUCKHOUT. 1977. Effect of malformin on the growth, development, and behavior of *Phaseolus vulgaris* L. Bot. Gaz. 138:153–58.
Curtis, R. W., and W. W. JOHN. 1975. Effect of divalent cations on malformin-induced efflux and biological activity. Plant Cell Physiol. 16:835–44.
DARBYSHIRE, B. 1971a. The effect of water stress on indoleacetic acid oxidase in pea plants. Plant Physiol. 47:65–67.
———. 1971b. Changes in indoleacetic acid oxidase activity associated with plant water potential. Physiol. Plant. 25:80–84.
DARROW, R. A. 1943. Vegetative and floral growth of *Fouquieria splendens*. Ecology 24:397–414.
DARWIN, C. 1888. The Different Forms of Flowers on Plants of the Same Species. 2nd ed., J. Murray, London.
DAUBENMIRE, R. F. 1967. Plants and Environment. Wiley, New York.
DAVENPORT, T. I., and N. G. MARINOS. 1971. Cell-separation in isolated abscission zones. Aust. J. Biol. Sci. 24:709–15.
DAVENPORT, T. L., P. W. MORGAN, and W. R. JORDAN. 1976. Stress-induced foliar abscission: no enhanced ethylene production by leaf petioles. Plant Physiol. 57(suppl):97.
———. 1977. Auxin transport as related to leaf abscission during water stress in cotton. Plant Physiol. 59:554–57.
DAVIS, L. A., and F. T. ADDICOTT. 1972. Abscisic acid: correlations with abscis-

sion and with development in the cotton fruit. Plant Physiol. 49:644–48.
DAY, F. P., JR., and C. D. MONK. 1977. Seasonal nutrient dynamics in the vegetation on a Southern Appalachian watershed. Am. J. Bot. 64:1126–39.
Dela FUENTE, R. K., and A. C. LEOPOLD. 1969. Kinetics of abscission in the bean petiole explant. Plant Physiol. 44:251–54.
DENNIS, F. G. 1973. Physiological control of fruit set and development with growth regulators. Acta Hortic. 34(1):251–59.
De WILDE, R. C. 1971. Practical applications of (2-chloroethyl)phosphonic acid in agricultural production. HortScience 6:364–70.
DOBBIE, H. B., and M. CROOKES. 1952. New Zealand Ferns. 5th ed. Whitcombe & Tombs, Auckland.
DÖRFFLING, K., M. BÖTTGER, D. MARTIN, V. SCHMIDT, and D. BOROWSKI. 1978. Physiology and chemistry of substances accelerating abscission in senescent petioles and fruit stalks. Physiol. Plant. 43:292–96.
DUGGER, W. M., and I. P. TING. 1970. Air pollution oxidants—their effects on metabolic processes in plants. Annu. Rev. Plant Physiol. 21:215–34.
DUNCAN, J. F. 1925. 'Pull roots' of *Oxalis esculenta*. Trans. Bot. Soc. Edinb. 29:192–96.
DURIEUX, A. J. B. 1975. Additional lighting of lilies (cv. "Enchantment") in the winter to prevent flower-bud abscission. Acta Hortic. 47:237–40.
DWELLE, R. B. 1975. Abscission of *Phaseolus* and *Impatiens* explants. Effects of ionizing radiation upon endogenous growth regulators and *de novo* enzyme synthesis. Plant Physiol. 56:529–34.
DYSON, J. G., and M. CHESSIN. 1961. Effect of auxins on virus-induced leaf abscission. Phytopathology 51:195.
EAGLES, C. F., and P. F. WAREING. 1964. The role of growth substances in the regulation of bud dormancy. Physiol. Plant. 17:697–709.
EAMES, A. J., and L. J. MacDANIELS. 1947. An Introduction to Plant Anatomy. 2nd ed. McGraw-Hill, New York.
EDGERTON, L. J. 1971. Apple abscission. HortScience 6:378–82.
———. 1973. Control of abscission of apples with emphasis on thinning and pre-harvest drop. Acta Hortic. 34(1):333–43.
EDWARDS, D. S. 1980. Evidence for the sporophytic status of the Lower Devonian plant *Rhynia gwynne-vaughanni* Kidston and Lang. Rev. Palaeobot. Palynol. 29:177–88.
El-ANTABLY, H. M. M., P. F. WAREING, and J. HILLMAN. 1967. Some physiological responses to d,l-abscisin (dormin). Planta 73:74–90.
ELIASSON, L. 1969. Growth regulators in *Populus tremula*. I. Distribution of auxin and growth inhibitors. Physiol. Plant. 22:1288–1301.
ELLIOTT, F. C. 1964. Defoliation with anhydrous ammonia and desiccation with arsenic acid in Texas. Pp. 8–10, Proc. 18th Cotton Defoliation-Physiol. Conf. (Nat. Cotton Council, Memphis).
Elliott, F. C., M. HOOVER, and W. K. PORTER, JR., eds. 1968. Advances in Production and Utilization of Quality Cotton: Principles and Practices. Iowa State Univ. Press, Ames.
El-ZIK, K. M., H. YAMADA, and V. T. WALHOOD. 1980. Effect of management on blooming, boll retention, and productivity of upland cotton, *Gossypium hirsutum* L. Pp. 1–4, 1980 Beltwide Cotton Prod. Res. Conf. Proc.

EMSWELLER, S. L. 1954. Plant regulators and plant breeding. Pp. 161–69 in H. B. Tukey (ed.), Plant Regulators in Agriculture. Wiley, New York.

ENGELBRECHT, L. 1971. Cytokinin activity in larval infected leaves. Biochem. Physiol. Pflanz. 162:9–27.

EPSTEIN, E. 1972. Mineral Nutrition of Plants: Principles and Perspectives. Wiley, New York.

EVANS, H. J., E. R. PURVIS, and F. E. BEAR. 1950. Molybdenum nutrition of alfalfa. Plant Physiol. 25:555–66.

EVANS, L. S., and K. F. LEWIN. 1980. Effects of simulated acid rain on growth and yield of soybeans and pinto beans. Pp. 299–308 in D. S. Shriner, C. R. Richmond, and S. E. Lindberg (eds.), Symposium on Potential Environmental and Health Effects of Atmospheric Deposition. Ann Arbor Science Pub., Ann Arbor, Mich.

EVENARI, M. 1949. Germination inhibitors. Bot. Rev. 15:153–94.

———. 1961. Chemical influences of other plants (allelopathy). Pp. 691–736 in W. Ruhland (ed.), Encyclopedia of Plant Physiology 16. Springer-Verlag, Berlin.

EWART, A. J. 1935. Disarticulation of the branches in *Eucalyptus*. Ann. Bot. 49:507–11.

FAHN, A., and E. WERKER. 1972. Anatomical mechanisms of seed dispersal. Pp. 151–221 in T. T. Kozlowski (ed.), Seed Biology. Vol. I. Academic Press, New York.

FARRAR, D. R. 1967. Gametophytes of four tropical fern genera reproducing independently of their sporophytes in the Southern Appalachians. Science 155:1266–67.

FARRINGTON, P., and J. S. PATE. 1981a. Fruit set in *Lupinus angustifolius* cv. Unicrop. I. Phenology and growth during flowering and early fruiting. Aust. J. Plant Physiol. 8:293–305.

———. 1981b. Fruit set in *Lupinus angustifolius* cv. Unicrop. II. Assimilate flow during flowering and early fruiting. Aust. J. Plant Physiol. 8:307–18.

FELICIANO, A., and R. H. DAINES. 1970. Factors influencing ingress of *Xanthomonas pruni* through peach leaf scars and subsequent development of spring cankers. Phytopathology 60:1720–26.

FENWICK, J. A., and M. F. MOSELEY. 1979. The splitting of the axes of *Eriogonum ovalifolium* Nutt. var. *nivale* (Canby) Jones. Bot. Soc. Amer. Misc. Series, Pub. 157, p. 10.

FERENBAUGH, R. W. 1976. Effects of simulated acid rain on *Phaseolus vulgaris* (Fabaceae). Am. J. Bot. 63:283–88.

FITTING, H. 1911. Untersuchungen über die vorzeitige Entblätterung von Blüten. Jahrb. f. wiss. Bot. 49:187–263.

FLETCHER, R. A., T. OEGEMA, and R. F. HORTON. 1969. Endogenous gibberellin levels and senescence in *Taraxacum officinale*. Planta 86:98–102.

FLICK, J. D., and L. HERMANN. 1978. Effects of gibberellic acid on fruit set of Passe Crassane pear. Acta Hortic. 80:143–47.

FOSTER, A. S., and E. M. GIFFORD, JR. 1974. Comparative Morphology of Vascular Plants. 2nd ed. Freeman, San Francisco.

FRANKENBERG, J. M., and D. A. EGGERT. 1969. Petrified *Stigmaria* from North America: Part I. *Stigmaria ficoides*, the underground portions of Lepidodendraceae. Palaeontogr., Abt. B, Paleophytol. 128:1–47.

FRIEND, J., and D. R. THRELFALL (eds.). 1976. Biochemical Aspects of Plant-Parasite Relationships. Annu. Proc. Phytochemical Soc. Academic Press, London.

FRITSCH, F. E. 1945. Structure and Reproduction of the Algae. Vols. I and II. Cambridge Univ. Press, London.

GA'ASH, D., and S. LAVEE. 1973. The effect of growth regulators on maturation, quality and pre-harvest drop of stone fruits. Acta Hortic. 34:449–55.

GAFF, D. F. 1971. Desiccation-tolerant flowering plants in Southern Africa. Science 174:1033–34.

GAHAGAN, H. E., R. E. HOLM, and F. B. ABELES. 1968. Effect of ethylene on peroxidase activity. Physiol. Plant. 21:1270–79.

GARDNER, F. E., P. C. MARTH, and L. P. BATJER. 1939. Spraying with plant growth substances to prevent apple fruit dropping. Science 90:208–9.

GARNER, W. W., and H. A. ALLARD. 1923. Further studies in photoperiodism, the response of the plant to relative length of day and night. J. Agr. Res. 23:871–920.

GAUR, B. K., and A. C. LEOPOLD. 1955. The promotion of abscission by auxin. Plant Physiol. 30:487–90.

GAWADI, A. G., and G. S. AVERY, JR. 1950. Leaf abscission and the so-called "abscission layer." Am. J. Bot. 37:172–80.

GERDTS, M. H., G. L. OBENAUF, J. H. LaRUE, and G. M. LEAVITT. 1977. Chemical defoliation of fruit trees. Calif. Agric. 31(4):19.

GERSANI, M., S. H. LIPS, and T. SACHS. 1980. The influence of shoots, roots, and hormones on sucrose distribution. J. Exp. Bot. 31:177–84.

GILL, A. M. 1976. Fire and the opening of *Banksia ornata* follicles. Aust. J. Bot. 24:329–35.

GILLILAND, M. G., C. H. BORNMAN, and F. T. ADDICOTT. 1976. Ultrastructure and acid phosphatase in pedicel abscission of *Hibiscus*. Am. J. Bot. 63:925–35.

GOEBEL, K. 1897. Organography of Plants. Especially of the Archegoniatae and Spermaphyta. Part I. General Organography. Authorized English Edition by I. B. Balfour. Hafner Publishing, New York. 1969.

GOLDSCHMIDT, E. E. 1974. Hormonal and molecular regulation of chloroplast senescence in *Citrus* peel. Pp. 1027–33 in Y. Sumiki (ed.), Plant Growth Substances 1973. Hirokawa Publishing, Tokyo.

———. 1980. Abscisic acid in citrus flower organs as related to floral development and function. Plant Cell Physiol. 21:193–95.

Goldschmidt, E. E., and B. LESHEM. 1971. Style abscission in the citron (*Citrus medica* L.) and other *Citrus* species: morphology, physiology, and chemical control with picloram. Am. J. Bot. 58:14–23.

Goldschmidt, E. E., S. K. EILATI, and R. GOREN. 1972. Increase in ABA-like growth inhibitors and decrease in gibberellin-like substances during ripening and senescence of *Citrus* fruits. Pp. 611–17 in D. J. Carr (ed.), Plant Growth Substances 1970. Springer-Verlag, Berlin.

Goldschmidt, E. E., R. GOREN, Z. EVEN-CHEN, and S. BITTNER. 1973. Increase in free and bound abscisic acid during natural and ethylene-induced senescence of citrus fruit peel. Plant Physiol. 51:879–82.

GOLDWIN, G. K. 1978. Improved fruit setting with plant growth hormones. Acta Hortic. 80:115–21.
GOREN, R., and M. HUBERMAN. 1976. Effects of ethylene and 2,4-D on the activity of cellulase isoenzymes in abscission zones of the developing orange fruit. Physiol. Plant. 37:123–30.
GORTER, C. J. 1964. Studies on abscission in explants of *Coleus*. Physiol. Plant. 17:331–45.
GOULD, F. W. 1968. Grass Systematics. McGraw-Hill, New York.
GOULD, R. E., and T. DELEVORYAS. 1977. The biology of *Glossopteris*: Evidence from petrified seed-bearing and pollen-bearing organs. Alcheringa 1:387–99.
GRAY, R., and J. BONNER. 1948a. An inhibitor of plant growth from the leaves of *Encelia farinosa*. Am. J. Bot. 35:52–57.
———. 1948b. Structure determination and synthesis of a plant growth inhibitor, 3-acetyl-6-methoxybenzaldehyde, found in the leaves of *Encelia farinosa*. J. Am. Chem. Soc. 70:1249–53.
GREENBERG, J., R. GOREN, and J. RIOV. 1975. The role of cellulase and polygalacturonase in abscission of young and mature Shamouti orange fruits. Physiol. Plant. 34:1–7.
GREENBLATT, G. A. 1965. Interactions of hormones and growth retardants in abscission. Ph.D. dissertation. Univ. Calif., Davis.
GREGORY, P. H. (ed.). 1974. *Phytophthora* Disease of Cocoa. Longman, London.
GRIESEL, W. O. 1954. Cytological changes accompanying abscission of perianth segments in *Magnolia grandiflora*. Phytomorphology 4:123–32.
GRUEN, H. E. 1959. Auxins and fungi. Annu. Rev. Plant Physiol. 10:405–40.
GUILLARMOD, A. J. 1976. Use of odourless carrier, a petroleum product, in preparing herbarium material. Taxon 25:219–21.
GUINN, G. 1976. Water deficit and ethylene evolution by young cotton bolls. Plant Physiol. 57:403–5.
GYURÓ, F., J. NYÉKI, M. SOLTÉSZ, and Z. TISZA. 1978. Effect of treatments with gibberellic acid on fruit setting in pear. Acta Hortic. 80:139–41.
HACKETT, W. P., R. M. SACHS, and J. DeBIE. 1972. Growing *Bougainvillea* as a flowering pot plant. Calif. Agric. 26(8):12–13.
HAGEMANN, P. 1971. Histochemische Muster beim Blattfall. Ber. Schweiz. Bot. Ges. 81:97–138.
HALEVY, A. H., and S. MAYAK, 1981. Senescence and postharvest physiology of cut flowers, Part 2. Hortic. Rev. 3:59–143.
HALEVY, G. 1974. Effects of gazelles and seed beetles (Bruchidae) on germination and establishment of *Acacia* species. Isr. J. Bot. 23:120–26.
HALL, J. L., and R. SEXTON. 1974. Fine structure and cytochemistry of the abscission zone cells of *Phaseolus* leaves. II. Localization of peroxidase and acid phosphatase in the separation zone cells. Ann. Bot. 38:855–58.
HALL, W. C., and J. L. LIVERMAN. 1956. Effect of radiation and growth regulators on leaf abscission in seedling cotton and bean. Plant Physiol. 31:471–76.
Hall, W. C., and P. W. MORGAN. 1964. Auxin-ethylene interrelationships. Pp. 727–45 in J. P. Nitsch (ed.), Régulateurs Naturels de la Croissance Végétale. Cent. Natl. Rech. Sci., Paris.
Hall, W. C., G. B. TRUCHELUT, C. L. LEINWEBER, and F. A. HERRERO. 1957.

Ethylene production by the cotton plant and its effects under experimental and field conditions. Physiol. Plant. 10:306–17.
HALLÉ, F., R. A. A. OLDEMAN, and P. B. TOMLINSON. 1978. Tropical Trees and Forests. Springer-Verlag, Berlin.
HÄNISCH ten CATE, C. H., and J. BRUINSMA. 1973a. Abscission of flower bud pedicels in *Begonia*. I. Effects of plant growth regulating substances on the abscission with intact plants and with explants. Acta Bot. Neerl. 22:666–74.
———. 1973b. Abscission of flower bud pedicels in *Begonia*. II. Interaction and time sequence of plant growth regulating substances on the abscission with explants. Acta Bot. Neerl. 22:675–80.
Hänisch ten Cate, C. H., J. Van NETTEN, J. F. DORTLAND, and J. BRUINSMA. 1975a. Cell wall solubilization in pedicel abscission of *Begonia* flower buds. Physiol. Plant. 33:276–79.
Hänisch ten Cate, C. H., J. BERGHOEF, A. M. H. Van der HOORN, and J. BRUINSMA. 1975b. Hormonal regulation of pedicel abscission in *Begonia* flower buds. Physiol. Plant. 33:280–84.
HARDIN, J. W. 1979. Patterns of variation in foliar trichomes of Eastern North American *Quercus*. Am. J. Bot. 66:576–85.
HARDWICK, R. C. 1979. Leaf abscission in varieties of *Phaseolus vulgaris* (L.) and *Glycine max* (L.) Merrill—a correlation with propensity to produce adventitious roots. J. Exp. Bot. 30:795–804.
HARRIS, T. M., W. MILLINGTON, and J. MILLER. 1974. The Yorkshire Jurassic Flora. Ginkgoales and Czekanowskiales. Vol. 14. British Museum (Natural History), London.
HARRISON, A. T., E. SMALL, and H. A. MOONEY. 1971. Drought relationships and distribution of two Mediterranean-climate California plant communities. Ecology 52:869–75.
HARTMANN, H. T., A. J. HESLOP, and J. WHISLER. 1968. Chemical induction of fruit abscission in olives. Calif. Agric. 22(7):14–16.
HARTUNG, W., and J. WITT. 1968. Über den Einfluss der Bodenfeuchtigkeit auf den Wuchsstoffgehalt von *Anastatica hierochuntica* und *Helianthus annuus*. Flora (Abt. B) 157:603–14.
HAYWARD, H. E., and C. H. WADLEIGH. 1949. Plant growth on saline and alkali soils. Adv. Agron. 1:1–38.
HEAD, G. C. 1973. Shedding of roots. Pp. 237–93 in T. T. Kozlowski (ed.), Shedding of Plant Parts. Academic Press, New York.
HEALD, F. D. 1933. Manual of Plant Diseases. 2nd ed. McGraw-Hill, New York.
HEMPHILL, D. D., JR., and H. B. TUKEY, JR. 1973. The effect of intermittent mist on abscisic acid content of *Euonymus alatus* Sieb. 'Compactus.' J. Am. Soc. Hortic. Sci. 98:416–20.
HENRICI, M. 1929. Structure of the cortex of grass roots in the more arid regions of South Africa. Sci. Bull. No. 85, Dept. Agric., Union of South Africa. Pp. 3–12.
HENRICKSON, J. 1972. A taxonomic revision of the Fouquieriaceae. Aliso 7:439–537.
HENRY, E. W. 1975. Peroxidases in tobacco abscission zone tissue. III. Ultrastructural localization in thylakoids and membrane-bound bodies of chloroplasts. J. Ultrastruct. Res. 52:289–99.

Henry, E. W., and T. E. JENSEN. 1973. Peroxidases in tobacco abscission zone tissue. I. Fine structural localization in cell walls during ethylene induced abscission. J. Cell Sci. 13:591–601.

Henry, E. W., J. G. VALDOVINOS, and T. E. JENSEN. 1974. Peroxidases in tobacco tissue. II. Time course studies of peroxidase activity during ethylene-induced abscission. Plant Physiol. 54:192–96.

HERRETT, R., H. H. HATFIELD, D. G. CROSBY, and A. J. VLITOS. 1962. Leaf abscission induced by the iodide ion. Plant Physiol. 37:358–63.

HEWITT, W. B. 1938. Leaf-scar infection in relation to the olive-knot disease. Hilgardia 12:41–71.

HIELD, H. Z., C. W. COGGINS, JR., and M. J. GARBER. 1958. Gibberellin tested on *Citrus*. Fruit set on Bearss lime, Eureka lemon, and Washington navel orange increased by treatments in preliminary investigations. Calif. Agric. 12(5):9–11.

Hield, H. Z., R. M. BURNS, and C. W. COGGINS, JR. 1964. Pre-harvest use of 2,4-D on *Citrus*. Univ. of Calif. Div. of Agr. Sci. Circular 528.

Hield, H. Z., C. W. COGGINS, JR., J. C. F. KNAPP, and R. M. BURNS. 1966. Chemical fruit thinning on mandarins and Valencia orange. Calif. Citrogr. 51:312–43.

HILDEBRAND, E. M. 1944. The mode of action of the pollenicide, Elgetol. Proc. Am. Soc. Hortic. Sci. 45:53–58.

HILLIARD, O. M., and B. L. BURTT. 1971. *Streptocarpus* an African Plant Study. Univ. Natal Press, Pietermaritzburg.

HINMAN, R. L., and J. LANG. 1965. Peroxidase-catalyzed oxidation of indole-3-acetic acid. Biochemistry 4:144–58.

HINTRINGER, A. 1927. Über die Ablösung der Samen von der Plazenta, beziehungsweise vom Perikarp. Sitz.-Ber. Akad. Wiss. Wien, math.-naturw. Kl., Abt. 1, 136:257–79.

HO, D. T-H., and J. E. VARNER. 1976. Response of barley aleurone layers to abscisic acid. Plant Physiol. 57:175–78.

HODA, M. N., and N. B. SYAMAL. 1975. Effect of zinc and growth regulators on sex, fruit formation and abscission layer in Litchi. Sci. Cult. 41:448–50.

HODGSON, R. W. 1918. An account of the mode of foliar abscission in *Citrus*. Univ. Calif. Publ. Bot. 6:417–28.

HOFFMAN, H. 1964. Morphogenesis of bacterial aggregations. Annu. Rev. Microbiol. 18:111–27.

HOLDEN, D. J. 1956. Factors in dehiscence of the flax fruit. Bot. Gaz. 117:294–309.

HOLLOWAY, J. E. 1918. The prothallus and young plant of *Tmesipteris*. Trans. N.Z. Inst. 50:1–44.

HOLM, R. E., and F. B. ABELES. 1967. Abscission: the role of RNA synthesis. Plant Physiol. 42:1094–1102.

Holm, R. E., and W. C. WILSON. 1977. Ethylene and fruit loosening from combinations of *Citrus* abscission chemicals. J. Am. Soc. Hortic. Sci. 102:576–79.

HOOVER, M., and V. T. WALHOOD. 1974. Chemical Harvest Aids for Cotton. Univ. Calif. Coop. Ext. Bull. No. AXT-208.

HORTON, R. F. 1976. The regulation of stem abscission in succulents. Pp. 152–53 in P. E. Pilet (ed.), Ninth Intl. Conf. Plant Growth Subst., Collected Abstracts. Lausanne.

Horton, R. F., and D. J. OSBORNE. 1967. Senescence, abscission and cellulase activity in *Phaseolus vulgaris*. Nature 214:1086–88.
HOSHAW, R. W., and A. T. GUARD. 1949. Abscission of marcescent leaves of *Quercus palustris* and *Q. coccinea*. Bot. Gaz. 110:587–93.
HÖSTER, H. R., W. LIESE, and P. BÖTTCHER. 1968. Untersuchungen zur Morphologie und Histologie der Zweigabwürfe von *Populus* 'Robusta.' Forstwiss. Centralbl. 87:356–68.
HOUGH, W. A. 1968. Carbohydrate reserves of saw-palmetto: seasonal variation and effects of burning. For. Sci. 14:399–405.
HUXLEY, P. A. 1970. Some aspects of the physiology of Arabica coffee—the central problems and the need for a synthesis. Pp. 255–68 in L. C. Luckwill and C. V. Cutting (eds.), Physiology of Tree Crops. Academic Press, New York.
HYODO, H. 1978. Ethylene production by wounded tissue of *Citrus* fruit. Plant Cell Physiol. 19:545–51.
INGOLD, C. T. 1965. Spore Liberation. Clarendon Press, Oxford.
———. 1971. Fungal Spores. Their Liberation and Dispersal. Clarendon Press, Oxford.
INMAN, T. 1848. The causes that determine the fall of leaves. Pp. 89–92, Proc. Literary and Philosophical Soc., Liverpool, 36th Session, No. IV.
ITAI, C., and Y. VAADIA. 1965. Kinetin-like activity in root exudate of water-stressed sunflower plants. Physiol. Plant. 18:941–44.
———. 1971. Cytokinin activity in water-stressed shoots. Plant Physiol. 47:87–90.
IWAHORI, S., and R. F. M. Van STEVENINCK. 1976. Ultrastructural observation of lemon fruit abscission. Sci. Hortic. 4:235–46.
JACKSON, J. M. 1952. Physiology of leaf abscission. Arkansas Acad. Sci. Proc. 5:73–76.
JACKSON, M. B., and D. J. OSBORNE. 1970. Ethylene, the natural regulator of leaf abscission. Nature 225:1019–22.
———. 1972. Abscisic acid, auxin, and ethylene in explant abscission. J. Exp. Bot. 23:849–62.
Jackson, M. B., I. B. MORROW, and D. J. OSBORNE. 1972. Abscission and dehiscence in the squirting cucumber, *Ecballium elaterium*. Can. J. Bot. 50:1465–71.
JACOBS, M. R. 1955. Growth Habits of the Eucalypts. Forestry and Timber Bureau (Aust.). Commonwealth Government Printer, Canberra.
JACOBS, W. P. 1955. Studies on abscission: the physiological basis of the abscission-speeding effect of intact leaves. Am. J. Bot. 42:594–604.
———. 1958. Further studies of the relation between auxin and abscission of *Coleus* leaves. Am. J. Bot. 45:673–75.
———. 1962. Longevity of plant organs: internal factors controlling abscission. Annu. Rev. Plant Physiol. 13:403–36.
Jacobs, W. P., and S. C. KIRK. 1966. Effect of gibberellic acid on elongation and longevity of *Coleus* petioles. Plant Physiol. 41:487–90.
Jacobs, W. P., J. A. SHIELD, JR., and D. A. OSBORNE. 1962. Senescence factor and abscission of *Coleus* leaves. Plant Physiol. 37:104–6.
JACOBSEN, J. V. 1977. Regulation of ribonucleic acid metabolism by plant hormones. Annu. Rev. Plant Physiol. 28:537–64.

JAHNS, H. M. 1973. Anatomy, morphology, and development. Pp. 3–58 in V. Ahmadjian and M. E. Hale (eds.), The Lichens. Academic Press, New York.

JANZEN, D. H. 1967. Synchronization of sexual reproduction of trees within the dry season in Central America. Evolution 21:620–37.

———. 1970. *Jacquinia pungens*, a heliophile from the understory of tropical deciduous forest. Biotropica 2:112–19.

———. 1974. Tropical blackwater rivers, animals, and mast fruiting by the Dipterocarpaceae. Biotropica 6:69–103.

———. 1975. Ecology of Plants in the Tropics. Edward Arnold, London.

Janzen, D. H., and D. E. WILSON. 1974. The cost of being dormant in the tropics. Biotropica 6:260–62.

JENSEN, T. E., and J. G. VALDOVINOS. 1967. Fine structure of abscission zones. I. Abscission zones of the pedicels of tobacco and tomato flowers at anthesis. Planta 77:298–318.

———. 1968a. Structure of microbodies in abscission cells. Plant Physiol. 43:2062–65.

———. 1968b. Fine structure of abscission zones. III. Cytoplasmic changes in abscising pedicels of tobacco and tomato flowers. Planta 83:303–13.

JEPSON, W. L. 1925. A Manual of the Flowering Plants of California. Associated Students' Store, Berkeley.

JERIE, P. H. 1976. The role of ethylene in abscission of Cling peach fruit. Aust. J. Plant Physiol. 3:747–54.

JOHNSON, R. E., and F. T. ADDICOTT. 1967. Boll retention in relation to leaf and boll development in cotton (*Gossypium hirsutum* L.). Crop Sci. 7:571–74.

JONES, R. G. W., and O. R. LUNT. 1967. The function of calcium in plants. Bot. Rev. 33:407–26.

JONKER, F. P. 1976. The Carboniferous "genera" *Ulodendron* and *Halonia*—an assessment. Palaeontogr. Abt. B, Palaeophytol. 157:97–111.

JOUBERT, T. G. La G. 1965. A third jointless gene in tomatoes. S. Afr. J. Agri. Sci. 8:571–82.

JUNIPER, B. 1972. Mechanisms of perception and patterns of organisation in root caps. Pp. 119–31 in M. W. Miller and C. C. Kuehnert (eds.), The Dynamics of Meristem Cell Populations. Plenum Press, New York.

JUNTTILA, O. 1976. Apical growth cessation and shoot tip abscission in *Salix*. Physiol. Plant. 38:278–86.

KAMERBEEK, G. A., and W. J. de MUNK. 1976. A review of ethylene effects in bulbous plants. Sci. Hortic. 4:101–15.

KAWASE, M. 1976. Ethylene accumulation in flooded plants. Physiol. Plant. 36:236–41.

———. 1978. Anaerobic elevation of ethylene concentration in waterlogged plants. Am. J. Bot. 65:736–40.

KAYS, S. J., C. A. JAWORSKI, and H. C. PRICE. 1976. Defoliation of pepper transplants in transit by endogenously-evolved ethylene. J. Am. Soc. Hortic. Sci. 101:449–51.

KENDALL, J. N. 1918. Abscission of flowers and fruits in the Solanaceae with special reference to *Nicotiana*. Univ. Calif. Publ. Bot. 5:347–428.

KENDRICK, W. B., and J. W. CARMICHAEL. 1973. Hyphomycetes. Pp. 323–509

in G. C. Ainsworth, F. K. Sparrow, and A. S. Sussman (eds.), The Fungi. Vol. IVA. Academic Press, New York.
KILLIAN, K. 1911. Beiträge zur Kenntnis der Laminarien. Zeitschr. Bot. 3: 433–94.
KIMMEY, J. W. 1945. The seasonal development and the defoliating effect of *Cronartium ribicola* on naturally infected *Ribes roezli* and *R. nevadense*. Phytopathology 35:406–16.
KING, E. E. 1973. Endo-polymethylgalacturonase of boll weevil larvae, *Anthonomus grandis*: an initiator of cotton flower bud abscission. J. Insect Physiol. 19:2433–37.
King, E. E., and H. C. LANE. 1969. Abscission of cotton flower buds and petioles caused by protein from boll weevil larvae. Plant Physiol. 44:903–6.
KNEE, M. 1978a. Properties of polygalacturonate and cell cohesion in apple fruit cortical tissue. Phytochemistry 17:1257–60.
———. 1978b. Metabolism of polymethylgalacturonate in apple fruit cortical tissue during ripening. Phytochemistry 17:1261–64.
KNOLL, A. H., and E. S. BARGHOORN. 1977. Archean microfossils showing cell division from the Swaziland System of South Africa. Science 198:396–98.
KOEK-NOORMAN, J., and B. J. H. ter WELLE. 1976. The anatomy of branch abscission layers in *Perebea mollis* and *Naucleopsis guianensis* (Castilleae, Moraceae). Leiden Bot. Ser. 3:196–203.
KÖKÉNDYNÉ-INÁNTSY, I., T. SZABÓ, and T. BUBÁN. 1978. Histochemical properties of the embryo and fruit set in sour cherry following treatments by growth regulators. Acta Hortic. 80:153–56.
KOMIYA, K., and T. MISAWA. 1976. Studies on the leaf abscission by fungicidal injury. Research for metabolic pathway related to abscission by use of organic inhibitor. Annu. Rep. Soc. Plant Protection of North Japan 27:113.
KONAR, R. N., and R. K. KAPOOR. 1972. Anatomical studies on *Azolla pinnata*. Phytomorphology 22:211–23.
KONDO, K., A. WATANABE, and H. IMASEKI. 1975. Relationships in actions of indoleacetic acid, benzyladenine and abscisic acid in ethylene production. Plant Cell Physiol. 16:1001–7.
KONDO, N., and K. SUGAHARA. 1978. Changes in transpiration rate of SO_2-resistant and -sensitive plants with SO_2 fumigation and the participation of abscisic acid. Plant Cell Physiol. 19:365–73.
KONTA, F. 1974. Frond articulation in the fern genus *Loxogramme*. Acta Phytotaxon. Geobot. 26:119–26.
KORF, R. P. 1973. Discomycetes and Tuberales. Pp. 249–319 in G. C. Ainsworth, F. K. Sparrow, and A. S. Sussman (eds.), The Fungi. Vol. IVA. Academic Press, New York.
KORIBA, K. 1958. On the periodicity of tree-growth in the tropics, with reference to the mode of branching, the leaf-fall, and the formation of the resting bud. The Gardens' Bull. (Singapore), 27:11–81.
KOSHIMIZU, K., M. INUI, H. FUKUI, and T. MITSUI. 1968. Isolation of (+)-abscisyl-β-D-glucopyranoside from immature fruit of *Lupinus luteus*. Agric. Biol. Chem. 32:789–91.
KOYAMA, T., M. YAMADA, and M. MATSUHASHI. 1977. Formation of regular packets of *Staphylococcus aureus* cells. J. Bacteriol. 129:1518–23.

KOZLOWSKI, T. T. 1960. Some problems in use of herbicides in forestry. Pp. 1–10, Proc. 17th N. Cent. Weed Control Conf., Milwaukee.

———. 1973. Extent and significance of shedding of plant parts. Pp. 1–44 in T. T. Kozlowski (ed.), Shedding of Plant Parts. Academic Press, New York.

KRAMER, P. J. 1951. Causes of injury to plants resulting from flooding of the soil. Plant Physiol. 26:722–36.

KREBS, A. T. 1965. Radiation-induced premature aging in leaves and autumnal events in nature. Beitr. Biol. Pflanz. 41:157–74.

KREZDORN, A. H. 1969. The use of growth regulators to improve fruit set in *Citrus*. Pp. 1113–19 in H. D. Chapman (ed.), Proc. 1st Int. Citrus Symposium. Vol. III. Univ. Calif., Riverside.

KUBART, B. 1906. Die organische Ablösung der Korolle nebst Bemerkungen über die MOHLsche Trennungsschichte. Sitzungsber. Akad. Wiss. Wien, Math.-Naturw. Kl. Abt. I. 115:1491–1518.

KUHN, D. A., and M. P. STARR. 1965. Clonal morphogenesis of *Lampropedia hyalina*. Arch. Mikrobiol. 52:360–75.

KUIJT, J. 1969. The Biology of Parasitic Flowering Plants. Univ. Calif. Press, Berkeley.

KULMAN, H. M. 1971. Effects of insect defoliation on growth and mortality of trees. Annu. Rev. Entomol. 16:289–324.

KURAISHI, S., and R. M. MUIR. 1962. Increase in diffusible auxin after treatment with gibberellin. Science 137:760–61.

KÜSTER, E. 1911. Die Gallen der Pflanzen. Verlag S. Hirzel, Leipzig.

———. 1916. Beiträge zur Kenntnis des Laubfalles. Ber. Dtsch. Bot. Ges. 24:184–93.

LADIGES, P. Y., and A. KELSO. 1977. The comparative effects of waterlogging on two populations of *Eucalyptus viminalis* and one population of *Eucalyptus ovata*. Aust. J. Bot. 25:159–69.

LAIBACH, F. 1932. Pollenhormon und Wuchsstoff. Ber. Dtsch. Bot. Ges. 50:383–90.

———. 1957. Der Einflusz von Auxin und Gibberellin auf die Abstoszung von Pflanzenorganen. Naturwissenschaften 44:594–95.

Laibach, F., and E. MASCHMANN. 1933. Über den Wuchsstoff der Orchideenpollinien. Jahrb. Wiss. Bot. 78:399–430.

LAMBERS, H., E. STEINGRÖVER, and G. SMAKMAN. 1978. The significance of oxygen transport and of metabolic adaptation in flood-tolerance of *Senecio* species. Physiol. Plant. 43:277–81.

LAMPREY, H. F. 1967. Notes on the dispersal and germination of some tree seeds through the agency of mammals and birds. East Afr. Wildl. J. 5:179–80.

Lamprey, H. F., G. HALEVY, and S. MAKACHA. 1974. Interactions between *Acacia*, bruchid seed beetles and large herbivores. East Afr. Wildl. J. 12:81–85.

LARSEN, F. E. 1969. Promotion of leaf abscission of deciduous tree fruit nursery stock with abscisic acid. HortScience 4:216–18.

———. 1973. Promotion of leaf abscission in fruit nursery stock. Acta Hortic. 34(1):129–33.

Larsen, F. E., and G. D. LOWELL. 1977. Tree fruit nursery stock defoliation with harvest aide chemical and surfactant mixtures. HortScience 12:580–82.

La RUE, C. D. 1936. The effect of auxin on the abscission of petioles. Proc. Natl. Acad. Sci. U.S.A. 22:254–59.

LAUDI, G. 1956. Studi sulla fisiologia dell'abscissione: influenza della presenza di rami normali, foglie ascellari sull'abscissione di piccioli privati del lembo. Nuovo Giorn. Bot. Ital. 63:204–12.

Laudi, G., and L. LAMBRI. 1956. Studi sulla fisiologia dell'abscissione. Effetti di trattamenti capaci di potenziare o di indebolire l'azione auxinica nella parte parennante o nella parte caduca. Nuovo Giorn. Bot. Ital. 63:324–35.

LEE, C. I., and H. B. TUKEY, JR. 1972. Effect of intermittent mist on development of fall color in foliage of *Euonymus alatus* Sieb. 'Compactus'. J. Am. Soc. Hortic. Sci. 97:97–101.

LEE, E. 1911. The morphology of leaf-fall. Ann. Bot. 25:51–105.

LEINWEBER, C. L., and W. C. HALL. 1959a. Foliar abscission in cotton. I. Effect of age and defoliants on the respiratory rate of blade, petiole, and tissues of the abscission zone. Bot. Gaz. 120:144–51.

———. 1959b. Foliar abscission in cotton. II. Influence of age and defoliants on chemical composition of blades and pulvinoids. Bot. Gaz. 120:183–86.

LEOPOLD, A. C. 1964. Kinins and the regulation of leaf senescence. Pp. 705–18 in J. P. Nitsch (ed.), Régulateurs Naturels de la Croissance Végétale. Cent. Natl. Rech. Sci., Paris.

———. 1977. Modification of growth regulatory action with inorganic solutes. Pp. 33–41 in C. A. Stutte (ed.), Plant Growth Regulators, No. 159. American Chemical Soc., Washington, D.C.

Le PELLEY, R. H. 1942. The food and feeding habits of *Antestia* in Kenya. Bull. Ent. Res. 33:71–89.

LESQUEREUX, L. 1891. The flora of the Dakota Group. U.S. Geol. Surv., Monogr. 17:1–400.

LETHAM, D. S. 1967. Chemistry and physiology of kinetin-like compounds. Annu. Rev. Plant Physiol. 18:349–64.

LEUTY, S. J., and M. J. BUKOVAC. 1968. A comparison of the growth and anatomical development of naturally abscising, non-abscising and naphthaleneacetic acid-treated peach fruits. Phytomorphology 18:372–79.

LEVINE, E., and F. R. HALL. 1977. Effect of feeding and oviposition by the plum curculio on apple and plum fruit abscission. J. Econ. Entomol. 70:603–7.

LEVITT. J. 1972. Responses of Plants to Environmental Stresses. Academic Press, New York.

LEWIS, L. N., and J. C. BAKHSHI. 1968a. Interactions of indoleacetic acid and gibberellic acid in leaf abscission control. Plant Physiol. 43:351–58.

———. 1968b. Protein synthesis in abscission: the distinctiveness of the abscission zone and its response to gibberellic acid and indoleacetic acid. Plant Physiol. 43:359–64.

Lewis, L. N., and J. E. VARNER. 1970. Synthesis of cellulase during abscission of *Phaseolus vulgaris* leaf explants. Plant Physiol. 46:194–99.

Lewis, L. N., R. L. PALMER, and H. Z. HIELD. 1968. Interactions of auxins, abscission accelerators, and ethylene in the abscission of *Citrus* fruit. Pp. 1303–13 in F. Wightman and G. Setterfield (eds.), Biochemistry and Physiology of Plant Growth Substances. Runge Press, Ottawa.

Lewis, L. N., F. T. LEW, P. D. REID, and J. E. BARNES. 1972. Isozymes of cellulase in the abscission zone of *Phaseolus vulgaris*. Pp. 234–39 in D. J. Carr (ed.), Plant Growth Substances 1970. Springer-Verlag, Berlin.
LICITIS-LINDBERGS, R. 1956. Branch abscission and disintegration of the female cones of *Agathis australis* Salisb. Phytomorphology 6:151–67.
LIEBERMAN, M. 1979. Biosynthesis and action of ethylene. Annu. Rev. Plant Physiol. 30:533–91.
LINSKENS, H. F. 1964. Pollen physiology. Annu. Rev. Plant Physiol. 15:255–70.
LIPE, J. A., and P. W. MORGAN. 1972. Ethylene: role in fruit abscission and dehiscence processes. Plant Physiol. 50:759–64.
Lipe, J. A., P. W. MORGAN, and J. B. STOREY. 1969. Growth substances and fruit shedding in the pecan, *Carya illinoensis*. J. Am. Soc. Hortic. Sci. 94:668–71.
LIPE, W. N. 1966. Physiological Studies of Peach Seed Dormancy with Reference to Growth Substances. Ph.D. dissertation, Univ. Calif., Davis.
LIVINGSTON, G. A. 1948. Contributions to knowledge of leaf abscission in the Valencia orange. Ph.D. dissertation, Univ. Calif., Los Angeles.
———. 1950. In vitro tests of abscission agents. Plant Physiol. 25:711–21.
LLOYD, F. E. 1914. Injury and abscission in *Impatiens sultani*. Pp. 72–79, Sixth Annu. Rep. Que. Soc. Prot. Plants.
———. 1916. Abscission in *Mirabilis jalapa*. Bot. Gaz. 61:213–30.
———. 1920a. Abscission in fruits in *Juglans californica quercina*. Trans. R. Soc. Can. (Series 3) 14:17–22.
———. 1920b. Environmental changes and their effect upon boll-shedding in cotton (*Gossypium herbaceum*). Ann. N.Y. Acad. Sci. 29:1–131.
———. 1926. Cell disjunction in *Spirogyra*. Pap. Mich. Acad. Sci. Arts Lett. 6: 275–87.
LOMINSKI, I., and M. RAFI SHAIKH. 1968. Long-chain mutants of *Streptococcus faecalis* induced by ultraviolet irradiation. J. Med. Microbiol. 1:219–20.
Lominski, I., J. CAMERON, and G. WYLLIE. 1958. Chaining and unchaining *Streptococcus faecalis*—a hypothesis of the mechanism of bacterial cell separation. Nature 181:1477.
LONGMAN, K. A., and J. JENÍK. 1974. Tropical Forest and Its Environment. Longman, London.
LOUIE, D. S., JR. 1963. Studies on abscission: I. Auxin in the physiology of abscission. II. Water relations and the defoliability of cotton. Ph.D. dissertation. Univ. Calif., Los Angeles.
Louie, D. S., Jr., and F. T. ADDICOTT. 1970. Applied auxin gradients and abscission in explants. Plant Physiol. 45:654–57.
LOVE, R. M. 1963. Mission veldtgrass. Calif. Agric. 17(10):2–3.
LÖWI, E. 1907. Untersuchungen über die Blattablösung und verwandte Erscheinungen. Sitzungsber. Akad. Wiss. Wien, Math.-Naturw. Kl. Abt. I. 116:983–1024.
LUCKWILL, L. C. 1953. Studies of fruit development in relation to plant hormones. I. Hormone production by the developing apple seed in relation to fruit drop. J. Hortic. Sci. 28:14–24.
———. 1957. Hormonal aspects of fruit development in higher plants. Pp. 63–85 in H. K. Porter (ed.), The Biological Action of Growth Substances. Symp. Soc. Exp. Biol. XI. Cambridge Univ. Press, Cambridge.

———. 1970. The control of growth and fruitfulness of apple trees. Pp. 237–54 in L. C. Luckwill and C. V. Cutting (eds.), Physiology of Tree Crops. Academic Press, New York.

Luckwill, L. C., H. JONKERS, and R. ANTOSZEWSKI, eds. 1978. Symposium on Growth Regulators in Fruit Production. Skierniewice, Poland. Acta Horticulturae 80.

LYON, J. L. 1964. Interactions of abscission regulators and the petiole in the control of abscission. M.Sc. thesis. Univ. Calif., Davis.

Lyon, J. L., and R. J. COFFELT. 1966. Rapid method for determining numerical indexes for time-course curves. Nature 211:330.

Lyon, J. L., and O. E. SMITH. 1966. Effects of gibberellins on abscission in cotton seedling explants. Planta 69:347–56.

Lyon, J. L., K. OHKUMA, and F. T. ADDICOTT. 1972. Current status of the abscission retardants and accelerants of young cotton fruit. Pp. 36–40, Proc. 1972 Beltwide Cotton Prod. Res. Conf. (Nat. Cotton Council, Memphis).

McAFEE, J. A., and P. W. MORGAN. 1971. Rates of production and internal levels of ethylene in the vegetative cotton plant. Plant Cell Physiol. 12:839–47.

McCALLA, D. R., M. K. GENTHE, and W. HOVANITZ. 1962. Chemical nature of an insect gall growth-factor. Plant Physiol. 37:98–103.

MacCONNELL, W. P., and L. KENERSON. 1964. Chemi-pruning northern hardwoods. J. For. 62:463–66.

MACKIE, W., and R. D. PRESTON. 1974. Cell wall and intercellular region polysacharides. Pp. 40–85 in W. D. P. Stewart (ed.), Algal Physiology and Biochemistry. Univ. Calif. Press, Berkeley.

MACLACHLAN, G. A. 1977. Cellulose metabolism and cell growth. Pp. 13–20 in P. E. Pilet (ed.), Plant Growth Regulation. Springer-Verlag, Berlin.

MacLEAN, D. C., D. C. McCUNE, L. H. WEINSTEIN, R. H. MANDLE, and G. N. WOODRUFF. 1968. Effects of acute hydrogen fluoride and nitrogen dioxide exposures on *Citrus* and ornamental plants of Central Florida. Environ. Sci. Technol. 2:444–49.

McLEMORE, B. F. 1973. Chemicals Fail to Induce Abscission of Loblolly and Slash Pine Cones. U.S. For. Serv. Res. Note SO-155.

McMICHAEL, B. L., W. R. JORDAN, and R. D. POWELL. 1973. Abscission processes in cotton: induction by plant water deficit. Agron. J. 65:202–4.

MAGNUS, P. 1893. Ueber *Synchytrium papillatum* Farl. Ber. Dtsch. Bot. Ges. 11: 538–42.

MAI, G. 1934. Korrelationsuntersuchungen an entspreiteten Blattstielen mittels lebender Orchideenpollinien als Wuchsstoffquelle. Jahrb. Wiss. Bot. 79:681–713.

MANI, M. S. 1964. Ecology of Plant Galls. Dr. W. Junk Publishers, The Hague.

MANNING, W. J., and W. A. FEDER. 1976. Effects of ozone on economic plants. Pp. 47–60 in T. A. Mansfield (ed.), Effects of Air Pollutants on Plants. Cambridge Univ. Press, Cambridge.

MANUEL, F. E. 1968. A Portrait of Isaac Newton. Belknap Press of Harvard Univ. Press, Cambridge, Mass.

MARTIN, G. C. 1973. Peach fruit-set and abscission. Acta Hortic. 34(1):345–52.

Martin, G. C., and W. H. GRIGGS. 1970. The effectiveness of succinic acid 2,2-

dimethylhydrazide in preventing preharvest drop of 'Bartlett' pears. HortScience 5:258–59.

Martin, G. C., R. J. ROMANI, S. A. WEINBAUM, C. NISHIJIMA, and J. MARSHACK. 1980. Measurement of abscisic acid and polysomes at anthesis and shortly after in pollinated and non-pollinated 'Winter Nelis' pear flowers. J. Am. Soc. Hortic. Sci. 105:318–21.

MARTIN, L. B. 1954. Abscission and starch distribution following application of sucrose and indoleacetic acid to excised abscission zones of *Phaseolus vulgaris*. Ph.D. dissertation. Univ. Calif., Los Angeles.

MARYNICK, M. C. 1976. Studies on abscission in cotton explants. Ph.D. dissertation. Univ. Calif., Davis.

———. 1977. Patterns of ethylene and carbon dioxide evolution during cotton explant abscission. Plant Physiol. 59:484–89.

Marynick, M. C., and F. T. ADDICOTT. 1976. Evidence for a dual role for oxygen in control of abscission. Nature 264:668–69.

MASON, T. G. 1922. Growth and abscission in sea island cotton. Ann. Bot. 36:458–84.

MATILE, P. 1975. The Lytic Compartment of Plant Cells. Springer-Verlag, New York.

———. 1978. Biochemistry and function of vacuoles. Annu. Rev. Plant Physiol. 29:193–213.

MATZKE, E. B. 1936. The effect of street lights in delaying leaf-fall in certain trees. Am. J. Bot. 23:446–52.

MAYAK, S., and A H. HALEVY. 1972. Interrelationships of ethylene and abscisic acid in the control of rose petal senescence. Plant Physiol. 50:341–46.

Mayak, S., A. H. Halevy, and M. KATZ. 1972. Correlative changes in phytohormones in relation to senescence processes in rose petals. Physiol. Plant. 27:1–4.

MEGURO, M., C. A. JOLY, and M. M. BITTENCOURT. 1977. Desiccation tolerant *Xerophyta plicata* Spreng.-Velloziaceae. Plant Physiol. 59(suppl):54.

MERRILL, E. D. 1945. Plant Life of the Pacific World. Macmillan, New York.

METCALFE, C. R. 1967. Distribution of latex in the plant kingdom. Econ. Bot. 21:115–27.

MEYERS, R. M. 1940. Effect of growth substances on the absciss layer of *Coleus*. Bot. Gaz. 102:323–38.

MIDDLETON, J. T. 1961. Photochemical air pollution damage to plants. Annu. Rev. Plant Physiol. 12:431–48.

MILBORROW, B. V. 1967. The identification of (+)-abscisin II [(+)-dormin] in plants and measurement of its concentrations. Planta 76:93–113.

———. 1974. Chemistry and biochemistry of abscisic acid. Pp. 57–91 in V. C. Runeckles, E. Sondheimer, and D. C. Walton (eds.), Recent Advances in Phytochemistry. Vol. 7. The Chemistry and Biochemistry of Plant Hormones. Academic Press, New York.

MILLER, C. S., L. H. WILKES, E. L. THAXTON, and J. L. HUBBARD. 1971. Cotton wilt-harvest and wiltant defoliation effectiveness in Texas. Texas Agric. Exp. Stn. Bull. MP-1010. College Station.

MILLER, P. R., J. R. PARMETER, JR., O. C. TAYLOR, and E. A. CARDIFF. 1963. Ozone injury to the foliage of *Pinus ponderosa*. Phytopathology 53:1072–76.

MILLINGTON, W. F. 1963. Shoot tip abortion in *Ulmus americana*. Am. J. Bot. 50:371–78.
Millington, W. F., and W. R. CHANEY. 1973. Shedding of shoots and branches. Pp. 149–204 in T. T. Kozlowski (ed.), Shedding of Plant Parts. Academic Press, New York.
MILNE-REDHEAD, E., and H. G. SCHWEICKERDT. 1939. A new conception of the genus *Ammocharis* Herb. J. Linn. Soc. (Bot.) 52:159–97.
MITCHELL, J. W., and G. A. LIVINGSTON. 1968. Methods of Studying Plant Hormones and Growth-Regulating Substances. U.S. Dep. Agric. Handbook No. 336.
MIZRAHI, Y., S. BLUMENFELD, S. BITTNER, and A. E. RICHMOND. 1971. Abscisic acid and cytokinin contents of leaves in relation to salinity and relative humidity. Plant Physiol. 48:752–55.
MOHAN RAM, H. Y., and I. V. RAMANUJA RAO. 1977. Prolongation of vase-life of *Lupinus hartwegii* Lindl. by chemical treatments. Sci. Hortic. 7:377–82.
MOLINE, H. E., C. E. LaMOTTE, C. GOCHNAUER, and A. McNAMER. 1972. Further comparative studies of pectin esterase in relation to leaf and flower abscission. Plant Physiol. 50:655–59.
MOLISCH, H. 1886. Untersuchungen über Laubfall. Sitzungsber. Akad. Wiss. Wien, Math.-Naturw. Kl., Abt. I. 93:148–84.
MOLLENHAUER, H. H., and D. J. MORRÉ. 1966. Golgi apparatus and plant secretion. Annu. Rev. Plant Physiol. 17:27–46.
MONK, C. D. 1966. An ecological significance of evergreenness. Ecology 47:504–5.
———. 1971. Leaf decomposition and loss of ^{45}Ca from deciduous and evergreen trees. Am. Midl. Nat. 86:379–84.
MONSELISE, S. P., R. GOREN, and J. COSTO. 1967. Hormone-inhibitor balance of some citrus tissues. Isr. J. Agric. Res. 17:35–45.
MOONEY, H. A., and E. L. DUNN. 1970. Convergent evolution of Mediterranean-climate evergreen sclerophyll shrubs. Evolution 24:292–303.
Mooney, H. A., E. L. Dunn, F. SHROPSHIRE, and L. SONG. 1970. Vegetation comparisons between the Mediterranean climatic areas of California and Chile. Flora (Jena) 159:480–96.
MORGAN, M. D. 1975. Ecology of aspen in Gunnison County, Colorado. M.Sc. thesis. Univ. Illinois, Urbana.
MORGAN, P. W. 1976. Gibberellic acid and IAA compete in ethylene-promoted abscission. Planta 129:275–76.
Morgan, P. W., and J. I. DURHAM. 1972. Abscission: potentiating action of auxin transport inhibitors. Plant Physiol. 50:313–18.
———. 1973. Morphactins enhance ethylene-induced leaf abscission. Planta 110:91–93.
———. 1975. Ethylene-induced leaf abscission is promoted by gibberellic acid. Plant Physiol. 55:308–11.
Morgan, P. W., and H. W. GAUSMAN. 1966. Effects of ethylene on auxin transport. Plant Physiol. 41:45–52.
Morgan, P. W., and W. C. HALL. 1964. Accelerated release of ethylene by cotton following application of indolyl-3-acetic acid. Nature 201:99.

Morgan, P. W., E. BEYER, JR., and H. W. GAUSMAN. 1968. Ethylene effects on auxin physiology. Pp. 1255–73 in F. Wightman and G. Setterfield (eds.), Biochemistry and Physiology of Plant Growth Substances. Runge Press, Ottawa.

Morgan, P. W., W. R. JORDAN, T. L. DAVENPORT, and J. I. DURHAM. 1977. Abscission responses to moisture stress, auxin transport inhibitors, and ethephon. Plant Physiol. 59:710–12.

MORRÉ, D. J. 1968. Cell wall dissolution and enzyme secretion during leaf abscission. Plant Physiol. 43:1545–59.

Morré, D. J., and J. H. CHERRY. 1977. Auxin hormone—plasma membrane interactions. Pp. 35–43 in P. E. Pilet (ed.), Plant Growth Regulation, Ninth Intl. Conf. on Plant Growth Substances. Springer-Verlag, Berlin.

MOTHES, K. 1964. The role of kinetin in plant regulation. Pp. 131–40 in J. P. Nitsch (ed.), Régulateurs Naturels de la Croissance Végétale. Cent. Natl. Rech. Sci., Paris.

MÜHLDORF, A. 1925. Über den Ablösungsmodus der Gallen von ihren Wirstpflanzen nebst einer kritischen übersicht über die Trennungserscheinungen im Pflanzenreiche. Beih. Bot. Centralbl. Abt. I. 42:1–110.

MUIR, R. M., and B. P. LANTICAN. 1968. Purification and properties of the enzyme system forming indoleacetic acid. Pp. 259–72 in F. Wightman and G. Setterfield (eds.), Biochemistry and Physiology of Plant Growth Substances. Runge Press, Ottawa.

Muir, R. M., and J. G. VALDOVINOS. 1970. Gibberellin and auxin relationships in abscission. Am. J. Bot. 57:288–91.

MUKHERJEE, A. K., and S. K. CHATTERJEE. 1969. Seasonal studies of ethylene effects on abscission of *Coleus* debladed petioles of different maturity. Indian J. Plant Physiol. 12(1&2):148–53.

MULLETT, J. H. 1966. The effect of a growth regulator on seed production and retention in *Phalaris tuberosa* L. J. Aust. Inst. Agric. Sci. 32:218–19.

MUNZ, P. A., and D. D. KECK. 1959. A California Flora. Univ. Calif. Press, Berkeley.

MURAMOTO, H. 1977. Caducous bracts. Pp. 8–9, Proc. 1977 Beltwide Cotton Prod. Improv. Conf. Atlanta, GA. (Nat. Cotton Council, Memphis).

MURASHIGE, T. 1966. The deciduous behavior of a tropical plant. Physiol. Plant. 19:348–55.

MUSSELL, H. W., and D. J. MORRÉ. 1969. A quantitative bioassay specific for polygalacturonases. Anal. Biochem. 28:353–60.

NAGY, L. A. 1974. Transvaal stromatolite: first evidence for the diversification of cells about 2.2×10^9 years ago. Science 183:514–16.

NAMIKAWA, I. 1926. Contributions to the knowledge of abscission and exfoliation in floral organs. J. Coll. Agric. Hokkaido Imp. Univ. 17(2):63–121.

NEGER, F. W., and J. FUCHS. 1915. Untersuchungen über den Nadelfall der Koniferen. Jahrb. f. wiss. Bot. 55:608–60.

NELJUBOW, D. 1901. Über die horizontale Nutation der Stengel von *Pisum sativum* und einiger anderer Pflanzen. Beih. Bot. Zentralbl. 10:128–38.

NEWTON, R. J., D. R. SHELTON, S. DISHAROON, and J. E. DUFFEY. 1978. Turion formation and germination in *Spirodela polyrhiza*. Am. J. Bot. 65:421–28.

NG, F. S. P. 1978. Strategies of establishment in Malayan forest trees. Pp. 129–62 in P. B. Tomlinson and M. H. Zimmermann (eds.), Tropical Trees as Living Systems. Cambridge Univ. Press, Cambridge.

NICKELL, L. G. 1978. Plant growth regulators. Controlling biological behavior with chemicals. Chem. Eng. News 56:18–34.

NIGHTINGALE, G. T., and R. B. FARNHAM. 1936. Effect of nutrient concentration on anatomy, metabolism, and bud abscission of sweet peas. Bot. Gaz. 97:477–517.

NITSCH, J. P. 1963. The mediation of climatic effects through endogenous regulating substances. Pp. 175–93 in L. T. Evans (ed.), Environmental Control of Plant Growth. Academic Press, New York.

NIXON, R. W., and R. E. GARDNER. 1939. Effect of certain growth substances on inflorescences of dates. Bot. Gaz. 100:866–71.

NOEL, A. R. A., and J. VAN STADEN. 1975. Phyllomorph senescence in *Streptocarpus molweniensis*. Ann. Bot. 39:921–29.

OBERLY, G. H., and D. BOYNTON. 1966. Apple nutrition. Pp. 1–50 in N. F. Childers (ed.), Temperate to Tropical Fruit Nutrition. Somerset Press, Somerville, N.J.

OGAWA, J. M., C. W. NICHOLS, and H. ENGLISH. 1955. Almond scab. Plant Dis. Rep. 39:504–8.

OJEHOMON, O. O., A. S. RATHJEN, and D. G. MORGAN. 1968. Effects of day length on the morphology and flowering of five determinate varieties of *Phaseolus vulgaris* L. J. Agric. Sci. 71:209–14.

OLMSTED, C. E. 1951. Experiments on photoperiodism, dormancy and leaf age and abscission in sugar maple. Bot. Gaz. 112:365–93.

OLSON, W. H., G. S. SIBBETT, G. L. CARNILL, and G. C. MARTIN. 1977. Lower ethephon rates effective in walnut harvest. Calif. Agric. 31(7):6–7.

ORMROD, D. P. 1978. Pollution in Horticulture. Elsevier Scientific Pub., Amsterdam.

ORSHAN, G. 1954. Surface reduction and its significance as a hydroecological factor. J. Ecol. 42:442–44.

OSBORNE, D. J. 1955. Acceleration of abscission by a factor produced in senescent leaves. Nature 176:1161–63.

———. 1958a. Changes in the distribution of pectin methylesterase across leaf abscission zones of *Phaseolus vulgaris*. J. Exp. Bot. 9:446–57.

———. 1958b. The role of 2,4,5-T butyl ester in the control of leaf abscission in some tropical woody species. Trop. Agric. 35:145–58.

———. 1959. Identity of the abscission accelerating substance in senescent leaves. Nature 183:1593.

———. 1968. Defoliation and defoliants. Nature 219:564–67.

———. 1973. Internal factors regulating abscission. Pp. 125–47 in T. T. Kozlowski (ed.), Shedding of Plant Parts. Academic Press, New York.

Osborne, D. J., and S. E. MOSS. 1963. Effect of kinetin on senescence and abscission in explants of *Phaseolus vulgaris*. Nature 200:1299–1301.

Osborne, D. J., and J. A. SARGENT. 1976a. The positional differentiation of ethylene-responsive cells in rachis abscission zones in leaves of *Sambucus nigra* and their growth and ultrastructural changes in senescence and separation. Planta 130:203–10.

———. 1976b. The positional differentiation of abscission zones during the development of leaves of *Sambucus nigra* and the response of the cells to auxin and ethylene. Planta 132:197–204.

Osborne, D. J., M. B. JACKSON, and B. V. MILBORROW. 1972. Physiological properties of abscission accelerator from senescent leaves. Nat. New Biol. 240: 98–101.

OVERBECK, F. 1924. Studien an den Turgeszenz-Schleudermechanismen von *Dorstenia contrayerva* and *Impatiens parviflora* D.C. Jahrb. f. wiss. Bot. 63:467–500.

OYEBADE, T. 1976. Influence of pre-harvest sprays of ethrel on ripening and abscission of coffee berries. Turrialba 26:86–89.

PALMER, R. L., H. Z. HIELD, and L. N. LEWIS. 1969. *Citrus* leaf and fruit abscission. Pp. 1135–43, Proc. 1st Int. Citrus Symp. Vol. 3. Univ. Calif., Riverside.

PARKIN, J. 1898. On some points in the histology of monocotyledons. Ann. Bot. 12:147–54.

———. 1900. Observations on latex and its functions. Ann. Bot. 14:193–214.

PATE, J. S. 1977. Functional biology of dinitrogen fixation by legumes. Pp. 473–517 in R. W. F. Hardy and W. S. Silver (eds.), A treatise on dinitrogen fixation, Section III: Biology. Wiley, New York.

PATHAK, S., and S. K. CHATTERJEE. 1976a. Promotive effects of NAA in relation to age. Geobios 3(3):93–94.

———. 1976b. Leaf abscission process of annual cotton in relation to some biochemical changes with progress of reproductive maturity. Indian J. Exp. Biol. 14:720–22.

PEDERSEN, M. W., G. E. BOHART, V. L. MARBLE, and E. C. KLOSTERMEYER. 1972. Seed production practices. Pp. 689–720 in C. H. Hanson (ed.), Alfalfa Science and Technology. Monogr. 15. Am. Soc. Agron., Madison, Wis.

PENFOLD, A. R., and J. L. WILLIS. 1961. The Eucalypts. Interscience Publishers, New York.

PERRY, T. O., and H. HELLMERS. 1973. Effects of abscisic acid on the growth and dormancy behavior of different races of red maple. Bot. Gaz. 134:283–89.

PESIS, E., Y. FUCHS, and G. ZAUBERMAN. 1978. Cellulase activity and fruit softening in avocado. Plant Physiol. 61:416–19.

PETERSON, R. L., M. G. SCOTT, and S. L. MILLER. 1979. Some aspects of carpel structure in *Caltha palustris* L. (Ranunculaceae). Am. J. Bot. 66:334–42.

PFEIFFER, H. 1928. Die pflanzlichen Trennungsgewebe. In K. Linsbauer (ed.), Handbuch der Pflanzenanatomie. Abt. I., Teil 2, Band V., Lief. 22. Borntraeger, Berlin.

PHAFF, H. J., M. W. MILLER, and E. M. MRAK. 1978. The Life of Yeasts. 2nd ed. Harvard Univ. Press, Cambridge, Mass.

PHILLIPS, D. A., and R. A. WHITE. 1967. Frond articulation in species of Polypodiaceae and Davalliaceae. Am. Fern J. 57:78–88.

PHILLIPS, I. D. J. 1964a. Root-shoot hormone relations. I. The importance of an aerated root system in the regulation of growth hormone levels in the shoot of *Helianthus annuus*. Ann. Bot. 28:18–35.

———. 1964b. Root-shoot hormone relations. II. Changes in endogenous auxin concentration produced by flooding of the root system in *Helianthus annuus*. Ann. Bot. 28:37–45.

PICKETT-HEAPS, J. D. 1975. Green Algae. Sinauer Assoc., Sunderland, Mass.

PIENIĄŻEK, J. 1971. Regulatory role of abscisic acid in the petiole abscission of apple explants. Acad. Polon. Sci. Bull. Ser. Sci. Biol. 19:125–29.

PIERIK, R. L. M. 1971. Auxin-induced secondary abscission in isolated apple flowers. Naturwissenschaften 58:568–69.

———. 1980. Hormonal regulation of secondary abscission in pear pedicels in vitro. Physiol. Plant. 48:5–8.

PLUMSTEAD, E. P. 1958. The habit of growth of Glossopteridae. Trans. Geol. Soc. S. Afr. 61:81–94.

———. 1966. The story of South Africa's coal. Optima, Dec. 1966:187–202.

———. 1967. A general review of the Devonian fossil plants found in the Cape System of South Africa. Palaeontol. Afr. 10:1–83.

———. 1975. A new assemblage of plant fossils from Milorgfjella, Dronning Maud Land. Br. Antarct. Surv. Sci. Rep. No. 83.

POELT, J. 1973. Systematic evaluation of morphological characters. Pp. 91–115 in V. Ahmadjian and M. E. Hale (eds.), The Lichens. Academic Press, New York.

POOVAIAH, B. W. 1974. Formation of callose and lignin during leaf abscission. Am. J. Bot. 61:829–34.

Poovaiah, B. W., and A. C. LEOPOLD. 1973. Inhibition of abscission by calcium. Plant Physiol. 51:848–51.

Poovaiah, B. W., H. P. RASMUSSEN, and M. J. BUKOVAC. 1973. Histochemical localization of enzymes in the abscission zones of maturing sour and sweet cherry fruit. J. Am. Soc. Hortic. Sci. 98:16–18.

PORTER, N. G. 1977. The role of abscisic acid in flower abscission of *Lupinus luteus*. Physiol. Plant. 40:50–54.

Porter, N. G., and R. F. M. VAN STEVENINCK. 1966. An abscission-promoting factor in *Lupinus luteus* (L). Life Sci. 5:2301–8.

PORTHEIM, L. 1941. Further studies on the action of heteroauxin on *Phaseolus vulgaris*. Ann. Bot. 5:35–46.

PRAKASH, G. 1974. Effect of growth substances on foliar abscission in *Catharanthus roseus*. J. Indian Bot. Soc. 53:124–27.

———. 1976. Abscission interaction between different lengths of petioles and growth regulators. Geobios (Jodhpur) 3:90–91.

PRANCE, G. T., and S. A. MORI. 1977. What is *Lecythis*? Taxon 26:209–22.

PRATT, H. K. 1974. The role of ethylene in fruit ripening. Pp. 153–60 in Facteurs et Régulation de la Maturation des Fruits. Colloques Int. Cent. Natl. Rech. Sci., No. 238, Paris.

Pratt, H. K., and J. D. GOESCHL. 1969. Physiological role of ethylene in plants. Annu. Rev. Plant Physiol. 20:541–84.

PRICE, S. R. 1911. The roots of some North African desert-grasses. New Phytol. 10:328–40.

PRYOR, L. D., and R. B. KNOX. 1971. Operculum development and evolution of Eucalypts. Aust. J. Bot. 19:143–71.

Pryor, L. D., M. M. CHATTAWAY, and N. H. KLOOT. 1956. The inheritance of wood and bark characters in *Eucalyptus*. Aust. J. Bot. 4:216–39.

PURI, V. 1951. The role of floral anatomy in the solution of morphological problems. Bot. Rev. 17:471–553.

PYATT, B. 1973. Lichen propagules. Pp. 117–45 in V. Ahmadjian and M. E. Hale (eds.), The Lichens. Academic Press, New York.
RABEY, G. G., and G. C. BATE. 1978. The effect of a period of darkness on the translocation of ^{14}C-labelled assimilates from leaves subtending five- to ten-day-old cotton bolls (*Gossypium hirsutum* L). Rhod. J. Agric. Res. 16:61–71.
RAPHAEL, H. J., A. J. PANSHIN, and M. W. DAY. 1954. "Chemical" bark peeling of aspen: 1952 and 1953 field tests. Q. Bull. Mich. Agric. Exp. Stn. 37: 230–40.
RAPPAPORT, L. 1957. Effect of gibberellin on growth, flowering and fruiting of the Earlypak tomato (*Lycopersicum esculentum*). Plant Physiol. 32:440–44.
RASMUSSEN, G. K. 1974. Cellulase activity in separation zones of citrus fruit treated with abscisic acid under normal and hypobaric atmospheres. J. Am. Soc. Hortic. Sci. 99:229–31
———. 1975. Cellulase activity, endogenous abscisic acid, and ethylene in four *Citrus* cultivars during maturation. Plant Physiol. 56:765–67.
RATNER, A., R. GOREN, and S. P. MONSELISE. 1969. Activity of pectin esterase and cellulase in the abscission zone of *Citrus* leaf explants. Plant Physiol. 44: 1717–23.
REED, H. S. 1907. The value of certain nutritive elements to the plant cell. Ann. Bot. 21:501–43.
REED, N. R., and B. A. BONNER. 1974. The effect of abscisic acid on the uptake of potassium and chloride into *Avena* coleoptile sections. Planta 116:173–85.
REICHE, C. 1885. Ueber anatomische Veränderungen welche in den Perianthkreisen der Blüten während der Entwicklung der Frucht vor sich gehen. Jahrb. Wiss. Bot. 16:638–87.
REID, D. M., and A. CROZIER. 1971. Effects of waterlogging on the gibberellin content and growth of tomato plants. J. Exp. Bot. 22:39–48.
REID, P. D., P. G. STRONG, F. LEW, and L. N. LEWIS. 1974. Cellulase and abscission in the red kidney bean *Phaseolus vulgaris*. Plant Physiol. 53:732–37.
REINERT, J., and H. URSPRUNG, eds. 1971. Origin and Continuity of Cell Organelles. Springer-Verlag, Berlin.
RETHKE, R. V. 1946. The anatomy of circumscissile dehiscence. Am. J. Bot. 33: 677–83.
REYNOLDS, D. R. 1971. Wall structure of a bitunicate ascus. Planta 98:244–57.
REYNOLDS, E. R. C. 1975. Tree rootlets and their distribution. Pp. 163–77 in J. G. Torrey and D. T. Clarkson (eds.), The Development and Function of Roots. Academic Press, London.
RICH, S. 1964. Ozone damage to plants. Annu. Rev. Phytopathol. 2:253–66.
RICHARDS, L. A., ed. 1954. Diagnosis and Improvement of Saline and Alkaline Soils. U.S. Dep. Agric., Agric. Handbook No. 60.
RICHARDS, P. W. 1952. The Tropical Rain Forest. Cambridge Univ. Press, Cambridge.
RICHARDSON, D. E. 1976. Diagnosis of potato leaf-roll virus in post harvest glasshouse tests. Plant Pathol. 25:141–43.
RICK, C. M. 1967. Fruit and pedicel characters derived from Galápagos tomatoes. Econ. Bot. 21:171–84.
RILEY, C. V. 1882. Jumping seeds and galls. Proc. U.S. Nat. Mus. Wash. 5:632–35.

RIOV, J. 1974. A polygalacturonase from *Citrus* leaf explants. Role in abscission. Plant Physiol. 53:312–16.
ROBERTS, A. N., and R. L. TICKNOR. 1970. Commercial production of English holly in the Pacific Northwest. Am. Hortic. Mag. 49:301–14.
RODGERS, J. P. 1977. Plant growth substances in relation to fruit development and fruit abscission in cotton. Ph.D. dissertation. Univ. Rhodesia, Salisbury.
———. 1980. Cotton fruit development and abscission: Growth and morphogenesis. S. Afr. J. Sci. 76:90–92.
ROMBERGER, J. A. 1963. Meristems, Growth and Development in Woody Plants. U.S. Dep. Agric. For. Serv. Tech. Bull. No. 1293.
ROSAS, C., J. H. COCK, and G. SANDOVAL. 1976. Leaf fall in cassava. Exp. Agric. 12:395–400.
ROTHWELL, K., and R. L. WAIN. 1964. Studies on a growth inhibitor in yellow lupin (*Lupinus luteus* L.). Pp. 363–75 in J. P. Nitsch (ed.), Régulateurs Naturels de la Croissance Végétale. Cent. Natl. Rech. Sci., Paris.
ROY, B. K., and S. K. CHATTERJEE. 1967. Analysis of leaf-abscission in cotton (*Gossypium barbadense* L.). Indian J. Plant Physiol. 10:119–29.
RUBINSTEIN, B., and A. C. LEOPOLD. 1962. Effects of amino acids on bean leaf abscission. Plant Physiol. 37:398–401.
———. 1963. Analysis of the auxin control of bean leaf abscission. Plant Physiol. 38:262–67.
RUDNICKI, R., and J. PIENIĄŻEK. 1970. The changes in concentration of abscisic acid (ABA) in developing and ripe apple fruits. Bull. Acad. Pol. Sci. Ser. Sci. Biol. 18:577–80.
Rudnicki, R., J. Pieniążek, and N. PIENIĄŻEK. 1968. Abscisin II in strawberry plants at two different stages of growth. Bull. Acad. Pol. Sci. Ser. Sci. Biol. 16:127–30.
Rudnicki, R., J. MACHNIK, and J. PIENIĄŻEK. 1968. Accumulation of abscisic acid during ripening of pears (Clapp's Favourite) at different storage conditions. Bull. Acad. Pol. Sci., Ser. V, Sci. Biol. 16:509–12.
RUTLAND, J. 1888. The fall of the leaf. Trans. N.Z. Inst. 21:110–20.
SACHER, J. A. 1973. Senescence and postharvest physiology. Annu. Rev. Plant Physiol. 24:197–224.
SAGEE, O., R. GOREN, and J. RIOV. 1980. Abscission of *Citrus* leaf explants. Interrelationships of abscisic acid, ethylene, and hydrolytic enzymes. Plant Physiol. 66:750–53.
SAMET, J. S., and T. R. SINCLAIR. 1980. Leaf senescence and abscisic acid in leaves of field-grown soybean. Plant Physiol. 66:1164–68.
SAMPSON, H. C. 1918. Chemical changes accompanying abscission in *Coleus blumei*. Bot. Gaz. 66:32–53.
SAMUEL, G. 1927. On the shot-hole disease caused by *Clasterosporium carpophilum* and on the "shot-hole" effect. Ann. Bot. 41:375–404.
SANDSTEDT, R. 1971. Cytokinin activity during development of cotton fruit. Physiol. Plant. 24:408–10.
SASTRY, K. K. S., and R. M. MUIR. 1963. Gibberellin: effect on diffusible auxin in fruit development. Science 140:494–95.
SAUNDERS, P. 1978. Phytohormones and bud dormancy. Pp. 423–45 in D. S. Letham, P. B. Goodwin, and T. J. V. Higgins (eds.), Phytohormones and Re-

lated Compounds—A Comprehensive Treatise, Vol. II. Elsevier/North-Holland Biomedical Press, Amsterdam.
SAUNDERS, P. F., M. A. HARRISON, and R. ALVIM. 1974. Abscisic acid and tree growth. Pp. 871–81 in Y. Sumiki (ed.), Plant Growth Substances 1973. Hirokawa Publishing, Tokyo.
SCHAFFALITZKY de MUCKADELL, M. 1961. Environmental factors in development stages of trees. Pp. 289–97 in T. T. Kozlowski (ed.), Tree Growth. Ronald Press, New York.
SCHNEIDER, G. W. 1975. Ethylene evolution and apple fruit thinning. J. Am. Soc. Hortic. Sci. 100:356–59.
———. 1977. Studies on the mechanism of fruit abscission in apple and peach. J. Am. Soc. Hortic. Sci. 102:179–81.
SCHNEIDER, H., J. W. CAMERON, R. K. SOOST, and E.C. CALAVAN. 1961. Classifying certain diseases as inherited. Pp. 15–21 in W. C. Price (ed.), Proc. of the 2nd Conf. of the Intl. Organization of Citrus Virologists. Univ. Florida Press, Gainesville.
Schneider, H., R. G. PLATT, W. P. BITTERS, and R. M. BURNS. 1978. Diseases and incompatibilities that cause decline in lemons. Citrograph 63:219–21.
SCHOPF, J. W. 1972. Evolutionary significance of the Bitter Springs (late Precambrian) microflora. Pp. 68–77 in Proc. XXIV Intl. Geol. Cong., Sect. 1, Montreal.
———. 1978. The evolution of the earliest cells. Sci. Am. 239:111–38.
Schopf, J. W., and D. Z. OEHLER. 1976. How old are the eukaryotes? Science 193:47–49.
SCHROEDER, C. A. 1945. Tree foliation affected by street lights. Arborists News 10:1–3.
SCHWEBEL, J. P. O. 1973. Hormonal Control of Growth and Development in *Lemna minor* L., with Special Emphasis on the Role of Abscisic Acid (ABA). Ph.D. dissertation. Texas A & M Univ., College Station, Tex.
SCHWERTNER, H. A., and P. W. MORGAN. 1966. Role of indoleacetic acid oxidase in abscission control in cotton. Plant Physiol. 41:1513–19.
SEELEY, J. G. 1950. Potassium deficiency of greenhouse roses. Proc. Am. Soc. Hortic. Sci. 56:466–70.
SEN, S. 1968. An induced mutant of *Corchorus olitorius* L. with enhanced abscission rate. Ann. Bot. 32:863–66.
SENTANDREU, R., and D. H. NORTHCOTE. 1969. The formation of buds in yeast. J. Gen. Microbiol. 55:393–98.
SEQUEIRA, L., and A. KELMAN. 1962. The accumulation of growth substances in plants infected by *Pseudomonas solanacearum*. Phytopathology 52:439–48.
Sequeira, L., and T. A. STEEVES. 1954. Auxin inactivation and its relation to leaf drop caused by the fungus *Omphalia flavida*. Plant Physiol. 29:11–16.
SEWARD, A. C. 1931. Plant Life Through the Ages. Cambridge Univ. Press, Cambridge.
SEXTON, R. 1976. Some ultrastructural observations on the nature of foliar abscission in *Impatiens sultani*. Planta 128:49–58.
———. 1979. Spatial and temporal aspects of cell separation in the foliar abscission zones of *Impatiens sultani* Hook. Protoplasma 99:53–66.
Sexton, R., and J. L. HALL. 1974. Fine structure and cytochemistry of the abscis-

sion zone cells of *Phaseolus* leaves. I. Ultrastructural changes occurring during abscission. Ann. Bot. 38:849–54.

Sexton, R., and A. J. REDSHAW. 1981. The role of cell expansion in the abscission of *Impatiens* leaves. Ann. Bot. vol. 45.

Sexton, R., and G. G. C. JAMIESON, and M. H. I. L. ALLAN. 1977. An ultrastructural study of abscission zone cells with special reference to the mechanism of enzyme secretion. Protoplasma 91:369–87.

Sexton, R., M. L. DURBIN, L. N. LEWIS, and W. W. THOMSON. 1980. Use of cellulase antibodies to study leaf abscission. Nature 283:873–74.

SHAYBANY, B., S. A. WEINBAUM, and G. C. MARTIN. 1977. Identification of ABA stereoisomers in French prune seeds and association of ABA with ethylene-enhanced prune abscission. J. Am. Soc. Hortic. Sci. 102:501–3.

SHELDRAKE, A. R. 1973. The production of hormones in higher plants. Biol. Rev. 48:561–96.

SHOJI, K., and F. T. ADDICOTT. 1954. Auxin physiology in bean leaf stalks. Plant Physiol. 29:377–82.

Shoji, K., F. T. Addicott, and W. A. SWETS. 1951. Auxin in relation to leaf blade abscission. Plant Physiol. 26:189–91.

SIEGELMAN, H. W. 1951. The respiratory metabolism of flowers. Ph.D. dissertation. Univ. Calif., Los Angeles.

SIFTON, H. B. 1963. On the hairs and cuticle of Labrador tea leaves. A developmental study. Can. J. Bot. 41:199–207.

———. 1965. On the abscission region in leaves of the Blue Spruce. Can. J. Bot. 43:985–93.

SKOOG, F. 1940. Relationships between zinc and auxin in the growth of higher plants. Am. J. Bot. 27:939–51.

SLOGER, C., and B. E. CALDWELL. 1970. Response to cultivars of soybean to synthetic abscisic acid. Plant Physiol. 46:634–35.

SMITH, G. M. 1938. Cryptogamic Botany. Vol. II. Bryophytes and Pteridophytes. McGraw-Hill, New York.

———. 1950. The Fresh-Water Algae of the United States. 2nd ed. McGraw-Hill, New York.

———. 1969. Marine Algae of the Monterey Peninsula, California. 2nd ed. Stanford Univ. Press, Stanford, Calif.

SMITH, O. E., J. L. LYON, F. T. ADDICOTT, and R. E. JOHNSON. 1968. Abscission physiology of abscisic acid. Pp. 1547–60 in F. Wightman and G. Setterfield (eds.), Biochemistry and Physiology of Plant Growth Substances. Runge Press, Ottawa.

SMITH, P. F., and W. REUTHER. 1949. Observations on boron deficiency in *Citrus*. Proc. Fla. State Hortic. Soc. 62:31–38.

SOLMS LAUBACH, H. Graf zu. 1884. Der Aufbau des Stockes von *Psilotum triquetrum* und dessen Entwicklung aus der Brutknospe. Ann. Jard. Bot. Buitenzorg 4:139–86.

SORBER, D. G., and M. H. KIMBALL. 1950. Use of Ethylene in Harvesting the Persian Walnut (*Juglans regia*) in California. U.S. Dep. Agric. Tech. Bull. No. 996.

SPENCER, P. W., and J. S. TITUS. 1972. Biochemical and enzymatic changes in apple leaf tissue during autumnal senescence. Plant Physiol. 49:746–50.

SPRAGUE, H. B., ed. 1964. Hunger Signs in Crops. 3rd ed. McKay, New York.
STANLEY, R. G., and E. G. KIRBY. 1973. Shedding of pollen and seeds. Pp. 295–340 in T. T. Kozlowski (ed.), Shedding of Plant Parts. Academic Press, New York.
STARK, J. 1876. On the shedding of branches and leaves in the Coniferae. Trans. R. Soc. Edinb. 27:651–60.
STEAD, A. D., and K. G. MOORE. 1979. Studies on flower longevity in *Digitalis*. Pollination induced corolla abscission in *Digitalis* flowers. Planta 146:406–14.
STEMBRIDGE, G. E., and C. E. GAMBRELL. 1972. Peach fruit abscission as influenced by applied gibberellin and seed development. J. Am. Soc. Hortic. Sci. 97:708–11.
STOKEY, A. G. 1948. Reproductive structures of the gametophytes of *Hymenophyllum* and *Trichomanes*. Bot. Gaz. 109:363–80.
Stokey, A. G., and L. R. ATKINSON. 1958. The gametophyte of the Grammitidaceae. Phytomorphology 8:391–403.
STOPES, M. C. 1910. Adventitious budding and branching in *Cycas*. New Phytol. 9:235–41.
STOREY, J. B. 1957. Auxin physiology of cotton in relation to leaf abscission. Ph.D. dissertation. Univ. Calif., Los Angeles.
STÖSSER, R. 1969. Histoautoradiographische Lokalisierung von ^{45}Calcium in der Trennzone der Früchte von Süss- und Sauerkirsche. Z. Pflanzenphysiol. 61:314–21.
———. 1971. Über Beziehungen zwischen Trenngewebeausbildung und Haltekräften bei einigen Süss- und Sauerkirschensorten. Gartenbauwissenschaft 36:105–14.
———. 1975. Die Wirkung von Calcium auf die Trenngewebeausbildung bei Explantaten von Sauerkirschenfrüchten. Gartenbauwissenschaft 40:113–16.
Stösser, R., H. P. RASMUSSEN, and M. J. BUKOVAC. 1969. A histological study of abscission layer formation in cherry fruits during maturation. J. Am. Soc. Hortic. Sci. 94:239–43.
STOWE, B. B., and T. YAMAKI. 1957. The history and physiological action of the gibberellins. Annu. Rev. Plant Physiol. 8:181–216.
STROBEL, G. A. 1976. Toxins of plant pathogenic bacteria and fungi. Pp. 135–59 in J. Friend and D. R. Threlfall (eds.), Biochemical Aspects of Plant-Parasite Relationships. Academic Press, London.
STROHL, W. R., and J. M. LARKIN. 1978. Cell division and trichome breakage in *Beggiatoa*. Curr. Microbiol. 1:151–55.
STRONG, F. E., and E. C. KRUITWAGEN. 1968. Polygalacturonase in the salivary apparatus of *Lygus hesperus* (Hemiptera). J. Insect Physiol. 14:1113–19.
SULLIVAN, J. W. N. 1938. Isaac Newton 1642–1727. Macmillan, New York.
SUTCLIFFE, J. F., P. D. ARCH, P. A. LEGGETT, B. J. PHILLIPS, and R. SEXTON. 1969. Enzymic changes occurring during the development of the abscission zone of *Coleus blumei*. Pp. 213, Abstracts. XI Int. Bot. Congr. Seattle, Wash.
SUTTLE, J. C., and H. KENDE. 1978. Ethylene and senescence in petals of *Tradescantia*. Plant Physiol. 62:267–71.
SWANSON, B. T., JR., H. F. WILKINS, C. F. WEISER, and I. KLEIN. 1975. Endogenous ethylene and abscisic acid relative to phytogerontology. Plant Physiol. 55:370–76.

SWEET, G. B. 1973. Shedding of reproductive structures in forest trees. Pp. 341–82 in T. T. Kozlowski (ed.), Shedding of Plant Parts. Academic Press, New York.
SWETS, W. A., and F. T. ADDICOTT. 1955. Experiments on the physiology of defoliation. Proc. Am. Soc. Hortic. Sci. 65:291–95.
TAKAHASHI, N., I. YAMAGUCHI. T. KŌNO, M. IGOSHI, K. HIROSE, and K. SUZUKI. 1975. Characterization of plant growth substances in *Citrus unshiu* and their change in fruit development. Plant Cell Physiol. 16:1101–11.
TAKAKI, H., and M. KUSHIZAKI. 1970. Accumulation of free tryptophan and tryptamine in zinc deficient maize seedlings. Plant Cell Physiol. 11:793–804.
TALBOYS, P. W. 1968. Water deficits in vascular disease. Pp. 255–331 in T. T. Kozlowski (ed.), Water Deficits and Plant Growth. Vol. II. Academic Press, New York.
TAMAS, I. A., D. H. WALLACE, P. M. LUDFORD, and J. L. OZBUN. 1979. Effect of older fruits on abortion and abscisic acid concentration of younger fruits in *Phaseolus vulgaris* L. Plant Physiol. 64:620–22.
TeBEEST, D., R. D. DURBIN, and J. E. KUNTZ. 1973. Anatomy of leaf abscission induced by oak wilt. Phytopathology 63:252–56.
TEICH, A. H. 1970. Cone serotiny and inbreeding in natural populations of *Pinus banksiana* and *Pinus contorta*. Can. J. Bot. 48:1805–9.
TERPSTRA, W. 1956. Some factors influencing abscission of debladed leaf petioles. Acta Bot. Neerl. 5:157–70.
TEUBNER, F. G., and A. E. MURNEEK. 1955. Embryo abortion as mechanism of "hormone" thinning of fruit. Univ. Mo. Agric. Exp. Stn. Bull. No. 590.
THEOPHRASTUS. 285 B.C. Enquiry into Plants. English translation by A. Hort. Putnam's, New York (1916).
THOMPSON, A. H. 1957. Chemical thinning of apples. Univ. Md. Agric. Exp. Stn. Bull. No. A-88.
THOMPSON, C. R., E. G. HENSEL, G. KATZ, and O. C. TAYLOR. 1970. Effects of continuous exposure of navel oranges to nitrogen dioxide. Atmos. Environ. 4:349.
TIFFANY, L. H. 1924. A physiological study of growth and reproduction among certain green algae. Ohio J. Sci. 24:65–98.
TISON, A. 1899. Sur la cicatrisation du système fasciculaire et celle de l'appareil sécréteur lors de la chute des feuilles. Compt. Rend. Acad. Paris 129:125–27.
———. 1900. Recherches sur la chute des feuilles chez les dicotyledonées. Mem. Soc. Linn. Normandie 20:121–37.
TOMASZEWSKA, E. 1964. Phenols and auxin as internal factors controlling leaf abscission. Bull. Acad. Pol. Sci., Ser. Sci. Biol. 12:541–45.
———. 1968. The naturally occurring regulators of leaf abscission in *Deutzia*. Arbor. Kornickie 13:173–215.
Tomaszewska, E., and M. TOMASZEWSKI. 1970. Endogenous growth regulators in fruit and leaf abscission. Nesz. Nauk. Uniw. Mikolaja Kopernika Toruniu Nauki Mat-Przyr. Biol. 23:45–53.
TRESHOW, M. 1970. Environment and Plant Response. McGraw-Hill, New York.
———. 1971. Fluorides as air pollutants affecting plants. Annu. Rev. Phytopathol. 9:21–44.
TRIPPI, V. S., and J. BONINSEGNA. 1966. Studies on ontogeny and senility in

plants. XIII. Effect of gibberellic acid on abscission and growth in young and adult plants of *Morus* and *Robinia*. Phyton 23:1–4.
TRONSMO, A., and J. RAA. 1977. Life cycle of the dry eye rot pathogen *Botrytis cinerea* on apple. Phytopathol. Z. 89:203–7.
UNRATH, C. R. 1978. The development of ethephon's thinning potential for spur 'Delicious' apples. Acta Hortic. 80:233–43.
VAADIA, Y., F. C. RANEY, and R. M. HAGAN. 1961. Plant water deficits and physiological processes. Annu. Rev. Plant Physiol. 12:265–92.
VALDOVINOS, J. G., and T. E. JENSEN. 1968. Fine structure of abscission zones. II. Cell-wall changes in abscising pedicels of tobacco and tomato flowers. Planta 83:295–302.
———. 1974. Abscission: A comparison of cell wall changes at the fine structural level during natural and ethylene-promoted abscission. Bull. R. Soc. N.Z. 12:855–61.
Valdovinos, J. G., T. E. Jensen, and L. M. SICKO. 1974. Abscission: cellular changes at the ultrastructural level. Pp. 1034–41 in Y. Sumiki (ed.), Plant Growth Substances 1973. Hirokawa Publishing, Tokyo.
Van der PIJL, L. 1952. Absciss-joints in the stems and leaves of tropical plants. Ned. Akad. Wetensch. Ser. C. 55:574–86.
———. 1953. The shedding of leaves and branches of some tropical trees. Indonesian J. Nat. Sci. Nos. 1–3. Pp. 11–25.
Van der VOSSEN, H. A. M., and G. BROWNING. 1978. Prospects of selecting genotypes of *Coffea arabica* L. which do not require tonic sprays of fungicide for increased leaf retention and yield. J. Hortic. Sci. 53:225–33.
Van der WOUDE, W. J., D. J. MORRÉ, and C. E. BRACKER. 1971. Isolation and characterization of secretory vesicles in germinated pollen of *Lilium longiflorum*. J. Cell Sci. 8:331–51.
Van SCHAIK, P. H., and A. H. PROBST. 1958. Effects of some environmental factors on flower production and reproductive efficiency in soybeans. Agron. J. 50:192–97.
Van STADEN, J. 1976. Seasonal changes in the cytokinin content of *Ginkgo biloba* leaves. Physiol. Plant. 38:1–5.
———. 1977. Seasonal changes in the cytokinin content of the leaves of *Salix babylonica*. Physiol. Plant. 40:296–99.
Van Staden, J., and J. E. DAVEY. 1978. Endogenous cytokinins in the laminae and galls of *Erythrina latissima* leaves. Bot. Gaz. 139:36–41.
Van STEVENINCK, R. F. M. 1957. Factors affecting the abscission of reproductive organs in yellow lupins (*Lupinus luteus* L.). I. The effect of different patterns of flower removal. J. Exp. Bot. 8:373–81.
———. 1959a. Abscission-accelerators in lupins (*Lupinus luteus* L.). Nature 183:1246–48.
———. 1959b. Factors affecting the abscission of reproductive organs of yellow lupins (*Lupinus luteus* L.). III. Endogenous growth substances in virus-infected and healthy plants and their effects on abscission. J. Exp. Bot. 10:367–76.
———. 1972. Abscisic acid stimulation of ion transport and alteration in K^+/Na^+ selectivity. Z. Pflanzenphysiol. 67:282–86.
VARMA, S. K. 1976. Role of abscisic acid in the phenomena of abscission of flower

buds and bolls of cotton (*Gossypium hirsutum*) and its reversal with other plant regulators. Biol. Plant. 18:421–28.

———. 1977. Variation in the endogenous abscisic acid, gibberellin, auxins, ascorbic acid and nitrogen content in retained and abscising bolls of cotton (*Gossypium hirsutum* L.). Plant Biochem. J. 4:19–27.

VARNER, J. E. 1961. Biochemistry of senescence. Annu. Rev. Plant Physiol. 12: 245–64.

Varner, J. E., and D. T.-H. HO. 1976. Hormones. Pp. 713–70 in J. Bonner and J. E. Varner (eds.), Plant Biochemistry, 3rd ed. Academic Press, New York.

———. 1977. Hormonal control of enzyme activity in higher plants. Pp. 83–92 in H. Smith (ed.), Regulation of Enzyme Synthesis and Activity in Higher Plants. Academic Press, London.

VAUGHAN, A. K. F., and G. C. BATE. 1977. Changes in the levels of ethylene, abscisic-acid-like substances and total non-structural carbohydrate in young cotton bolls in relation to abscission induced by a dark period. Rhod. J. Agric. Res. 15:51–63.

VENDRIG. J. C. 1960. On the abscission of debladed petioles in *Coleus rhenaltianus* especially in relation to the effect of gravity. Wentia 3:1–96.

VENKATESH, C. S. 1955. The structure and dehiscence of the anther in *Memecylon* and *Mouriria*. Phytomorphology 5:435–40.

VERBOOM, W. C. 1966. *Brachiaria dura*, a promising new forage grass. J. Range Manage. 19:91–93.

VIEIRA da SILVA, J. B. 1973. Influence de la sécheresse sur la photosynthèse et la croissance du cotonnier. Pp. 213–20 in R. O. Slatyer (ed.), Plant Response to Climatic Factors. Proc. Uppsala Symposium, UNESCO, Paris.

Von BRETFELD, H. 1880. Ueber Vernarbung und Blattfall. Jahrb. Wiss. Bot. 12: 133–60.

Von GUTTENBERG, H. 1926. Die Bewegungsgewebe. *In* K. Linsbauer (ed.), Handbuch der Pflanzenanatomie. Abt. I., Teil 2., Band V., Lief. 18. Borntraeger, Berlin.

Von HÖHNEL, F. R. 1880. Weitere Untersuchungen über den Ablösungsvorgang von verholzten Zweigen. Mitt. Forstl. Versuchswesen Oesterr. 2:247–56.

Von MOHL, H. 1860a. Ueber die anatomischen Veränderungen des Blattgelenkes, welche das Abfallen der Blätter herbeiführen. Bot. Zeitg. 18:1–7; 10–17.

———. 1860b. Einige nachträgliche Bemerkungen zu meinem Aufsatze über den Blattfall. Bot. Zeitg. 18:132–33.

———. 1860c. Ueber den Ablösungsprocess saftiger Pflanzenorgane. Bot. Zeitg. 18:273–77.

WALHOOD, V. T. 1955. Abscission in cotton. Ph.D. dissertation. Univ. Calif., Los Angeles.

———. 1957. The effect of gibberellins on boll retention and cut-out in cotton. Pp. 24–31, Proc. 12th Annu. Beltwide Cotton Defoliation Physiol. Conf. (Nat. Cotton Council, Memphis).

WALKER, D. C. 1980. Sorus Abscission from Laminae of *Nereocystis luetkeana* (Mert.) Post. and Rup. Ph.D. dissertation. Univ. British Columbia, Vancouver.

WALSH, C. S. 1977. The relationship between endogenous ethylene and abscission of mature apple fruits. J. Am. Soc. Hortic. Sci. 102:615–19.

WAREING, P. F. 1956. Photoperiodism in woody plants. Annu. Rev. Plant Physiol. 7:191–214.
———. 1957. Endogenous inhibitors in seed germination and dormancy. Physiol. Plant. 10:266–80.
Wareing, P. F., and A. G. THOMPSON. 1975. Rapid effects of red light on hormone levels. Pp. 285–94 in H. Smith (ed.), Light and Plant Development. Butterworths, London.
Wareing, P. F., C. F. EAGLES, and P. M. ROBINSON. 1964. Natural inhibitors as dormancy agents. Pp. 377–86 in J. P. Nitsch (ed.), Régulateurs Naturels de la Croissance Végétale. Cent. Natl. Rech. Sci., Paris.
WARNE, L. G. G. 1947. Bud and flower dropping in lupins. J. R. Hortic. Soc. 62:193–95.
WASSCHER, J. 1947. The prevention of bud- and flower-drop in begonias by spraying with growth substance solutions. Meded. Direct. Tuinbouw 10:547–55.
WAY, J. M., and R. J. CHANCELLOR. 1977. Herbicides and higher plant ecology. Pp. 345–72 in L. J. Audus (ed.), Herbicides. Physiology, Biochemistry, Ecology. Vol. 2. Academic Press, London.
WEAVER, G. M., and H. O. JACKSON. 1968. Relationship between bronzing in white beans and phytotoxic levels of atmospheric ozone in Ontario. Can. J. Plant Sci. 48:561.
WEAVER, R. J. 1973. Altering set and size of grapes with growth regulators. Acta Hortic. 34(1):275–78.
Weaver, R. J., and R. M. POOL. 1969. Effect of ethrel, abscisic acid, and a morphactin on flower and berry abscission and shoot growth in *Vitis vinifera*. J. Am. Soc. Hortic. Sci. 94:474–78.
WEBSTER, B. D. 1968. Anatomical aspects of abscission. Plant Physiol. 43:1512–44.
———. 1973. Ultrastructural studies of abscission in *Phaseolus*: ethylene effects on cell walls. Am. J. Bot. 60:436–47.
———. 1974. Characteristics of abscission in *Phaseolus* plants treated with 2-chloroethylphosphonic acid. R. Soc. N.Z. Bull. 12:863–69.
Webster, B. D., and H. W. CHIU. 1975. Ultrastructural studies of abscission in *Phaseolus*: characteristics of the floral abscission zone. J. Am. Soc. Hortic. Sci. 100:613–18.
Webster, B. D., and A. C. LEOPOLD. 1972. Stem abscission in *Phaseolus vulgaris* explants. Bot. Gaz. 133:292–98.
Webster, B. D., T. W. DUNLAP, and M. E. CRAIG. 1976. Ultrastructural studies of abscission in *Phaseolus*: localization of peroxidase. Am. J. Bot. 63:759–70.
WEIER, T. E., C. R. STOCKING, and M. G. BARBOUR. 1974. Botany. 5th Edition. Wiley, New York.
WEINBAUM, S. A., and L. E. POWELL. 1975. Diffusible abscisic acid and its relationship to leaf age in tea crabapple. J. Am. Soc. Hortic. Sci. 100:583–86.
Weinbaum, S. A., J. M. LABAVITCH, and Z. WEINBAUM, 1979. The influence of ethylene treatment of immature fruit of prune (*Prunus domestica* L.) on the enzyme-mediated isolation of mesocarp cells and protoplasts. J. Am. Soc. Hortic. Sci. 104:278–80.
WEINHEIMER, W. H., and G. W. WOODBURY. 1967. The anatomy and morphol-

ogy of the abscission layer in two cultivars of *Solanum tuberosum* L. Am. Potato J. 44:402–8.

WEINSTEIN, L. H. 1957. Senescence of roses. I. Chemical changes associated with senescence of cut Better Times roses. Contrib. Boyce Thompson Inst. 19:33–48.

———. 1977. Fluoride and plant life. J. Occup. Med. 19:49–78.

Weinstein, L. H., and D. C. McCUNE. 1971. Effects of fluoride on agriculture. J. Air. Pollut. Control Assoc. 21:410–13.

WELD, L. H. 1957. Cynipid Galls of the Pacific Slope. Weld, Ann Arbor, Mich.

WELLENSIEK, S. J., ed. 1973. Symposium on Growth Regulators in Fruit Production. Acta Hortic. 34, Vol. I (pp. 1–507) and Vol. II (pp. 1–85).

WENT, F. W. 1942. The dependence of certain annual plants on shrubs in Southern California deserts. Bull. Torrey Bot. Club. 69:100–14.

Went, F. W., and K. V. THIMANN. 1937. Phytohormones. Macmillan, New York.

WERTHEIM, S. J. 1973. Chemical control of flower and fruit abscission in apple and pear. Acta Hortic. 34(1):321–31.

WESTER, H. V., and P. C. MARTH. 1950. Growth regulators prolong the bloom of oriental flowering cherries and dogwood. Science 111:611.

WETMORE, R. H., and W. P. JACOBS. 1953. Studies on abscission: the inhibiting effect of auxin. Am. J. Bot. 40:272–76.

WHITHAM, T. G. 1978. Habitat selection by *Pemphigus* aphids in response to resource limitation and competition. Ecology 59:1164–76.

WHITING, A. G., and M. A. MURRAY. 1948. Abscission and other responses induced by 2,3,5-triiodobenzoic acid in bean plants. Bot. Gaz. 109:447–73.

WHITNEY, P. J. 1977. Microbial Plant Pathology. Pica Press, New York.

WHITTICK, A., and R. G. HOOPER. 1977. The reproduction and phenology of *Antithamnion cruciatum* (Rhodophyta: Ceramiaceae) in insular Newfoundland. Can. J. Bot. 55:520–24.

WIATR, S. M. 1978. Physiology and ultrastructure of petal abscission in western blue flax (*Linum lewisii*). Ph.D. dissertation. Univ. Calif., Davis.

WIESE, M. V., and J. E. DeVAY. 1970. Growth regulator changes in cotton associated with defoliation caused by *Verticillium albo-atrum*. Plant Physiol. 45:304–9.

WIESNER, J. 1871. Untersuchung über die herbstliche Entblätterung der Holzgewäsche. Sitzungsber. Akad. Wiss. Wien, Math.-Naturwiss. Kl. Abt. I. 64:465–509.

———. 1904a. Über Laubfall infolge Sinkens des absoluten Lichtgenusses (Sommerlaubfall). Ber. Dtsch. Bot. Ges. 22:64–72.

———. 1904b. Über den Treiblaubfall und über Ombrophilie immergrüner Holzgewächse. Ber. Dtsch. Bot. Ges. 22:316–23.

———. 1904c. Über den Hitzlaubfall. Ber. Dtsch. Bot. Ges. 22:501–4.

———. 1905. Über Frostlaubfall nebst Bemerkungen über die Mechanik der Blattablösung. Ber Dtsch. Bot. Ges. 23:49–60.

WILCOX, H. E., F. J. CZABATOR, G. GIROLAMI, D. E. MORELAND, and R. F. SMITH. 1956. Chemical debarking of some pulpwood species. Syracuse Univ., Coll. For. Tech. Publ. 77.

WILLIAMS, M. W., and L. P. BATJER. 1964. Site and mode of action of 1-naphthyl N-methylcarbamate (Sevin) in thinning apples. Proc. Am. Soc. Hortic. Sci. 85:1–10.
WILLIAMS, R. F. 1955. Redistribution of mineral elements during development. Annu. Rev. Plant Physiol. 6:25–42.
WILLIAMSON, C. E., and A. W. DIMOCK. 1953. Ethylene from diseased plants. Pp. 881–86 in Plant Diseases. The Yearbook of Agriculture 1953. U.S. Dep. Agric., Washington, D.C.
WILLIAMSON, G. B. 1976. A note on the abscission of spiny leaves of palms. Principes 20:116–18.
WILLIS, J. C. 1973. A Dictionary of the Flowering Plants and Ferns. 8th ed. Cambridge Univ. Press, London.
WILSON, W. C. 1978. The mode of action of growth regulators and other chemicals in loosening *Citrus* fruit. Acta Hortic. 80:265–70.
WITTWER, S. H. 1954. Control of flowering and fruit setting by plant regulators. Pp. 62–80 in H. B. Tukey (ed.), Plant Regulators in Agriculture. Wiley, New York.
Wittwer, S. H., and M. J. BUKOVAC. 1958. The effects of gibberellin on economic crops. Econ. Bot. 12:213–55.
WITZTUM, A. 1974. Abscission and the axillary frond in *Spirodela oligorhiza* (Lemnaceae). Am. J. Bot. 61:805–8.
Witztum, A., and O. KEREN. 1978a. Factors affecting abscission in *Spirodela oligorhiza* (Lemnaceae). I. Ultraviolet radiation. New Phytol. 80:107–10.
———. 1978b. Factors affecting abscission in *Spirodela oligorhiza* (Lemnaceae). II. Sucrose. New Phytol. 80:111–15.
WOOD, R. K. S. 1967. Physiological Plant Pathology. Blackwell Scientific Publications, Oxford.
WOODS, D. L., and K. W. CLARK. 1976. Preliminary observations on the inheritance of non-shattering habit in wild rice. Can. J. Plant Sci. 56:197–98.
WOODWELL, G. M. 1974. Variation in the nutrient content of leaves of *Quercus alba*, *Quercus coccinea*, and *Pinus rigida* in the Brookhaven Forest from budbreak to abscission. Am. J. Bot. 61:749–53.
WORLEY, C. L., and R. G. GROGAN. 1941. Defoliation of certain species as affected by α-naphthaleneacetic acid treatment. J. Tenn. Acad. Sci. 16:326–28.
WRIGHT, S. T. C. 1956. Studies of fruit development in relation to plant hormones. III. Auxins in relation to fruit morphogenesis and fruit drop in the black currant *Ribes nigrum*. J. Hortic. Sci. 31:196–211.
———. 1968. Multiple and sequential roles of plant growth regulators. Pp. 521–42 in F. Wightman and G. Setterfield (eds.), Biochemistry and Physiology of Plant Growth Substances. Runge Press, Ottawa.
———. 1978. Phytohormones and stress phenomena. Pp. 495–536 in D. S. Letham, P. B. Goodwin, and T. J. V. Higgins (eds.), Phytohormones and Related Compounds—A Comprehensive Treatise. Vol. II. Elsevier/North-Holland Biomedical Press, Amsterdam.
WULLSTEIN, L. H., and S. A. PRATT. 1981. Scanning electron microscopy of rhizosheaths of *Oryzopsis hymenoides*. Am. J. Bot. 68:408–19.

Wullstein, L. H., M. L. BRUENING, and W. B. BOLLEN. 1979. Nitrogen fixation associated with sand grain root sheaths (rhizosheaths) of certain xeric grasses. Physiol. Plant. 46:1–4.

YAGER, R. E. 1959. Effect of removal of leaves at various developmental stages upon floral abscission in tobacco. Phyton 13:125–131.

———. 1960a. Possible role of pectic enzymes in abscission. Plant Physiol. 35:157–62.

———. 1960b. A comparison of the mode of floral abscission in two varieties of *Nicotiana tabacum*. Proc. Iowa Acad. Sci. 67:86–91.

Yager, R. E., and R. M. MUIR. 1958a. Amino acid factor in control of abscission. Science 127:82–83.

———. 1958b. Interaction of methionine and indoleacetic acid in the control of abscission in *Nicotiana*. Proc. Soc. Exp. Biol. Med. 99:321–23.

YAMAGUCHI, I., and N. TAKAHASHI. 1976. Change of gibberellin and abscisic acid content during fruit development of *Prunus persica*. Plant Cell Physiol. 17:611–14.

YAMAGUCHI, S. 1954. Some interrelations of oxygen, carbon dioxide, sucrose and ethylene in abscission. Ph.D. dissertation. Univ. Calif., Los Angeles.

YAMAMURA, H., and R. NAITO. 1975. Mechanism of the thinning action of NAA in kaki fruits. I. Relation between NAA-induced fruit abscission and endogenous growth substances in fruit tissues. J. Jpn. Soc. Hortic. Sci. 43:406–14.

YEN, T. K. 1932. Carpel dehiscence in *Firmiana simplex*. Bot. Gaz. 93:205–12.

YOKOTA, T., M. OKABAYASHI, N. TAKAHASHI, I. SHIMURA, and K. UMEYA. 1974. Plant growth regulators in chestnut gall tissue and wasps. Pp. 28–38 in Y. Sumiki (ed.), Plant Growth Substances 1973. Hirokawa Publishing, Tokyo.

ZAUBERMAN, G., and M. SCHIFFMANN-NADEL. 1972. Pectin methylesterase and polygalacturonase in avocado fruit at various stages of development. Plant Physiol. 49:864–65.

ZEEVAART, J. A. D. 1971. (+)-Abscisic acid content of spinach in relation to photoperiod and water stress. Plant Physiol. 48:86–90.

———. 1976. Physiology of flower formation. Annu. Rev. Plant Physiol. 27:321–48.

ZIEGLER, H. 1959/60. Die Rhizothamnien bei *Comptonia peregrina* (L.) Coult. Mitt. Dtsch. Dendrol. Ges. 61:28–31.

ZOHARY, M., and G. ORSHAN. 1954. Ecological studies in the vegetation of the near eastern deserts. V. The *Zygophylletum dumosi* and its hydroecology in the Negev of Israel. Vegetatio 5–6:340–50.

ZUCCONI, F., R. STÖSSER, and M. J. BUKOVAC. 1969. Promotion of fruit abscission with abscisic acid. BioScience 19:815–17.

Index

Abortion in abscission, 13
　of shoot-tips, 52–53 (fig.)
ABA. *See* Abscisic acid
Abscise, definition, synonyms, 20
Abscised plant parts (list), 8
Abscisic acid (ABA). *See also* Hormone actions
　abscission hormone, 126
　abscission responses, 128–29 (figs. 122, 124), 313–14
　correlative occurrence, 127–28 (figs.), 311–13
　discovery, 126
　inhibition of α-amylase, 155
　interaction with ETH, GA, IAA, 123, 140–43 (fig. 124)
　mechanism of action, 129–30, 314
Abscisin I, 146
Abscisin II. *See* Abscisic acid
Abscission
　definition, 20
　ecological effects, 201
　evolution, 281–85
　plasticity, 283
　scope, 7–8
　time-course curves, 104–05
Abscission habits
　of branches, 10–11
　of flowers, fruits and seeds, 10
　of leaves, 9–10
Abscission indices, 104–05
Abscission zone, (figs. 5, 23, 24, 26)
　anatomy, 24–26
　definition, 20
　differentiation (ontogeny), 25
　morphology, 23–25
　number per leaf, 47
Abscissors (testing devices), 103–04
Absolute requirements for abscission, 19
Acacia
　coevolution with herbivores, 282
　number of abscising leaflets, 47
Acanthaceae fruit dehiscence, 41
Accelerators, naturally occurring, 146
　senescence factor, 147
Acer
　ABA and abscission, 126

abscission of marcescent leaves, 51
allelopathy, 201
photoperiod and leaf abscission, 187
Acid phosphatases, 170
　localization, 172–73 (figs.)
Acid rain, 200
Adansonia (baobab) architecture, 89 (figs. 90)
Adventitious abscission, 89–96
　definition, 20
　induced by ETH, GA, IAA, TIBA, 95–96
　induced by injury and disease, 92–95
　in leaves, 89, 91–94 (figs.)
　in stems, 92
Adventitious-bud abscission, 40, 54
Aegilops disseminule, 72
Aesculus, drought defoliation, 203
Agarum perforation development, 234
Agathis branch abscission, 56, 59 (fig. 57)
Agave
　adventitious-bud abscission, 54
　architecture, 85
Agricultural chemical-regulators
　hazards, 288
　limitations, 288
　uses, 290–306 (figs.)
Agropyron rhizosheaths, 78 (fig.)
Ailanthus, allelopathy, 201
Albizia, facultative deciduous habit, 206
Algal abscission, 224–34
Alkalinity, soil, and abscission, 198
Allelopathic factors in abscised parts, 201
Aloë
　adventitious leaf abscission, 91 (fig. 92)
　corolla abscission, 67 (fig.)
　dichotomous branching, 86
Alstonia, sympodial growth, 54
Alternaria, conidial abscission, 238 (fig. 240)
Ambrosiella, conidial abscission, 238 (fig. 239)
Amino acids as abscission substances, 145
Ammocharis leaf abscission, 47
Anagallis fruit dehiscence, 34 (fig. 35)
Anastatica, responses to flooding, 194
Andreaea capsule dehiscence, 224
Andricus, gall abscission, 212 (fig.)

Index

Animal abscission, 285–86
  similarity to plant abscission, 286
Annual habit, 280
  loss of leaf abscission, 280
Anomolous abscission, 96
  definition, 20
Antestia induction of coffee berry abscission, 211
Anther
  abscission, 67–68 (figs. 64–66)
  dehiscence, 37–38 (figs. 37, 38)
Anthocephalus branch abscission, 57–58
Anthonomus
  induction of flower bud abscission, 211–13
  production of PGU, 211–13
Anthracnose abscission in plane trees, 210
Antithamnion fragmentation, 234
Apple (*Malus* spp.)
  adventitious pedicel abscission, 96
  auxin correlation with fruit abscission, 114
  genetic aspects of fruit abscission, 258–59
  and Isaac Newton, 2
  preharvest drop prevention, 302 (fig.)
Araucaria, cladoptosis, 54
Arbutus
  bark abscission, 74 (fig. 76)
  failure of mistletoe establishment, 77
Arceuthobium (dwarf mistletoe) explosive dehiscence, 42
Architecture, plant, 81–89 (figs.)
  involvement of abscission, 81–83
Arctostaphylos bark abscission, 74
Aroid leaf scars, 30 (fig.)
Artemisia
  allelopathy, 201
  drought deciduous habit, 206
Artifacts of microtechnique, 174
Ascorbic acid and abscission, 146
Ascus dehiscence, 242–44 (figs. 246)
Aspergillus, source of malformin, 209
Asplenium, separation of plantlets, 219
Astragalus spine development, 47
Attritional abscission, 83 (figs. 84, 85)
Aulacomnium gemma abscission, 222
Autumnal defoliation, 202–03
  controlling factors, 202–03
Auxin and abscission. *See also* Hormone actions; IAA; NAA; 2,4-D
  acceleration, 116–22 (figs.)
  auxin-auxin balance, 123
  auxin gradient, 122–23
  auxin-regulators, 294
  correlative occurrence, 114 (figs. 115)
  delayed applications, 118 (fig. 119)
  history, 113–14
  mechanism of action, 123–26 (figs. 124)
  response curves, 120–21 (figs. 117, 119, 121, 122, 124)
  retardation, 116–22 (figs.)
  transport inhibitors, 122
Avocado (*Persea*)
  axillary bud abscission, 57
  vernal leaf abscission, 205
Azolla
  branch abscission, 218
  root abscission, 79, 218 (fig. 80)

Bacterial abscission (separation), 254–56
  in the Precambrian, 267
Bacterium savastanoi infections of olive leaf scars, 208
Ballistospore abscission and discharge, 244–47 (fig.)
Banksia, fire-induced dehiscence, 190
Baobab tree (*Adansonia*) architecture, 89 (figs. 90)
Bark abscission, 72–77 (figs.)
  chemical bark removal, 306
  root bark, 77
Basidiomycetous yeast, bud abscission, 253 (fig.)
Basidiospore discharge, 244–47 (fig.)
Bean (*Phaseolus vulgaris*)
  adventitious abscission induced by ETH, NaClO, TIBA, 95
  auxin occurrence and leaf abscission, 114 (fig. 115)
  explants, 99
  fruit abscission prevention, 299
  genetic aspects of leaf abscission, 259 (fig. 261)
  leaflet abscission, 47
  leaflet abscission zone, 24
  ozone-induced fruit abscission, 200
  photoperiod and flower bud abscission, 187
  ultrastructural changes, 168 (fig. 163)
Beet (*Beta vulgaris*) seed disseminule, 71
Beggiatoa fragmentation, 254
Benefits of abscission
  of bark, 18
  of branches, 17–18
  of flowers, fruits and seeds, 18
  of leaves, 17
Betula leaf abscission zone, 26 (fig.)
Biochemistry and abscission, 153–61
  cellulases, 156–58 (figs. 157–59)
  cell wall dynamics, 161
  lignases, 160–61
  pectinases, 158–60 (fig.)
Biotic factors in abscission, 207–16
  Anthonomus, 211–13
  general effects of insects, 210–13

insect galls, 214–16
Lygus, 211
mites, 213
pathogenic microorganisms, 207–10
Bird's nest fungi, peridiole abscission, 247–50 (figs. 248–49)
Bitunicate-ascus dehiscence, 244 (fig. 246)
Black currant (*Ribes nigrum*), auxin occurrence and fruit abscission, 114
Blue-green algae (Cyanophyta), 231–32
colony fragmentation, 231
hormogonia, 231 (fig. 232)
abscission in the Precambrian, 267 (fig. 268)
Boll weevil. See *Anthonomus*
Bombax, photoperiod and leaf abscission, 187
Boron deficiency and abscission, 196
Bothrodendron branch abscission, 58, 271
Botrychium, sporangial dehiscence, 220
Botrytis induction of apple abscission, 208
Bougainvillea
HF-induced leaf abscission, 200
NAA delay of bract abscission, 294 (fig. 295)
Brachystegia, deciduous habit, 206
Branch abscission, 56–61 (figs.)
action of wind, 194
anatomy, 56–57
chemical pruning, 305
cladoptosis, 54–56
of dead branches, 59 (fig. 60)
fire pruning, 305
of very large branches, 60
range of behavior, 8
self-pruning, 10
size of abscising branches, 57–58
and tree architecture, 10, 84–90
Broussonetia laticifer plugging, 29
Brown algae (Phaeophyta)
blade abscission, 232
blade splitting, 232 (fig. 234)
perforation development, 234
sorus abscission, 232 (fig. 233)
Brugiera (mangrove)
disseminule abscission, 72 (fig.)
abscission ultrastructure, 164–65 (figs.)
Bryophyllum
adventitious bud abscission, 54
adventitious internode abscission, 92
Bryopsis branchlet abscission, 229
Bud abscission
of higher plants, 52–54
of yeasts, 251–53 (figs.)
Bulblet abscission in ferns, 219
Bursera bark abscission, 74

Buxus, vernal leaf abscission, 203–04
Byrsocrypta gall dehiscence, 215

Cacataceae, disarticulation, 62
Calamites branch abscission, 271 (fig. 273)
Calcium
bridges in pectic substances, 177
deficiency and abscission, 112, 196
retardation of abscission, 112, 303, 305
Calocera ballistospore discharge, 247 (fig.)
Calochortus, septicidal dehiscence, 262 (fig. 263)
Caltha, dehiscence ontogeny, 33 (fig.)
Calyptra abscission of *Eucalyptus*, 63 (fig. 65)
Calyx abscission, 63 (figs. 64–66)
Candelabrum tree architecture, 86 (fig. 87)
Canthium, tree architecture, 85
Carbohydrate
abscission physiology, 110–11
and cell wall deposition, 110–11
and light intensity, 186
Carica (papaya)
architecture, 85 (fig. 86)
leaf scars, 30 (fig.)
Carpinus, abscission of marcescent leaves, 51, 203
Carya
bark abscission, 74
dehiscence promoted by ETH, 138
Cassia, ETH production by explants, 140
Castanea leaf abscission zone anatomy, 5 (fig.)
Castilla, tree architecture, 86 (fig.)
Catalpa leaf scar, 30 (fig.)
Catharanthus, influence of season on leaf abscission, 103
Cecidoses gall operculum dehiscence, 215
Cecropia, anther abscission, 68
Cedrus
branch abscission, 60
short shoot abscission, 56
Cellulase
ABA promotion of activity, 130 (fig.)
distribution on abscission zone, 157 (fig. 158)
in plant growth, 158
synthesis in abscission, 130 (fig.), 156–59 (figs.)
Cell wall
and carbohydrate nutrition, 110–11
changes in abscission, 170–84 (figs.)
dynamics, 161, 168–70
primary cell wall, 177 (figs. 173, 178)
retightening, 161
secondary cell wall, 177–84 (figs.)

Cenozoic era, 279–80
  abscission diversification, 280
  annual habit, 280
Cephaleuros, zoosporangial abscission, 230 (fig.)
Cephaliophora, conidial abscission, 240 (fig.)
Cephalozia capsule dehiscence, 255 (fig.)
Ceratocystis, induction of leaf abscission in *Quercus*, 208
Chaetosphaeria, conidial abscission, 237 (fig. 238)
Chemical defoliation, 290–93
  to control diseases, 291
  military use, 291–92
  physiology, 292–93
Cherry (*Prunus* spp.)
  floral cup abscission, 68 (figs. 69)
  fruit abscission, 71 (fig.)
  genetic aspects of fruit abscission, 258 (fig.)
CHI (cycloheximide)
  inhibition of PGU, 160
  promotion of abscission, 140, 305
  wound ETH induction, 140
Chlorophyta. *See* Green algae
Chloroplasts, 167–68
Christmas trees, control of needle abscission, 293
Chroococcus colony fragmentation, 231
Cinnamomum, vernal defoliation, 205
Cistus corolla abscission, 67, 194
Citron, persistent style, 68
Citrus
  drought-induced leaf abscission, 190, 207
  ETH responses, 136–37
  explants, 99
  flower-parts abscission, 63–68 (fig. 66)
  heat-induced leaf abscission, 189
  hypobaric abscission, 142
  leaf abscission habits, 45, 47, 190, 207
  leaflet abscission zone, 25
  persistent calyx, 63
  prevention of fruit abscission, 302
  promotion of fruit abscission, 305 (fig. 304)
  response to CHI, 140; to HF, 200; to NO_2, 200
  response to frost, 188–89
  shoot-tip abortion, 52 (fig. 53)
  style abscission, 68
  wall changes in abscission, 171 (fig. 173)
CK. *See* Cytokinin
Cladoptosis, 54–56 (fig.), 83
  definition, 20
Cladosporium and leaf abscission of almond, 208

Clastosporium, shot-hole diseases, 92 (fig. 93)
Clematis bark abscission, 74
Climacteric respiration, 109 (fig.)
Clostridium separation, 256
Cocos
  leaf abscission zone, 23
  leaf scars, 23 (fig. 31)
  tree architecture, 85 (fig. 86)
Codium propagule abscission, 229
Coevolution of abscission behavior, 280–83
Coffee (*Coffea* spp.)
  Antestia-induced abscission, 211
  Omphalia-induced abscission, 209
  responses to fruit thinning, 296
  tonic effect of copper, 293–94
Cold, 188–89
  in autumnal defoliation, 203, 281
Coleus
  as abscission research material, 99
  auxin occurrence and leaf abscission, 114 (fig. 115)
  ETH production by explants, 140 (fig. 141)
  explants, 99
  GA responses, 131–32
  micro-explants, 100
  seasonal influences on abscission, 103
  ultrastructural changes, 166 (fig.)
Combretum spine development, 47
Commiphora bark abscission, 74
Compensatory growth, 16
Competition in abscission, 13–15
  of flowers, 14–15
  of fruits, 14–15 (fig.)
  of leaves, 14
  in source-sink relations, 15
Compound leaves, development of, 48
Conidial abscission, 236–42 (figs.)
  ultrastructure, 238–41 (figs. 240–44)
Conidiobolus, conidial abscission and discharge, 241 (fig. 245)
Conotrachelus, induction of fruit abscission, 211
Constantinea blade abscission, 234 (fig. 235)
Cordia, tree architecture, 85
Coremiella, conidial abscission, 237 (fig. 238)
Corky bark disease, prevention of leaf abscission, 210
Cornus
  leaf scar tissues, 26 (fig.)
  dichotomous branching, 86
Correlative aspects of abscission, 11–15
  in homeostasis, 15–16
  of leaves, 11–13

in organ competition, 13–15
of petals, 12
Coryneum and leaf abscission of almond, 208
Cotton (*Gossypium* spp.)
adventitous stem abscission, 95
bud abscission from boll weevil, 211
corolla abscission, 63
dark-induced abscission, 128, 186
defoliation, 200, 290 (fig.)
ETH production by explants, 138 (fig.), 140
explants, 99–100 (fig. 101)
explant responses (*See under* Abscisic acid, Auxin, Gibberellin)
fruit dehiscence, 40–41 (fig.)
genetic aspects of fruit abscission, 259
genetics of caducous floral bracts, 264
heat-induced leaf abscission, 189
increased fruit-set, 300
leaf abscission from mites, 213
nitrogen and leaf abscission, 195–96
promotion of dehiscence, 301
responses to flooding, 194
responses to frost, 188–89
seasonal fruit abscission, 14 (fig.)
Cotyledon abscission, 47 (fig. 48)
Coumarin accelerated abscission, 137
Cowpea (*Vigna unguiculata*), genetic aspects of fruit abscission, 259
Crassula disarticulation, 62
Crataegus, ascorbic acid and leaf abscission, 146
Crenaticaulis rhizophore scars, 274
Cretaceous deciduous flora, 277 (fig. 279)
Crinum leaf abscission, 47
Critical period of response, 118
Cronartium and premature leaf abscission of *Ribes*, 208
Cryptantha calyx abscission, 63
Ctenopteris gemma abscission, 220 (fig.)
Cupressaceae bark abscission, 74
Cupressus, cladoptosis, 54
Cutin, 184
Cyanophyta. *See* Blue-green algae
Cyathea
architecture, 85
leaf scar development, 29, 218
Cyathus peridiole abscission, 247–50 (figs.)
Cybistetes leaf abscission, 47
Cycadaceae, leaf-base separation layers, 48, 91
Cycloheximide. *See* CHI
Cyclostigma leaf scars, 268 (figs. 269–70)
Cyphostemma branch abscission, 83 (fig. 82)
Cystopteris bulblet abscission, 219

Cytokinin (CK). *See also* Hormone actions
abscission responses, 134–35, 299
adventitious abscission induction, 96
correlative occurrence, 134
mechanism of action, 135
Cytology of abscission, 162–84
cell wall, 170–84
cytoplasmic, 162–70
Czekanowskiales, abscission in the Mesozoic, 277 (fig. 278)

Dactylosporium, conidial abscission, 238 (fig. 240)
Dawsonites, sporangial dehiscence, 268
Deciduous habits
definition, 20
drought deciduous, 190, 205–07
facultative deciduous habit, 206
vernal defoliation, 202–03
Definitions and terminology, 20–21
Defoliants, chemical, 290
physiological action, 292
Defoliator insects, abscission effects, 210
Defoliation
autumnal, 202–03
chemical-induced, 290–93 (fig.)
drought-induced, 190
vernal, 205
wiltants, use of, 291
Dehiscence
anatomy, 31–34
of anthers, 37–38 (figs.)
circumscissile, 34–37 (figs.)
definition, 21
of fern sporangia, 220–21 (figs. 222)
of fruits, 32–37 (figs. 34, 263, 265)
genetics, 262–65
of liverwort sporophyte, 225 (fig.)
loculicidal, 32, 262–63 (figs. 41, 263)
mechanical factors in, 40–43 (figs.)
ontogeny, 32–34 (figs.)
prevention, 301
promotion, 301
septicidal, 32, 262–63 (fig.)
Deightoniella, conidial abscission and discharge, 242 (fig. 245)
Deutzia, phenolics and abscission, 146
Diaspore disseminules of lichens, 253
Dichotomous branching, 86
Dicksonia architecture, 85
Dictyosomes, 162–68 (figs.)
and vesicles, 179 (fig.)
Dictyuchus, sporangial abscission, 236
Differentiation (ontogeny) of abscission zones, 25
Digitalis corolla abscission, 63

Dipteris, sporangial dehiscence, 222 (fig.)
Dipterocarpus cotyledon abscission, 47
Disarticulation
  of inflorescences, 68
  of stems, 62, 83
Disseminules, 71–72 (fig.)
Divided leaves, development of, 48
DNA and abscission, 154–55
Domingoella, conidial abscission, 238 (fig. 239)
Dracaena, dichotomous branching, 86
Drought
  in autumnal defoliation, 202–03
  in summer leaf abscission, 205–07
  variations in species patterns, 203
Drumopama, conidial abscission, 238 (fig. 239)
Drying in dehiscence, 40–41
Durian germination, 47 (fig. 48)
Durio cotyledon abscission, 47
Durvillaea blade splitting, 232

Ecballium fruit abscission, 39 (figs. 39, 40)
Ecology of abscission, 185–216
Elephants, seed dispersal by, 282
Encelia
  allelopathic substances from leaves, 201
  drought deciduous habit, 206
Endomembrane system, 162–68 (figs.)
Endoplasmic reticulum (ER), 162–68 (figs.)
  vesicles and plasmalemma, 179 (fig.)
Enzyme localization, 170
  acid phosphatase, 170 (figs. 172–73)
  peroxidase, 170 (figs. 171–72)
Ephedra stem disarticulation, 62
Eriogonum, anomalous axis abscission, 96
Erodium
  epidermal gall abscission, 216
  petal abscission, 63
Ervatamia, seasonal aspects of abscission, 103
Erythrina
  deciduous habit, 206
  photoperiod and leaf abscission, 187
Eschscholzia
  calyx abscission, 63 (fig. 65)
  petal abscission, 66 (fig. 65)
Essential factors for abscission
  energy source, 19
  low auxin, 19
  oxygen, 19
ETH. *See* Ethylene
Ethylene (ETH). *See also* Hormone actions
  abscission responses, 138–39, 141 (fig. 124)
  adverse side effects, 303–05
  atmospheric pollutant, 198

IAA counteraction, 140–41
and IAA-oxidase, 142
and IAA transport, 142
and IAA transport-inhibitors, 142
interrelations with hormones, 140–43 (figs. 124, 141)
mechanism of action, 143–45
occurrence, 135–38 (figs. 128, 137, 138, 141)
in plant diseases, 209
a secondary hormone, 144
wound ETH, 137 (fig. 138)
Eucalyptus
  allelopathy, 201
  bark abscission, 73–74 (fig. 75)
  branch abscission, 60 (fig. 85)
  branch stump separation, 59, 182 (fig. 60)
  calyptra abscission, 63 (fig. 65)
  dead branch abscission, 59 (fig. 60)
  drought-induced leaf abscission, 207
  genetics of bark abscission, 265
  heat-induced leaf abscission, 189
Euonymus, mist delay of leaf abscission, 193
Euphorbia
  ephemeral leaves, 46 (fig.)
  stem disarticulation, 62
  tree architecture, 86 (fig. 87)
Evergreen trees, vernal defoliation, 205
Evolution
  of abscission behavior, 280–85
  of protective layers, 280
  of separation layer, 280
  a theory of the evolution of abscission, 283–85
Excised abscission zones. *See* Explants
Explants
  holders, 101
  physiology of, 100–03
  research methods with, 99–103 (fig.)
  size, 100

Facultative deciduous habit, 206
Fagus
  leaf abscission habit, 46
  marcescent leaves, abscission, 51, 203
  pouch gall abscission, 213 (fig.)
Fern abscission, 218–21 (figs. 220–22)
  dehiscence, 220
  reproductive structures, 219
  vegetative organs, 218
Ficus
  branch abscission, 60
  facultative deciduous habit, 206
  laticifer cross-walls, 29
Fire, 189–90
  defoliation, 189
  dehiscence, 190

pruning, 189
and tree architecture, 190
Firmiana dehiscence ontogeny, 32 (figs. 33, 34)
Flax (*Linum* spp.)
dehiscence, 262
petal abscission, 63, 67
ultrastructural changes, 166–67, 175, 183 (figs.)
Flooding
abscission responses to, 193
hormonal responses to, 193–94
physiological effects of, 193–94
Flower abscission, 10–11, 63–68 (figs.), 294–96
calyx and sepals, 63 (fig. 65)
corolla and petals, 63–68 (figs.)
increased vase-life, 294–95
prevention by auxin-regulators, 294 (fig. 295)
prevention in herbarium specimens, 295–96
promotion by chemicals, 139, 295
stamens, 67–68 (figs. 64–66)
Fluorides, atmospheric pollutants, 200
Flushing and leaf abscission, 204
Forest communities, abscission habits and nutrient recycling, 201
Fouquieria
drought deciduous habit, 207 (figs. 191–92)
Fragmentation
of algae, 226, 231
of bacteria, 254–55
of fungi, 235
Fraxinus, ABA-induced leaf abscission, 126
Frost, 188–89
Fruit abscission
induced by plum curculio, 211
of mature fruit, 70 (figs. 70, 258), 301–05 (figs.)
ozone promotion, 200
prevention, 299–303
promotion, 296–99, 303–05 (fig.)
of young fruit, 68–70 (fig.), 296–300
Fruit thinning, 296–99 (fig.)
physiological action of chemicals, 298–99
to prevent biennial bearing, 296
Fuchsia bark abscission, 74
Fungi
abscission and dehiscence, 235–53
ascus dehiscence, 242–44 (figs. 246)
conidial abscission, 236–42 (figs.)
fragmentation, 235
sporangial abscission and dehiscence, 237–37 (figs.)
Fusarium and leaf abscission, 208

GA. *See* Gibberellin
Galactose promotion of abscission, 145–46
Galanthus leaf abscission, 47
Gall abscission and dehiscence, 212–16 (figs.)
Gamma radiation, 188
Gangamopteris leaf abscission, 272 (fig.)
Gardenia, HF-induced leaf abscission, 200
Gemma abscission
of *Ctenopteris*, 220 (fig.)
of *Hymenophyllum*, 219
of *Lycopodium*, 219
of *Marchantia*, 222 (fig. 223)
of *Psilotum*, 219 (fig. 220)
of *Trichomanes*, 221 (fig.)
of *Xiphopteris*, 220 (fig.)
Genetics
of abscission, 257–62 (figs.), 264–65
of dehiscence, 262–65 (figs.)
Geotrichum, conidial abscission, 237 (fig. 238)
Geranium petal abscission, 63
promotion by ETH, 139
Geum style abscission, 68
Gibberellin (GA). *See also* Hormone actions
abscission responses, 95, 131–33 (fig.), 137, 299–300
and α-amylase, 155
correlative occurrence, 130
interaction with ETH, with IAA, 132 (fig. 133), 140, 142
mechanism of action, 134
Ginkgo, CK in leaves, 134
Ginkgoales, abscission in the Mesozoic, 277 (fig. 278)
Gleocapsa colony fragmentation, 231
Glossopteris leaf abscission, 270 (fig. 272)
Glycocalyx
of bacteria, 255
similarity to pectic substances, 255
Gnomonium-induced abscission in plane trees, 210
Gosslingia rhizophore scars, 274
Gossypium. *See* Cotton
Grape (*Vitis* spp.)
bark abscission, 74
corolla abscission, 63 (fig. 66)
leaf abscission inhibited by corky bark disease, 210
Green algae (Chlorophyta)
abscission, 224–31
disarticulation, 225–26 (fig. 227)
dissemination, 224
fragmentation, 226

Heat and water stress abscission, 189
Helianthus response to flooding, 193–94

Heterocysts in fragmentation of blue-green algae, 231
Hevea
  climate and leaf abscission, 206
  laticifer blockage, 29
  photoperiod and leaf abscission, 187
Hibiscus mucilage canals, 24
Homeostasis, role of abscission in, 15–16
  and compensatory growth, 16
Hormogonial separation, 231
Hormone(s)
  changes with photoperiod, 187
  definition, 21
Hormone actions. *See also* Auxin; Abscisic acid; Ethylene; Cytokinin; Gibberellin
  in the abscission zone, 148–50 (fig.)
  responses to water stress, 191
  in source-sink relations, 151–52 (fig.)
  in the subtended organ, 150–51
Hormosira fragmentation, 232
Hormospore separation, 231 (fig. 232)
Horticultural practices that affect abscission, 288–89. *See also* Ecology; Physiology
Hyacinthus, adventitious leaf abscission, 91
Hydrodictyon, flagellar resorption, 229
Hygrophobic leaf abscission, 207
Hymenophyllum gemma abscission, 219
Hyoscyamus fruit dehiscence, 34
Hypobaric atmosphere and abscission, 142
Hypodermella prevention of *Larix* needle abscission, 210

IAA (indoleacetic acid). *See also* Auxin
  acceleration of abscission, 116–22 (figs.)
  ETH inhibition of transport, 142–43
  inactivation by ETH, 143
  adventitious abscission induction, 96
  interaction with ABA, with ETH, 123–26 (fig.)
  response curves, 120–22 (figs. 117, 119, 121–22)
  retardation of abscission, 116–22 (figs.)
IAA-oxidase, 108
  stimulation by ETH, 142
Impatiens
  adventitious internode abscission, 62
  fruit dehiscence, 41–42 (fig. 43)
  internode (stem) abscission, 62
  leaf separation pattern, 28
  micro-explants, 101
  ultrastructure of leaf abscission, 174, 176 (figs.)
Inhibition of abscission, 12. *See also* Physiology
Inman, T., 5
Insects, abscission effects of feeding, 210–13

Insect galls, 214–16 (figs. 212–14)
  abscission behavior, 214–16 (fig. 212)
  dehiscence, 215 (figs. 212–15)
  hormone production, 215
  jumping galls, 215–16 (fig. 214)
Internode (stem) abscission
  following injury, 62–63
  responses to hormones, 63
Iodoacetic acid, ETH production, 136
Ipomoea batatas (sweet potato), mid-petiole abscission, 93 (fig. 94)
Isidia, lichen disseminules, 253

Jacquinia, hygrophobic leaf abscission, 207
Juglans
  allelopathy, 201
  dehiscence promotion by ETH, 138
June drop, 70
Juniperus, cladoptosis, 54
Jute (*Cochorus olitorius*) leaf abscission, genetics and physiology, 264

Kalanchoë
  abscission of adventitious buds, 54
  adventitious leaf abscission, 91
  dispersal by raindrops, 40
Kochia stem abscission, 61

Laminaria
  blade abscission, 232
  blade splitting, 232 (fig. 234)
Lampropedia separation, 255
Larix (larch) cladoptosis, 54
  needle abscission prevented by *Hypodermella*, 210
Larrea, lack of leaf abscission, 23
Laticifer, 24, 29
Leaf longevity, 45 (fig. 46)
Leaf abscission
  acceleration, 289–92
  chemical induction, 290–92
  propagation by, 49
  retardation, 193, 289, 293
Leaf abscission habits, 9–10, 45–52
  autumnal deciduous, 9, 202–03
  competition effects, 9
  drought deciduous, 9, 190, 205–07
  vernal deciduous, 9–10, 203–05
Leaf scars, 29–31 (figs. 5, 30, 31)
Leaflets, number abscising, 47
Lecythis (sapucaia) fruit operculum dehiscence, 34–37 (fig.)
Ledum trichome abscission, 52
Legends
  of the *Moringa* tree, 1
  of Newton and the apple, 2
  of Siegfried (and the linden leaf), 2

Legume dehiscence, 41 (fig. 42)
Lemnaceae, branch abscission, 57
Lemon (*Citrus limon*) defoliation from sieve-tube decline, 259
Lepidocarpon, megasporangial abscission, 277
Lepidodendron
  branch abscission, 271 (figs. 274–75)
  leaf abscission and scars, 270 (fig. 271)
  stigmarian appendage abscission, 274 (figs. 276–77)
Lessonia bladed splitting, 232
Leucojum leaf abscission, 47
Libocedrus, cladoptosis, 54
Lichens, separation of disseminules, 253–54
Light and abscission, 186–88
  and carbohydrate accumulation, 186
  intensity, 186
  photoperiod, 186–87, 281
Lignase, 160–61, 184
Lignin, 25
  false reactions, 178–80
  positive identification, 179
  in protective layers, 177
  reagents, 179
Lignification, 25
  separation of lignified elements, 180–84 (figs.)
Ligno-suberization, 178
Lilium flower bud abscission from low light, 186
Limacinula ascus dehiscence, 244 (fig. 246)
Linum. See Flax
Liriodendron, response to flooding, 194
Lithocarpus trichome abscission, 52
Liverwort abscission, 221–23
  gemmae, 222 (fig. 223)
Lloyd, F. E., 5
  adventitious stem abscission in *Impatiens*, 92
  end-wall separation in *Spirogyra*, 226–27 (fig)
Localization of enzymes, 170
Lunularia gemma abscission, 222
Lupin (*Lupinus*)
  ABA and abscission, 126
  fruit dehiscence, 42 (fig.)
  prolonged flower vase-life, 294
Lycophyta leaf abscission in the Paleozoic, 268
Lycopodium
  gemma abscission, 219
  sporangial dehiscence, 220
Lygus
  induction of abscission, 211
  salivary PGU, 211
Lyngbya, hormongonial separation, 231

Macrocystis blade splitting, 232
Magnolia
  flower abscission, 63–67 (fig.)
  petal abscission, 66–67 (fig. 64)
  sympodial branching, 87 (fig. 88)
  vernal leaf abscission, 205
Main stem abscission, 61–63
Malformin, induction of leaf abscission, 209
Mangrove (*Brugiera* spp., *Rhizophora* spp.)
  cotyledon abscission, 47
  disseminule abscission, 72 (fig.)
  ultrastructure of abscission, 164–65 (figs.)
Manihot, photoperiod and leaf abscission, 187
Marchantia gemma abscission, 222 (fig. 223)
Marcescent leaves, abscission, 49–51 (fig.), 203–04
Mature fruit abscission, 70 (figs. 70, 258)
  prevention, 301–03 (fig.)
  promotion, 303–05 (fig.)
Mechanical factors in abscission, 38–40
  physical factors, 39–40
  tissue tensions, 38
  turgor, 38–39
Mediterranean climate, drought-induced abscission, 205–06
Melaleuca, HF-induced leaf abscission, 200
Melanorrhoea, hygrophobic leaf abscission, 207
Memecylon anther dehiscence, 37 (fig.)
Mercuric perchlorate and abscission, 142
Merospore abscission, 243–44 (figs.)
Mesozoic era, 277–78
  climatic changes and abscission, 277
  deciduous floras, 277 (figs. 278–79)
Metasequoia
  cladoptosis, 54 (fig. 55)
  paleontology, 280
Methionine, promotion of abscission, 145
Methods, physiological research, 98–105
  abscission indices, 104–05
  explant methods, 99–103 (figs.)
  field and greenhouse, 98–99
  testing devices, 103–04
Microbodies, 168 (fig. 169)
Microtubules, 167–68
Mid-petiole abscission, 93 (fig. 94)
Middle lamella, 170–77 (figs.)
  abscission changes, 179 (fig.)
  calcium bridges, 177
  pectic substances, 177
Mikiola gall abscission, 213 (fig.)
Mineral deficiencies, 111–12, 195–97
  boron, 196
  calcium, 112, 196
  molybdenum, 197

nitrogen, 195
zinc, 112, 195, 197
Mineral nutrition, 111–12
  deficiency-induced abscission, 111–12, 195–97
  return of nutrients to soil, 201
Mineral toxicities, 197
Mirabilis internode abscission, 28, 62
Mist and abscission, 193
Mistletoe (*Arceuthobium*, *Phoradendron*)
  and bark abscission, 77
  explosive dehiscence, 42
Mites, induction of leaf abscission, 213
Mitochondria, 167–68
Moisture and fruit dehiscence in Acanthaceae, 41
Moisture stress, 190–93
  drought defoliation, 190
  hormonal changes, 191 (fig. 192)
Molybdenum deficiency and abscission, 197
Monophenols, abscission promotion, 142, 146
  and IAA-oxidase, 142, 146
Moringa
  legend, 1
  branch abscission, 83
Morus
  GA responses, 132
  laticifer plugging, 29
Moss abscission, 221–24
  gemmae, 222 (fig. 223)
  opercula, 223–24 (fig.)
Mougeotia end wall separation, 226
Mucilage canals, 24
Multivesicular bodies, 163
Myrothamnus, leaf retention, 191

NAA (naphthaleneacetic acid). See also Auxin; IAA
  in abscission retardation, 293 (fig. 295), 299–300, 302
  in fruit thinning, 296–98 (fig.)
  toxic responses, 114–16, 298
Narcissus
  leaf abscission, 47
  vascular separation, 181 (fig.)
Naucleopsis branch abscission anatomy, 56
Necridia (separation discs), 231
Nereocystis sorus abscission, 232 (fig. 233)
Neuroterus, jumping gall abscission, 215–16 (fig. 214)
Newton, Sir Isaac and the apple, 2
Newtonia gall operculum dehiscence, 215
NH_3, atmospheric pollutant, 200
  as a defoliant, 200
Nicotiana
  corolla abscission, 4 (fig.)

pedicel abscission zone explants, 6 (fig.), 99
Nidulariaceae, periodiole abscission, 247–50 (figs.)
Nitrogen physiology, 111
  and IAA levels, 195
  deficiency, 195
  and defoliability, 195–96
NO_2, atmosphere pollutant, 200
Nuclear envelope, 163
Nucleolus, 162–63
Nutritional factors, 110–13
  carbohydrates, 110–11
  minerals, 111–12
  nitrogen, 111
  preabscission changes in, 112–13

Oedogonium
  cell division, 226 (figs. 227–29)
  division ring, 226 (figs. 227–28)
  flagellar abscission, 229
Oil palm, retardation of fruit abscission, 139, 302
Olive (*Olea*) leaf scar infection by *Bacterium*, 208
Omphalia-induced leaf abscission of coffee, 209
Ononis, drought abscission habit, 206
Ontogeny (differentiation)
  of abscission zones, 24 (fig.)
  of dehiscence zones, 32–34 (figs.)
Operculum abscission
  of asci, 242–44 (fig. 246)
  of insect galls, 215 (fig.)
  of moss capsule, 223 (fig. 224)
  of sapucaia (*Lecythis*) fruit, 34–37 (fig.)
Ophioglossum, sporangial abscission, 220
Opuntia, adventitious abscission, 93
Organelles and abscission, 162–70
Oryzopsis rhizosheath abscission, 76 (fig.)
Oscillatoria, hormogonial separation, 231
Osmunda, sporangial dehiscence, 221 (fig. 222)
Ostrya, marcescent leaf abscission, 51, 203
Oxygen
  acceleration of abscission, 107–08 (figs.)
  requirement, 107
Ozone, atmospheric pollutant, 198–200 (fig.)

Paleostigma bud scars, 268 (fig. 269)
Paleozoic era, evidence of abscission, 268–77
Palms, coevolution with herbivores, 282
Panicum disseminule, 72
Parthenium
  absence of leaf separation, 23

marcescent leaves, 51
resin canals, 24, 29
Parthenocissus
  leaflet abscission, 47
  petiole ETH production, 136, 138
Pathogenic microorganisms, 207–10
Patterns of separation, 27–28
  in *Impatiens* leaf, 28
  in *Mirabilis* stem, 28
Peach (*Prunus persica*)
  genetic aspects of fruit abscission, 258–59
  leaf scar infection by *Xanthomonas*, 208
  vascular separation, 181 (fig.)
  time-course of fruit abscission, 70 (fig.)
Pear (*Pyrus communis*)
  adventitious pedicel abscission, 96
  preharvest drop prevention, 302
Pecan (*Carya illinoensis*), promotion of dehiscence, 301
Pectic substances
  in bacterial glycocalyx, 255
  calcium bridges, 177
  in middle lamella, 177
  polygalacturonides, 177
Pectinases, 158–60
  pectin methylesterase (PME), 159
  polygalacturonase (PGU), 160
  role in xylem separation, 182, 184
  secretion by pathogens, 209, 211–13
Pectin methylesterase (PME), 159
Pemphigus galls delay leaf abscission, 214 (fig. 212)
Penicillium, conidial abscission, 239 (fig.)
Perebea branch abscission anatomy, 56
Periderm, 28
Peridiole abscission, 247–50 (figs.)
Peroxidase and abscission, 155 (fig. 156)
  localization, 170
Persea (avocado)
  axillary bud abscission, 57
  vernal leaf abscission, 205
Petal abscission, 63–68 (figs.)
Petraphera abscission, 267
Pfeiffer, H., 3
Phaeophyta. *See* Brown algae
Phaffia budding, 253 (fig.)
Phase I and II, 118–19
Phaselous vulgaris. See Bean
Phenolic substances and abscission, 146
Phenology, 202
  autumnal defoliation, 202–03
  drought-induced (summer) leaf abscission, 205–07
  hygrophobic leaf abscission, 207
  vernal leaf abscission, 203–05
Phoenix, toxicity of NAA, 114–16
Phoradendron and bark abscission, 77

Photoperiod and abscission, 186–87
  in autumnal defoliation, 202
  hormonal effects, 187
Phyllidia, lichen disseminules, 254
Phyllopodium and fern leaf abscission, 218
Physalis leaf abscission from tobacco mosaic virus, 209
Physical factors and abscission, 186–94
  light, 186–87
  radiation, 188
  temperature, 188
  wind, 194
Physiology
  of abscission, 97–152
  of chemical defoliants, 292
  of fruit-thinning chemicals, 298
Phytochrome in abscission, 187
Phytophthora, leaf scar infection, 31
Picea
  allelopathy, 201
  vernal leaf abscission, 203–04
Pilobolus sporangiophore dehiscence, 236 (fig. 237)
Pinus spp.
  bark abscission, 73–74 (fig. 75)
  branch abscission, 60, 83
  cone opening, fire-induced, 190, 264
  fire pruning, 190
  genetics of serotiny, 264
  needle abscission, 56, 188, 198
  needle retention, 45
  ozone-induced needle abscission, 198 (fig. 199)
Plantago fruit dehiscence, 34 (fig. 35)
Plasmalemma, 162–65 (figs.)
  vesicle fusion, 167 (fig.)
Platanus
  abscission from *Gnomonium*, 210
  bark abscission, 74 (fig. 76)
  branch abscission, 60
  leaf scars, 30 (fig.)
Plum curculio (*Conotrachelus*), induction of fruit abscission, 211
Plumeria
  laticifer blockage, 29
  photoperiod and leaf abscission, 187
Podocarpus branch abscission, 59
Pollutants, atmospheric
  acid rain, 200
  ETH, 198
  fluorides, 200
  NH_3, 200
  NO_2, 200
  ozone, 198 (fig. 199)
  SO_2, 200
Polygalacturonase (PGU), 160
  secretion by insects, 211–13

Polygalacturonides, 177
Polysaccharide degrading enzymes, 161
Populus
   branch abscission, 3 (fig.), 56, 59–60, 259
   galls and leaf abscission, 214 (fig. 212)
   genetic aspects of leaf abscission, 259
Pore, in ascus dehiscence, 244 (fig. 246)
Potato (*Solanum tuberosum*), flower and fruit abscission, 259
Poterium, drought abscission habit, 206
Preabscission changes in organs, 112–13
   biochemical, 112
   hormonal, 113
   mineral nutrients, 112
Precambrian era, evidence of abscission, 267–68 (fig.)
Preharvest drop
   genetic aspects, 258
   prevention, 301–03 (fig.)
Prickle abscission, 77
Primary cell wall
   lack of breakdown, 177 (fig. 179)
   swelling in abscission, 177 (figs. 173, 178)
Propagation by leaf abscission, 49
Protein synthesis in abscission, 154–55, 162
Protective layers, 28–31 (figs.)
   definition, 21
   evolution, 280
   leaf scars, 29–31 (figs. 5, 30–31)
   lignification, 177–79
   primary, 28
   secondary, 28–31
   and starch deposits, 168
Protosiphon bud abscission, 229
Prunus. *See also* Cherry
   abscission of leaf shot-holes, 92 (fig. 93)
   petiole ETH production, 136
Prunus cerasus. *See* Cherry
Prunus persica. *See* Peach
Psilotum gemma abscission, 219 (fig. 220)
Psoralea stem abscission, 61 (fig.)
Pteridium, sporangial dehiscence, 222 (fig.)
Pterygophera blade abscission, 232
Pyracantha, HF-induced leaf abscission, 200

Quercus
   branch abscission, 57, 59–60 (fig. 55)
   jumping-gall abscission, 215–16 (fig.)
   leaf abscission from *Ceratocystis*, 208
   marcescent leaf abscission, 51 (fig. 50), 203–04
   trichome abscission, 52
   vernal leaf abscission, 205

Radiation
   gamma radiation, 188
   UV radiation, 188
Rain and abscission, 193
Raindrops in dispersal
   of peridioloes, 247 (fig. 249)
   of plantlets, 40
   of seeds, 40
Recycling of minerals, 201
Red algae (Rhodophyta)
   blade abscission, 234 (fig. 235)
   fragmentation, 234
Requirements for abscission
   energy source, 19
   low auxin, 19
   oxygen, 19
Research methods, 98–105
   abscission index, 104–05
   abscissors, 103–04
   explant methods, 99–103
   field methods, 98–99
   greenhouse methods, 99
Resin canal, 24
   plugging, 29
Respiration and abscission, 109
   climacteric respiration, 109 (fig.)
   respiratory enzymes in abscission, 154
Resurrection plants, resistance to leaf abscission, 191
Retardants, naturally occurring, 146
Retightening phenomena, 161
Rhizoctonia, in stem abscission, 61
Rhizophora (mangrove)
   cotyledon abscission, 47
   disseminule, 72
Rhizosheath abscission, 77 (figs. 76, 78)
Rhododendron (Ericaceae)
   anther dehiscence, 37 (fig. 38)
   HF-induced leaf abscission, 200
Rhodophyta. *See* Red algae
Rhus, dichotomous branching, 86
Rhynia
   sporangial abscission, 268
   sporangial scar and branching, 268
Ribes
   auxin occurrence and fruit abscission, 114
   leaf abscission from *Cronartium*, 208
Riccardia, gemma abscission, 222–23
Ricinus, tree architecture, 88 (fig.)
RNA in abscission, 154–55, 162
Robinia, GA responses, 132
Root
   abscission, 77–80 (fig.)
   Azolla root abscission, 79 (fig. 80)
   cortex and bark abscission, 77 (fig. 76)
   nodule abscission, 79
   rhizosheath separation, 77 (figs. 76, 78)
Root cap abscission, 80–81
Roystonea leaf abscission, 31 (fig.)

Saccharomyces, bud abscission, 251 (figs. 251–52)
Salicornia disseminule, 72
Salinity, soil, and abscission, 197–98
Salix
 branch abscission, 59
 branch abscission zone, 3 (fig.)
 CK changes, 134
 leaf galls, 214
 shoot-tip abortion, 52
Salsola stem abscission, 61 (fig. 62)
Salvia, drought deciduous habit, 206
Salvinia branch abscission, 218
Sambucus leaflet separation layer, 27
Sapium, sympodial growth, 54
Saprolegnia, sporangial dehiscence, 236 (fig.)
Sapucaia (*Lecythis*) fruit operculum dehiscence, 34–37 (fig.)
Sarcocaulon spine development, 47
Sargassum fragmentation, 232
Scars
 leaf, 29–31 (figs.)
 branch, 59–60 (fig.), 271–75 (figs.)
Schinus leaf gall dehiscence, 215
Schizaea, sporangial abscission, 221 (fig. 222)
Schizidia, lichen disseminules, 254
Schizogoniales, abscission of shoots, 227
Schizosaccharomyces cell separation, 253 (fig. 254)
Scopulariopsis, conidial abscission, 238 (figs. 239, 241)
Secondary hormone, ETH role, 144
Seeds
 abscission, 71
 dispersal, 11
 fossil seeds, 275
 prevention of shattering, 301
Senecio, tolerance to flooding, 193
Senescence
 and abscission, 13
 definition, 21
Senescence factor, 147
Separation
 of bacteria, 254–55
 of lignified elements, 180–84 (figs.)
 patterns, 27–28
 ultrastructure, 174–76 (figs.)
Separation discs (necridia), 231
Separation layer
 anatomy, 26–27 (figs. 4, 26)
 cell shape changes, 39, 172
 definition, 21
 evolution, 280
 positions, 27, 47–49, 89–96
Septum, in budding yeast, 251–53 (figs.)
Sequoia cladoptosis, 54 (fig. 55)

Sequoiadendron
 cladoptosis, 54
 fire pruning and architecture, 189–90
Sesame (*Sesamum indicum*)
 genetics of dehiscence, 264 (fig. 265)
 shattering, 262
Shattering of seeds, 262
 prevention, 300–01
Shoot-tip abortion and abscission, 52–54 (fig.)
Shot-hole disease, abscission, 92–93 (fig.)
Siegfried and the linden leaf, 2
Sigillaria leaf scars, 271 (fig.)
Sinks, 15, 151 (fig.)
Siphonales, abscission of branchlets, 227
Snapdragon (*Antirrhinum*) corolla abscission, 295
Soil factors
 flooding, 193–94
 mineral deficiencies, 195–97
 mineral toxicities, 197
 salinity and alkalinity, 197–98
Soredia, lichen disseminules, 253–54
SO_2 (sulfur dioxide)
 ABA in resistant species, 200
 atmospheric pollutant, 200
Source-sink relations, 15, 151 (fig.)
Soybean (*Glycine max*)
 genetic aspects of leaf abscission, 259, 264
 photoperiod and young fruit abscission, 187
Sphagnum operculum abscission, 224 (fig.)
Sphaerobolus peridiole abscission, 250 (fig.)
Spilocaea, conidial abscission, 238 (figs. 239–40)
Spine development, abscission in, 47
Spirodela branch abscission, 57 (fig. 58), 188
Spirogyra
 calcium requirement for end walls, 231
 end wall separation, 226 (fig. 227)
Spondias, hygrophobic leaf abscission, 207
Sporangial abscission
 in the Devonian, 268
 in fungi, 235–36
Sporangial dehiscence
 in the Devonian, 268
 in ferns, 220–22 (figs.)
 in fungi, 235 (figs. 236–37)
Squirting cucumber (*Ecballium*), fruit abscission, 39 (figs. 39–40)
Stage I and II, 118–20
Staphylococcus separation (abscission), 255
Starch, preabscission deposits, 168 (fig. 169)
Stichococcus fragmentation, 226
Stigeoclonium, flagellar resorption, 229
Stigmella, salivary CK and abscission, 210

Stigmarian appendages, abscission anatomy, 274–77 (figs.)
Streptocarpus, adventitious leaf abscission, 89 (fig. 91)
  CK changes, 134
Streptococcus
  chain length, 255
  separation, 255
Stress-strain analyzer, 104
Strobilanthes leaf separation layer, 4 (fig.)
Strombosia cotyledon abscission, 47 (fig. 48)
Suberin, 184
Substrate requirement, 108–09
Sugars as abscission substances, 145
Summer (drought) -induced abscission, 205–07
Sweet potato (*Ipomoea batatas*), mid-petiole abscission, 93 (fig. 94)
Sympodial growth habits, 53–54, 83
  of trees, 86–89 (figs.)
Syncephalis merospore abscission, 243–44 (figs.)
Synchytrium gall abscission, 216

Tamarindus, facultative deciduous habit, 206
Taraxacum root phloem abscission, 77
Taxodium cladoptosis, 54 (fig. 55)
  GA response, 131
Taxonomy
  use of abscission characters, 259–62 (fig. 278)
  use of dehiscence characters, 262–64 (fig.)
Taxus flushing and leaf abscission, 203–04
Temperature and abscission
  cold, 188–89
  fire, 189–90
  frost, 188–89
  heat, 189
  response curves, 105–07 (fig. 106)
Terminology and definitions, 20–21
Tetramelis, hygrophobic leaf abscission, 207
Tetranychus induction of leaf abscission, 213
Tetraphis gemma abscission, 222
Thamnosa, allelopathy, 201
Theobroma
  flushing and leaf abscission, 204
  leaf scar infection, 31
  tree architecture, 88 (fig.)
Theophrastus, comments on abscission, 2, 98
Thuja cladoptosis, 54
TIBA abscission effects, 95, 142
Tissue tensions in abscission, 38
Tmesipteris shoot abscission, 218
Tobacco mosaic virus in leaf abscission of *Physalis*, 209

Tomato (*Lycopersicon esculentum*)
  genetic aspects of fruit abscission, 259 (fig. 260)
  increased fruit-set, 299–300
  mid-petiole abscission, 93
  ozone-induced fruit abscission, 200
Tonic effect of copper on coffee, 293–94
Toxicities, mineral, 197
Toxins of plant pathogens, 209
Transport inhibitors, interaction with ETH and IAA, 142–43
Tree fern
  architecture, 85
  leaf scars, 29
Trichomanes gemma abscission, 221 (fig.)
Trichome abscission, 52
Trichurus, conidial abscission, 242 (fig.)
Triplochiton, lower branch abscission, 84 (fig.)
Tulip, loculicidal dehiscence, 263 (fig.)
Tumbleweeds, 61–62 (figs.)
Turgor
  in abscission, 38–39, 241, 244–45 (fig.)
  in dehiscence, 41–43 (fig.)
2,4-D (2,4-dichlorophenoxyacetic acid). *See also* Auxin
  use in forest tree pruning, 305
Tyloses, 29
  in xylem separation, 182 (fig.), 184

Ulmus
  dehiscence of leaf galls, 215
  large branch abscission, 60
Ulothrix, flagellar resorption, 229
Ulvales, abscission of shoots, 227
Umbrella trees, 85 (fig. 86)
UV radiation, 188

Vascular anastomosis, 24–25
Vascular separation, 180–84 (figs.)
Vase-life of cut flowers, increased, 294
Vaucheria, flagellar resorption, 229
Verbascum corolla abscission, 67, 194
Vernal leaf abscission
  complete defoliation, 205
  of marcescent leaves, 203–04
  partial defoliation pattern, 204
  physiology, 205
Veronica corolla abscission, 67, 194
Verticillium and leaf abscission, 208
Vesicles, 162–70 (figs.)
  fusion with plasmalemma, 163, 165 (figs. 165–67)
  relation to abscission, 179 (fig.)
Vitis. *See* Grape
Von Mohl, H., separation and protection in abscission, 2, 22

Walnut (*Juglans regia*), promotion of dehiscence, 301
Water relations
  desiccation tolerant plants, 191
  flooding, 193–94
  mist, 193
  moisture stress, 190–93 (figs.)
  rain, 193
Westiella, hormogonial separation, 232 (fig.)
Wiesner, J., 4, 98
  role of heat in abscission, 189
  vernal leaf abscission, 203–04
Wild rice (*Zizania aquatica*) shattering, 262
Wind, abscission effects, 194

Xanthomonas, leaf scar infection of peach, 208
Xerophyta leaf retention, 191
Xiphopteris gemma abscission, 220 (fig.)

Yeast
  bud abscission, 250–53 (figs.)
  cell separation (fission), 253 (fig. 254)
  septum formation, 251 (fig. 252)
Young-fruit abscission, 68–70 (fig.)
  fruit thinning, 296–99 (fig.)
  genetic aspects, 258–59
  physiological aspects, 298–300
  prevention, 299–300

Zelkova branch abscission, 59, 83 (fig. 55)
Zinc deficiency
  and abscission, 112, 195
  and auxin levels, 195–96
  and fruit set, 299
Zosterophyllum, sporangial dehiscence, 268
Zygophyllum leaf abscission habit, 206

Designer:	Laurie Anderson
Compositor:	G&S Typesetters, Inc.
Printer:	Braun-Brumfield, Inc.
Binder:	Braun-Brumfield, Inc.
Text:	10/12 Sabon
Display:	Sabon

DATE DUE			
GAYLORD			PRINTED IN U.S.A.